Further praise for *A Silent Fire*

"Shilpa Ravella paints a captivating portrait of one of our body's most ingenious and intimate acts, sharing the formidable force of inflammation within us all. The way she charts forward is spot-on: curbing the rampant plague of modern maladies calls for diets and doctoring that protect the human body and its intrinsic power to keep us well."

—Anne Biklé, coauthor of *The Hidden Half of Nature*
and *What Your Food Ate*

"Shilpa Ravella's fascinating, poetic exploration of the body, food, and history will sweep you up sentence by sentence. A book that could not only reshape readers' understanding of their own systems and choices but possibly medicine itself."

—Lauren Sandler, author of *This Is All I Got*

"An enlightening dive into the medical science at the root of the modern epidemic of chronic diseases—and what it all means for healthy eating."

—David R. Montgomery, coauthor of *The Hidden Half of Nature*
and *What Your Food Ate*

A Silent Fire

The Story of Inflammation,
Diet, and Disease

SHILPA RAVELLA

W. W. NORTON & COMPANY

Independent Publishers Since 1923

This book is intended as a general information resource. It is not a substitute for professional medical advice. Seek immediate medical care if you experience symptoms like those described in this book. Patient names have been changed and identifying details changed or omitted; some patients are composites. The publisher is not responsible for, and should not be deemed to endorse or recommend, any website other than its own or any content that it did not create. The author, also, is not responsible for any third-party material.

For FGR

Humanity has but three great enemies: fever, famine and war. Of these by far the greatest, by far the most terrible, is fever.

—Sir Willam Osler

Disease usually results from inconclusive negotiations for symbiosis, an overstepping of the line by one side or the other, a biologic misinterpretation of borders.

—Lewis Thomas, *Lives of a Cell*

They were nothing more than people, by themselves. Even paired, any pairing, they would have been nothing more than people by themselves. But all together, they have become the heart and muscles and mind of something perilous and new, something strange and growing and great. Together, all together, they are the instruments of change.

—Keri Hulme, *The Bone People*

Contents

A Silent Fire

Introduction

When we were medical students, my best friend—we'll call him Jay—gave me a copy of Keri Hulme's *The Bone People*. Hulme's work is filled with isolation, fear, and violence. But at the core of this novel lies a love story, albeit one that distorts the notion of love so that it lives outside the margins of orthodoxy. The characters are figuratively stripped to the bone, their raw emotions bleeding through the page, juxtaposing the beauty and horrors of human nature. As we dissected cadavers that year, our surgical tools sliced through shallow and deep tissue, splitting soft gobs of yellow fat and tough, stringy muscle in an attempt to unveil and learn. The corpses on our tables, however, remained a mystery.

Nine years later, by the summer of 2012, Jay and I were living in Chicago. That July, a North American heat wave spread east from the Rocky Mountains, searing through the country with a ferocity unmatched since the 1930s. Highways in Illinois buckled under the pressure. In Grant Park, a large section of the pavement cracked and rose 3 feet into the air. I am haunted now by the audacity of that heat, and my own indifference to it, driving to and from work in a cool air-conditioned car. I wonder if, had I been dreaming less and watching more, things would have turned out differently.

It started insidiously, on a sweltering Friday evening. Jay had just returned from a workout at the gym. He made us a quick dinner of saucy linguine neatly nested in white plates. I remember how he escorted a spider from the windowsill out to the patio, setting it free. How he talked about escaping Chicago's sultry summers and frosty winters. Then, suddenly, he was tired. His fingers were cupped around his neck like he was going to choke himself. "Something feels off," he told me. "My neck is sore. I probably overdid it at the gym."

In the decade that I had known him, Jay was not one to complain about his health or even to ask for help when he was sick. I did not see any external injuries, and he could move his head freely. We chalked it up to muscle strain, a frequent sports-related injury that can cause local inflammation. He took some ibuprofen and expected to feel better soon.

A few weeks later, he was in worse shape. "My head and neck feel heavy, like someone draped bulky blankets over them," he said. The muscles in the back of his neck had gotten weaker. Doctors ordered a magnetic resonance imaging (MRI) scan of his head and neck, which came back normal.

Two more weeks passed. As time flitted by, we watched in a slowly rising panic as Jay's muscles continued to deteriorate. While driving home from work one day, he was barely able to lift his head up in order to gaze over the steering wheel. Soon, he developed a complete head drop and could not raise his chin off his chest. The moment at which this happened was hard to pinpoint. It was lost in a medley of moments both mundane and critical that we tried to sort through retrospectively.

At that point, Jay needed a body brace that extended to his waist. The brace was attached to a Philadelphia collar, a neck support used to prevent head and neck movement after a spinal cord injury. The contraption was uncomfortable, but it held his head up, redistributing its weight to the muscles in his middle and lower back. He took it off only if he needed to shower or sleep.

The average human head, which weighs about 10 pounds, is held up by a seemingly effortless, intricate balance among the muscles of the neck, an equilibrium so fluid and adept in its functions that one ceases to think about its mechanics. For Jay, that balance had been destroyed, suddenly and perhaps irrevocably. His condition continued to worsen. He was barely able to walk a few blocks before getting fatigued, and he started to have trouble swallowing food. The culpable force, whatever it was, had struck with terrifying speed and intention.

Doctors were baffled by the case. Jay was a young guy in his early thirties and had always been healthy. At first, neurologists considered a rare form of Parkinson's disease or the beginnings of amyotrophic lateral sclerosis, a rapidly progressive and fatal neurological condition in which nerve cells controlling the muscles degenerate. Patients lose the ability to move their arms and legs, swallow food, or speak. Muscles in the diaphragm and chest wall eventually fail, causing an inability to breathe. Within a few years, death ensues from respiratory failure.

But a clue floated in Jay's blood: unusually high levels of the enzyme creatine kinase, an indicator of muscle injury. This suggested that even in the absence of outer trauma, something was actively harming Jay's tissues, chewing up muscle. Over half of the muscles in his neck would be permanently destroyed, with the wreckage gleaming on a follow-up MRI scan.

Rheumatologists, specialists in treating autoimmune diseases, had another idea: Jay was badly *inflamed*. They guessed that an atypical autoimmune disease was responsible for his condition. Even if his symptoms and blood tests did not fit any known pattern. Even if we could not initially see the inflammation. Doctors sharpened an arsenal of anti-inflammatory drugs, ready to be deployed in rapid succession, but could give us no sense as to how Jay would fare. "Wait and see," they said. "Wait and see."

The word *inflammation* comes from the Latin verb *inflammare*, "to ignite"—a kindling, a setting on fire, as the ancient Romans described it. It is an ancestral response that evolved to protect the body from threats and contain damage, be it from a microbe, chemical, or trauma—the same defense employed by animals as primitive as starfish. Flaring up, tackling a problem, and dying away, inflammation is a fundamental immune response that has served us well through most of human history.

But in modern times, the threats we face are more insidious than those our ancestors faced. Today we find that inflammation can persist, with or without a known trigger, destroying healthy tissue. Autoimmune diseases like arthritis or lupus, which turn inflammation against the body, can be devastating and some-times fatal.

As medical students, Jay and I had learned about these and other inflammatory diseases. But inflammation as an entity did not cap-ture my imagination then. Pathological inflammation could be neatly packaged into distinct categories and aptly named. It was omnipresent, an indelible part of health and most diseases, essen-tial yet unassuming. But Jay's illness triggered a shift. All of a sud-den, inflammation morphed into something whole and consuming. It nested in the center of my consciousness, the thing my mind and eyes grappled with first when confronted with sickness.

As I spent over a decade in training and graduated to my first years as an attending gastroenterologist, many patients came to my office inflamed. Some suffered from inflammatory bowel disease, an autoimmune condition in which severe intestinal inflammation can result in surgery to remove most or all of the intestines. Others coped with inflammation from acid reflux, food sensitivities, celiac disease, irritable bowel syndrome, and more. I also treated patients with intestinal or multiple-organ transplants, whose immune sys-tems threatened to mount inflammatory attacks against their new organs. I prescribed anti-inflammatory drugs in my practice, from concoctions to treat generalized inflammation, pain, and fever—like aspirin—to a wide range of agents with new therapeutic targets, including potent immune-modulating drugs used for autoimmune diseases and in transplant medicine. Evidence supports these drugs for specific inflammatory disorders.

But what had happened to Jay eluded me. It had no name or color, no beginning or end. It was absent from medical textbooks. The inflammation came, took, and went, initially escaping detection. It could return as it wished. Treating Jay with anti-inflammatory drugs was both logical and whimsical. It meant believing in a fury despite the maddening dearth of evidence to support its existence. I started to become fascinated by what had not been emphasized in medical school: hidden inflammation.

There is a yin and yang to inflammation that is analogous to using a fire hose. Too little water pressure in the hose and the fire—be it a germ or another intruder—wins. Too much and the body can turn on itself, drowning in autoimmunity. But sometimes the hose simply leaks, and a low-level inflammation simmers quietly in the body. Doctors do not regularly test patients for this type of inflammation. It is an amorphous foe that often lacks routine treatments. Fighting it means moving in the darkness—the same problem Jay's doctors had initially grappled with. We cannot see hidden inflammation with our naked eye—as we can the swollen joints of patients with rheumatoid arthritis and the rashes of those with lupus—or with the tools we typically use to diagnose inflammatory diseases. Otherwise healthy people walking around with this type of inflammation are entirely unaware that it exists within them: there may be no obvious signs or symptoms.

But hidden inflammation, which once lived in the margins of medical literature, is far from benign, and uncovering it—seeing what was once unseen—has been a process as slow and sinuous as the disease itself, one that demanded the life's work of numerous scientists, a few of whom I portray in this book. Scientists' quest to comprehend inflammation began with milestone discoveries in the nineteenth century, and it is still unfolding. In the 1850s, German scientist Rudolf Virchow became the first to identify the cellular characteristics of inflammation, seeing what could not be seen with the naked eye—a radical departure from medicine's past. His work inspired Russian zoologist Elie Metchnikoff's serendipitous discovery of macrophages, one of the cells central to our present understanding of inflammation. A vicious battle to uncover the inner workings of the inflammatory response ensued.

A century later, when scientists stumbled upon the neglected works of these historic figures, it set them on a subversive journey that would revive ancient theories and help to redefine inflammation and disease in modern medicine. Hidden inflammation, silent and sinister, lurks in heart disease and smolders beneath developing tumors. It is tied to many other chronic conditions as well, including obesity, diabetes, and neurodegenerative and psychiatric diseases. It affects aging, the germs in our gut, and the function of our intestines. It weakens immunity, paradoxically predisposing us to infections. Even worse, it increases the likelihood that our immune system will mount an overwhelming, inappropriate attack against infections, leading to grim outcomes. In fact, hidden inflammation may shed light on why ostensibly healthy individuals can succumb to severe illness during epidemics and pandemics. Hidden inflammation can be found adorning specific spots in organs or traveling through the body's vessels—usually it does both. It rests on a wide continuum and can be as simple as an activation of certain inflammatory genes.

We are not accustomed to routinely diagnosing and treating most cases of hidden inflammation despite the damage it is known to cause, and this is perhaps its salient unifying thread at present. In the modern age, tools to capture inflammation move beyond the naked eye or even microscopes. Doctors can probe and poke at every part of the body with fine instruments, getting a close look inside organs and tissues, or use imaging studies and blood tests to elicit even more information. They can pinpoint loss of function, a cardinal sign of inflammation. They name what they see, often using the suffix -*itis* to denote inflammation. Only twenty examples of -*itis* nouns existed before 1800 (the earliest recorded use was *arthritis*, described in 1543 as "weakness in all the joints of the body, a noughty humor flowing to the same"), after which the form exploded. Inflammation of the brain is *encephalitis*, and that of the bowels, *colitis*. In the liver, *hepatitis*, a *nephritis* in the kidneys, a *myocarditis* of the heart muscle. The number of conditions ending in -*itis* are legion in medical dictionaries, numbering in the hundreds, and many of them are familiar to the general public: appendicitis, tonsillitis, bronchitis, dermatitis. Acute or chronic, inflammation links all kinds of doctors and diseases, bridges clinics and inpatient wards. But the pioneering work of scientists embroiled in the biology of hidden inflammation initially remained nameless. The inflammation was persistent, or chronic, but did not fall into known divisions. It stood outside the gates of traditional immunology, attempting to wrangle its way in.

In the twenty-first century, we have reached a tipping point of sorts when, for the first time, scientists have been able to staunchly declare that hidden inflammation can be both a consequence *and* a cause of disease, colluding with our genes and environment to evoke disaster. In fact, it may be a common thread running through nearly all disease. The force that fought our historic top killers, healing wounds and keeping microbes in check, now marches alongside diseases of modernity, insidious and imbalanced, brooding quietly before it explodes in a sudden rage.

But if one is not diagnosed with a typical inflammatory disease, what exactly does it mean to be "inflamed" in the modern age? What tests can catch hidden inflammation? Where does it come from—is it a response to underlying disease, or is it triggered by something in the environment, like an unhealthy diet or pollution or stress? How much evidence ties it to our present-day chronic calamities? And how can we prevent, suppress, or even reverse it?

Beyond drug therapy, two developing narratives have begun to take hold in the struggle to understand and combat inflammation, both grounded in rapidly expanding science and poised to disrupt modern medicine.

First, the story of food. A growing body of research points to the power of our diets to cause, avert, or treat inflammation. I have studied nutrition for many years, both as a gastroenterologist taking care of patients with special dietary needs—including the sickest ones fed with liquid formula running through their veins or stomach tubes—and as a consumer poring over the gamut of nutrition science data, hoping to understand how best to address the questions that popped up in my clinics. Patients want to know about the discussions they are watching on television or reading about in magazines, the things they are hearing about from friends and family and almost everyone but their doctors. Is fat the problem, or carbohydrates? Are eggs back in? What about sugar or the conflicting information on gluten and grains? They want to know about one of the most coveted yet confusing topics in nutrition: the "anti-inflammatory" diet. Is there evidence for a *real* anti-inflammatory diet, one that could help to hinder or heal the chronic, fatal disease of modernity? And what makes certain foods "inflammatory," when eating itself is an inflammatory act?

Second, the story of germs. The gut microbiome, comprising

trillions of bugs living in our intestines, was thrust into the scientific spotlight at the turn of the twenty-first century. Germs are now understood to be symbionts crucial to human health, not just pathogens responsible for disease. Intestinal germs play a crucial role in immune function and inflammation. In the fourth century, "yellow soup" made of dried stool was used to treat diarrhea. Modern versions of this treatment include fecal transplants and probiotics. Manipulating gut germs to prevent or reverse disease is a work in progress, a rising diagnostic and therapeutic market with remarkable potential. We can leverage their power to foster an optimal inflammatory state, an immune system that does not overreact or underreact to challenges, strengthening immunity while staving off autoimmunity and other chronic diseases.

Inflammation is at once pervasive and abstruse, spreading through sickness and health with infinite, scrappy resolve. The aim of this book is to expose its new depths, which will help to shape medicine's future: its links to common, fatal diseases of modernity and the interlaced roles of food and germs. Modern medicine, replete with specialties of every sort, addresses inflammation piecemeal. If inflammation is a beast akin to the elephant in an old Buddhist parable, each of us examines a portion of the elephant, never the whole, describing its shape and form with limited information and drawing disparate conclusions. But it is this elephant that binds doctors in a common fight—rheumatologists, cardiologists, gastroenterologists, oncologists, endocrinologists, neurologists, and more—against an adversary both good and evil, seen and unseen, coming and going as it pleases under a multitude of guises.

Today, as nutrition science grows, as gut microbiome research explodes, and as diseases linked to inflammation—both old and new—are becoming commonplace, deciphering the science is vital. Jay's strange illness could have struck any seemingly healthy individual. Uncontrolled inflammation, which may lie at the root of his condition, is tied to a modern plague of illnesses.

In the decades to follow, more of the mystery will be unraveled, with new names, diseases, and iterations of inflammation. But the facts detailing the birth of this story will remain static, trapped in time, softened by hindsight. This book explores the works of modern scientists but also seeks glimpses from history. We owe, in large part, the contemporary understanding of inflammation, food, and germs to centuries-old discoveries. The resulting constellation of stories, including those from my own life and work, are an attempt to give the elephant skeletal shape and form.

CHAPTER 1

Metamorphosis

O n May 3, 1845, a large crowd gathered at the Friedrich-Wilhelms University to celebrate the birthday of its late founder. An equestrian statue of King Frederick II of Prussia, surrounded by a cast-iron ring fence, graced the large courtyard. The university, located in an eighteenth-century palace, would host many of Germany's greatest thinkers over the years, becoming one of Europe's most prestigious institutions. In the crowd stood a young man with humble origins in rural Pomerania. His grandparents were butchers, his father a shopkeeper. He had arrived in Berlin only a few years prior to attend medical school. Awed by the city's glitter and sophistication and anxious not to appear shabby, he had written to his father asking for money with which to buy fashionable stockings.

That day, when he was called forth to give his prepared talk, the audience did not know what to expect from the short, thin, and blond boy who appeared before them as an apparition, seemingly fragile enough to be blown asunder by the wind. He was known to patients as *der kleine doctor*, or "the little doctor." As he spoke, Rudolf Virchow's dark, nearly lashless eyes bored into the audience. He shone on stage, with an eloquence that belied his youth and diminutive physique.

Virchow tore into the ideas that held fast in the minds of Berlin's most prominent physicians and scientists. Many of these men were mired in an intellectual movement called Romanticism, which swept western Europe to counter the late-eighteenth-century Enlightenment and ruled German medicine at the turn of the nineteenth century. Romanticism rejected analytical methods, holding that the natural scientist could deduce truths from a priori first principles, without observation or experimentation. Organic matter possessed a vital spark that could not be explained by the laws of physics or chemistry. It was the age of humorism, a theory with roots in ancient Greek and Roman medicine that dominated Western medicine for more than two thousand years. Human bodies were filled with a quartet of liquids, or humors: red blood, black bile, yellow bile, and white phlegm. Each resembled fire, earth, water, or wind—the four elements responsible for creating everything in the universe. An imbalance in these fluids, or dyscrasia, was believed to be responsible for all the diseases known to humankind.

Virchow laughed at these ideas. Life, he said, was nothing but a sum of phenomena at the mercy of ordinary physical and chemical laws. Scientific medicine based on a mechanistic approach was the new way forward, and research should include clinical observation, animal experimentation, and autopsy. His views were radical for the time. No one had publicly challenged age-old doctrines so aggressively. The audience was angry and shaken. Discordant chatter filled the room. "Did you hear it? We do not know anything anymore," an elderly physician said.

Despite the heated reactions to his speech, Rudolf Virchow would go on to become world-famous in his time. And with his steadfast belief that close observation of nature was the fundamental means of arriving at scientific truths, he laid the foundation for our modern understanding of inflammation. The story of hidden inflammation begins with him.

At the Charité, the largest hospital in Berlin, Virchow made rounds twice daily on suffering patients. He changed bandages, administered laxatives, and wrote prescriptions. He was also tasked with balancing the four humors according to salient practices of the day, including cupping and bloodletting—therapies he did not believe in. He tied patients' arms to make their veins swell before slicing through them, letting supposedly toxic blood flow out of the body, a process called "breathing the vein." Or he applied leeches to the skin. The segmented worms with suckers at both ends and hundreds of teeth in three sets of jaws would latch onto tissue and drink a volume of blood up to ten times their body weight.

But in his research laboratory, there was no room for the abstract and irrational. Troubled by a world in which much was yet unseen, Virchow turned to methodical scientific experimentation. A year before his speech, in 1844, he was asked to verify claims made by leading pathologists that phlebitis, or inflammation of the veins, was responsible for most diseases. This popular theory made sense at the time. Cadavers routinely filled morgues after common surgical operations—or even childbirth—went awry, and the average autopsy uncovered abscess-ridden bodies and veins filled with blood clots. Virchow, who had been interested in inflammation since his student days, welcomed the challenge.

Toiling in his research laboratory, Virchow coined the term *thrombus* to describe blood clots, delving into the mechanisms behind their development. He refused to accept, as many had, that blood clots were simply a "vital substance," the "stuff of life." And he showed that phlebitis did indeed exist. But this type of inflammation was not the cause of all diseases, as pathologists assumed. In most cases the inflammation was reactionary; it was the body's *response* to the formation of a blood clot.

Nearly five thousand years ago, writings on Egyptian papyri pointed to heat and redness as markers of disease. In 25 CE, the Roman Aulus Cornelius Celsus described an illness he could see with the naked eye: "Now the signs of an inflammation are four: redness and swelling with heat and pain." He recorded his findings in his medical treatise *De Medicina*, naming the remedies for inflammation as "rest, abstinence, a belt of sulphurated wool and the wormwood draught upon an empty stomach," as well as poultices or ointments to "repress and soothe."

A century and a half later, the Greek physician Galen believed

that a noxious buildup of one of the four humors in the body created the four cardinal symptoms of inflammation that Celsus first
described: *rubor* (redness), *tumor* (swelling), *calor* (heat), and *dolor*
(pain). This idea lingered well into the nineteenth century. Although
the ancients first saw, felt, and recorded the expression of one of the
oldest diseases known to humankind—if not the oldest—they could
not grasp its dimensions through the limits of the human senses.
Beneath heat and color and deformed flesh lay a fragmented, shape-
shifting wilderness.

At the time, the idea that the natural world consisted of tiny components invisible to the human eye was unthinkable for most people,
masking the true nature of inflammation for centuries. But by the
early seventeenth century, Dutch spectacle makers started experimenting with magnifying glasses. They put several glasses in a tube
and noticed something strange: the object at the end of the tube was
larger than what any single glass could magnify. Scientist Antony
Van Leeuwenhoek developed new methods for making and using a
real microscope, a handheld device with a single lens made of a small
glass ball ground and polished to yield magnifications of around
270×, the finest at the time.

The microscope was a breakthrough for understanding inflammation. For the first time, scientists could see minute changes in
blood vessels and blood flow around inflamed tissues. The invention
of the achromatic lens in the 1750s, which allowed even greater mag-
nification levels, would set Virchow on his life's mission to understand disease on a cellular level.

By the time the revolution of 1848 broke out in the streets of Berlin, disrupting long years of stability in central Europe, Virchow
was an established physician and gave frequent, well-attended lectures. Loyal to the lower classes from which he had risen, he fought
for improved medical care for the poor, secular nursing schools, and
physician autonomy. "If medicine is to fulfill its vast duty, it must
influence political and social life," he wrote. But after the uprising
was squashed by a conservative aristocracy, many liberals were
forced into exile to escape political persecution. Virchow was dismissed from the Charité and faced the possibility that there would
be no place left in Germany for him to work. Then a letter came

from the small Bavarian University of Würzburg, in south Germany, offering him the country's first chair in pathological anatomy. Virchow packed his bags, leaving behind the loud voices and bitter disillusionment of Berlin.

In Würzburg, Virchow was happy with his new wife Rose, who understood him "better than anybody else." He spent seven of the most creative years of his life there, enjoying clinical work, research, and teaching. He pushed his students to "learn to see microscopically." In lectures, he would pass around a small table-railroad, a special contraption that held microscopes. During these years he came up with his still-most famous notion: *omnis cellula e cellula*, or "all cells arise only from preexisting cells." Cell theory was initially developed by scientists Theodor Schwann and Matthias Schleiden in 1839. Using a microscope, they identified cells to be the most basic units of life. Virchow completed classical cell theory with his explanation of cell reproduction.

In 1858, a year before Charles Darwin published *Origin of Species*, Virchow published his landmark *Cellular Pathology*, which swept away speculative theories of disease and laid the foundation for modern medicine. He described a new way of understanding inflammation, shifting the focus from vessels visible upon autopsy to the microscopic world of cells, a radical departure from medicine's past.

Virchow's studies were the first to look at cellular changes underlying the tissue damage that accompanies inflammation, explaining the four cardinal signs apparent to the naked senses—redness, swelling, heat, and pain—based on microscopic observations. Redness and heat are due to increased blood flow, while swelling is connected to what scientists call exudation: the walls of inflamed blood vessels become more permeable, and inflammatory cells, protein, and fluid ooze out into injured tissues, working to heal the injury. Pain follows. He saw white blood cells clustered at sites of inflammation and guessed, correctly, that they play an important role, but the details were a mystery.

To the four historic marks of inflammation, Virchow added a fifth cardinal sign, *functio laesa*, or "loss of function." "Nobody would expect a muscle which is inflamed to perform its functions normally," he wrote. "Nobody would expect an inflamed gland-cell to secrete normally." He reasoned that "changes must have occurred in the composition of the cellular elements, altering their natural functional power." He also stressed the importance of an inflammatory trigger, writing that "we cannot imagine inflammation to take place without an irritating stimulus." Nearly two centuries later, when scientists' understanding of inflammation would become deeply nuanced and sophisticated, the hunt for irritating stimuli would hold the key to preventing and treating inflammation.

Virchow was one of the first scientists to truly *see* inflammation, characterizing it in ways previously unimaginable. He wrestled with the definition of inflammation for the duration of his life—and contained the beast with characteristic intellectual humility, a harbinger of the trials to be faced by generations of physicians.

"Inflammation is thus an active and passive process," he wrote, ". . . not a particular process, unique and uniform, but a group of processes that become particular only through their special arrangement in time and space . . . those states of irritation which we witness in the course of the severer forms of disease—the really *inflammatory* kinds of irritation—never in any case admit of a simple explanation."

Virchow was eventually called back to Berlin to succeed his aging teacher, Johannes Müller. On his own terms, he returned to a city that had driven him out only a few years before, asking for the creation of a special pathological institute at the university. From that building, near the Charité, the poor boy from Prussia would train generations of physicians.

Virchow's medical research would amount to more than two thousand books and papers, a phenomenal level of productivity that would help bring about Berlin's golden age in science, transforming a coarse intellectual frontier into a dynamic metropolis. His insights came at a time when microscopes were rudimentary, when fixing and staining methods and tissue cultures did not exist, when only the skeletons of present-day chemistry and physiology were at his disposal. Yet his name is only mentioned briefly in medical schools today. Students memorize "Virchow's node," a large lymph node near the left clavicle bone, one of the earliest signs of gastrointestinal cancer, and "Virchow's triad,"*three factors thought to contribute to blood clot formation. The Nazi government tarnished Virchow's reputation after his death, feeling he had been a dangerous man due to a liberal political bent that included championing racial equality and health care for the impoverished. Most of his personal archives were destroyed, and what remained were held in Soviet Germany, out of reach of Western scholars. Moreover, Germany's unpopularity after World War II left Virchow's name besmirched.

Virchow's story, however, is essential to inflammation's historical emergence and to its resurrection in modern science. His early research had identified inflammation as a consequence of disease. He would soon hypothesize, however, that it was also a root cause, one involved in chronic conditions like heart disease and cancer—ideas ignored in his time but revived by modern scientists. His work would inspire many individuals, including an impassioned zoologist, setting the stage for a broad portrait of the immune system and its inflammatory response—in all its intricacies—to fully emerge.

Immunology, the branch of medicine that deals with the immune system, was a burgeoning and still unnamed field in Virchow's day. (The scientific journal *Cellular Immunology* would first appear in 1970.) Immunology, at its core, is an unveiling of inflammation's inner machinery and a look at the interplay between the complex forces that hold its reins tight or slack. Inflammation, which had captivated Virchow for a lifetime, is the immune system's imprint, its concrete impression on our bodies and on the germs and other matter with which we interact. The immune system gives rise to diverse types of inflammation—acute or chronic, overt or obscure. These may be classified according to the immune system's intent. For example, immunity, an essential function of the immune system, is our ability to defend against harmful germs and other intruders.

---

* Virchow's triad describes the presence of blood vessel damage, the slowing or disruption of blood flow, and a hypercoagulable state as central to the formation of blood clots.

A pathogen or poison that makes its way into the body may cause inflammation as the immune system attempts to rid the body of the interloper. Autoimmunity, on the other hand, occurs when the immune system employs inflammation against our own body, as with the inflamed joints of an arthritis patient.

By the mid-1870s, French chemist Louis Pasteur and physician Robert Koch established germ theory, identifying tiny organisms that could invade the human body and cause disease. Future concepts of immunity would hinge on the acceptance of germ theory: that infectious diseases were specific and reproducible, each caused by a unique germ. The medieval notions of ill humors and miasmas slowly gave way to this recognition.* Pasteur went on to create the world's first successful anthrax vaccine, weakening anthrax bacteria with heat and using them to protect patients against the disease. But he did not know *why* the vaccine worked. The idea of immunity—a primeval force with benevolent intent—had not yet emerged.

Immune cells, which Virchow first glimpsed clustered at sites of inflammation, are born in the marrow of the bone and mature in the thymus, a butterfly-shaped organ sitting behind the breastbone that is the stuff of sweetbreads. They congregate in immune organs like the spleen, lymph nodes, tonsils, and linings of body cavities like the bowels and lungs, where they are constantly conversing with foreigners. Our understanding of the immune system, as Eula Biss writes, "is remarkably dependent upon metaphor, even at the most technical level. Immunologists describe the activities of our cells with terms like interpretation and communication, imbuing them with essentially human characteristics." Immune cells, like characters in literary fiction, are complex and nuanced, transforming with the passage of time, developing new names and personalities as science wrests nature's secrets. Macrophages are one of the key white blood cells in inflammation, playing a part not only in immunity but also in a variety of modern chronic diseases. In the story of

---

* The miasma theory held that poisonous air from decomposing matter was responsible for a variety of diseases.

their discovery lies the origins of what contemporary science is just beginning to reveal.

———————

"We are going to the circus. You must come!" the children pleaded. It was the winter of 1882, and Russian zoologist Elie Metchnikoff had rented a seaside cottage near the port of Messina, a harbor city situated at the base of the jagged Sicilian hills. Metchnikoff loved the five children, who were his wife's orphaned younger siblings. He had taken them in after the deaths of Olga's parents. But he declined the invitation to see the circus apes. He had come to Italy to focus on smaller, and to him more interesting, creatures.

From the cottage he had a broad view of the cerulean waters of the Messina strait. If he made his way past the dirty quay, lined with dilapidated buildings, fishermen would provide him with the marine animals he coveted for his research. That morning, Metchnikoff's family left him alone in a small living room furnished with a desk and a microscope, surrounded by flasks full of fresh seawater. His unruly dark hair fell into bespectacled eyes as he hunched over the desk. He was only thirty-seven years old.

Through the lens of his microscope, Metchnikoff examined the starfish larva *Bipinnaria asterigera*. These minuscule, mesmerizing baby starfish, covered with thousands of hairy bands on their sinuous edges, were eerily spectral, as transparent as the water they came from. Watching them was like peering into a glass house. When Metchnikoff added a few drops of carmine dye to the slides, some "wandering cells" in the starfish gobbled up the dye and turned deep red. He had seen these types of cells in other spineless, primordial animals, such as worms, sea anemones, and bottom-dwelling sponges that had inhabited the seas for nearly a billion years. He had seen them in medusae, comb jellies, and siphonophores.

Metchnikoff thought the wandering cells were nature's earliest digestive mechanism. But why, he wondered, did these cells wander about in organisms like the starfish larvae, which already secreted digestive juices to break down food? And if they ate bits of dye, what else would they eat? Did they have a greater purpose? He was so excited he started pacing up and down in the room. He left the cottage and ran to the seashore, walking along the water's edge. The cells devoured not only food but also waste. They were scavengers, he thought. They got rid of unwanted things. Maybe even *harmful* things. It could be, he mused, that the special cells protected the organism against intruders, forming a rudimentary type of self-defense. This simple idea was so overwhelming in its implications that it took his breath away.

He raced back home, gait heavy and beard unkempt, making his way through cobblestone alleys and drawing stares from curious Sicilian housewives. The garden in the back of his cottage held a tangerine tree that had been decorated for Christmas. He plucked some rose thorns from the tree and took them to the living room, where he inserted the thorns into the skin of the starfish larvae. Metchnikoff was restless that night as he awaited the results of his experiment.

The next morning, when he looked through his microscope, a strange sight raged: a multitude of wandering cells within the larvae had encircled the rose thorns from every side, walling them off, preventing them from affecting larval functions. He repeated the experiment using other materials—goat milk, cooked peas, sea urchin eggs, and even drops of human blood—and each time the wandering cells ingested or fenced out the foreign objects. He then examined water fleas (*Daphnia*) that were infested with fungus. In these animals, which were just as transparent as the starfish larvae, he watched wandering cells ingest needle-shaped fungus spores. This was no mere digestion, Metchnikoff thought, but rather a deadly struggle between species occurring on a tiny scale, a battlefield teeming with soldiers delivering concerted, deliberate attacks.

That spring, in March of 1883, the great pathologist Rudolf Virchow came to Messina to watch Mount Etna erupt. Virchow was an admirer of the German literary luminary Johann Wolfgang von Goethe, who had scaled the mountain in the late eighteenth century. Metchnikoff met Virchow through a mutual acquaintance, a Messina university professor. As a boy, he had idolized Virchow. Cell theory had inspired him, sparking a desire to create his own grand theory in medicine. At the time of their chance meeting in Messina, cell theory was only about twenty years old.

The wandering cells in his starfish larvae, Metchnikoff explained to Virchow, were akin to white blood cells in human beings. A zoologist at heart, Metchnikoff reasoned that if his ideas about the wandering cells were correct, they would manifest themselves throughout the phylogenetic ladder of the animal kingdom. Thrusting a rose thorn into a starfish was like shoving a splinter into a human finger, creating redness, heat, swelling, pain, and loss of function, those iconic features of inflammation. White blood cells, miniature warriors that accrued at sites of inflammation and made up yellowish pus, rushed to defend the body from invaders, be it a foreign object or a microbe.

Virchow had glimpsed white blood cells teeming at sites of inflammation, but he had not understood their function. "We pathologists think and teach the exact opposite," he told Metchnikoff. "Microbes are well off *within* the white blood cells and use them as means of transport to disseminate through the host." Most scientists of the time agreed with this notion. Inflammation was harmful, a force to be fought and suppressed (medical dictionaries of the time described inflammation as "morbid"). Metchnikoff was arguing the opposite, that the core force behind inflammation was beneficial. The idea was wild and subversive, reminiscent of the old "inner curative force" from Hippocratic times, the vitalistic belief that the body had magical powers and could purge itself of disease. Virchow was impressed. Here was a novel theory of immunological defense that held cells central to the process. Despite a seemingly outlandish veneer, it was fashioned by a true scientist who could provide hard data for his ideas.

The world's first modern theory of immunity drew together disparate intellectual streams, including pathologists' cellular basis of inflammation and microbiologists' germ theory. It emerged from evolutionary underpinnings, as befitted a zoologist inching his way into medicine. Metchnikoff was eighteen when he read Darwin's *Origin of Species*, which had left him in awe. Darwin's ideas on natural selection, like Virchow's cell theory, pushed an understanding of organisms away from the four humors, or a balance of nebulous forces existing in perfect harmony to promote health. Rather, human beings were the products of imperfect structures and functions slowly molded by evolutionary forces and biological needs. Perfection was an elusive goal shaped by earthy processes playing out in competitive, hostile environments, forcing adaptation for survival.

Metchnikoff's notion of immunity responded to this new understanding of how an organism is defined by evolutionary challenges. In higher animals with complex digestive tracts, he correctly posited that the wandering cells armed themselves against new threats, taking on roles beyond their vestigial digestive capabilities. He named the cells *phagocytes*, from the Greek *phago*, "to devour," and *cytos*, "cell," calling the process *phagocytosis*. Later, in a study published in a journal established by Virchow and now known as *Virchows archiv*, he separated the phagocytes into two classes. He named the larger phagocytes *macrophages*, or "big eaters," and the smaller ones *microphages*, or "small eaters," known today as neutrophils.

Both types of phagocytes are white blood cells that move swiftly to sites of infection or injury, controlling damage and congealing into crusty pus as they die off. Of the two, neutrophils are short-lived and rush in more quickly, dominating in sites of acute inflammation. "The larger and less mobile macrophages play an important role in devouring weakened or dead elements," Metchnikoff wrote, noting that these cells not only fight infection but also maintain tissue. Studying the metamorphosis of frogs, he noted that macrophages digest the muscles in a tadpole's tail, getting rid of the tail and other useless larval organs as the animal matures.

On a warm summer day in August of 1883, Metchnikoff gave a historic talk in a large assembly hall at Novorossiya University, situated in his native Ukraine. The talk, "On the Curative Forces of the Organism," was the first time he elaborated on his theory of immunity in a public forum. He was nervous at first as he stood before the packed audience, but he became increasingly animated, gesturing with his hands. "Protection against disease is one of the most crucial issues to have ever preoccupied humanity," he noted, "so it is only natural that great attention should have been devoted to it from the most remote times." He pointed out that germs had been invading lower organisms like plants and insects for much longer than they had humans. How did these organisms protect themselves? "Whether the bacteria penetrate [us] through the lung vesicles, the wall of the digestive tract, or a wound in the skin," he said, "everywhere they risk being captured by mobile cells, capable of consuming and destroying them." He went on to say that human beings possessed "an entire system of the organs of curative digestion," naming key players in immunity, including the spleen, lymph glands, and bone marrow. Extraordinary cells within these organs, he said, fought germs. Metchnikoff's words, uttered in the late nineteenth century, were arrestingly prescient. With his razor intuition, stemming from only a few early experiments, he had come up with the concept of an immune system.

Frustrated that his efforts to conduct research in Russia were being obstructed by political and other barriers, he turned down an offer to head a laboratory in St. Petersburg and instead sought refuge abroad. He dreamed of a "peaceful little university town" in which to work. Metchnikoff looked to Germany first, that scientific superpower he had admired since he was a child, but was badly snubbed when he visited microbiologist Robert Koch. Satisfied with studying the germs that caused disease, Koch was not curious about how organisms responded to an invasion. He believed that bacteria, rather than being attacked and devoured by phagocytes, used them as incubators in which they could multiply,* and brushed off Metchnikoff's ideas.

Metchnikoff had better luck in Paris, where he met with another scientist he had long admired. "I saw a frail elderly man of a short stature, the left side of his body semi paralyzed, with penetrating gray eyes and a gray mustache and beard," he recalled around thirty years later of his first encounter with Louis Pasteur, who had by then suffered a stroke. "He received me very kindly, and immediately spoke to me of the question which interested me most, the struggle of the organism against microbes." A swift friendship developed between the two men.

Both were foreigners in the world of medicine—a zoologist and a chemist—and polymaths, whose varied interests proved fertile ground for cross-pollination, bringing radical ideas into the field. The

---

* Pathogens, including bacteria and viruses, can indeed replicate inside macrophages and other cells.

Franco-Prussian War had ended in 1871, but in its aftermath a bitter duel ensued between the two dominant schools of microbiology—one headed by Koch in Germany and the other by Pasteur in France. Pasteur was happy to support a young scientist whose ideas were being assailed in Germany and had spoken of Metchnikoff's phagocyte theory as "most original and so creative," telling him, "I at once placed myself on your side, for I have for many years been struck by the struggle between the diverse microorganisms which I have had occasion to observe. I believe you are on the right road."

Though he dreaded living in a large and noisy city, Metchnikoff was taken by Pasteur's warm welcome and generosity. The older scientist offered to make him the head of a laboratory in the newly constructed Pasteur Institute on Rue Dutot in the outskirts of Paris, which had opened in 1888 to continue Pasteur's successful research on germs and vaccines. Metchnikoff did not know that he would one day become the institute's most celebrated researcher or that he would, quite literally, never leave the place. (By his request, his ashes would be kept in a marble urn on a bookshelf in the library.)

Paris of the late nineteenth century was alight with the *joie de vivre* spirit of the Belle Epoque period and yet untouched by two impending world wars. When Metchnikoff arrived, it was one of the first cities to decorate its streets with electric lights, and people sought glimpses of the towering steel edifice recently unveiled by French engineer Gustav Eiffel—the world's tallest building, the product of a brief but golden equilibrium in Parisian history. The city was a cultural mecca, doused in physical beauty and replete with endless sources of entertainment, igniting creative pursuits. Guy de Maupassant feverishly wrote short stories, entering the most fertile period of his life, and Émile Zola penned *La Bete Humaine*. Cafes multiplied, and the Moulin Rouge first opened its doors to the public. The *flaneurs*, fashionably dressed aesthetes who roamed the streets in leisure, drank in their surroundings with a languid ease. Metchnikoff, on the other hand, in his worn suit, barely paid attention to the temptations of the city as he rushed to and from the laboratory, wholly consumed by his work.

As he spent the next quarter century defending his theories, Metchnikoff repeatedly came under attack. A famous French scientist called his theory of immunity "an oriental fairy tale." German pathologist Paul Baumgarten pointed out that most patients with relapsing fever recovered even though microbes floated freely in their blood, untouched by the phagocytes that were able to—as Metchnikoff claimed—combat germs. Metchnikoff responded by showing that phagocytosis in these patients *was* indeed occurring, but in the spleen, not in the blood.

Then came a new study that seemed to deliver a devastating blow to the phagocyte theory. In 1890, the German physician Emil von Behring, who joined the Koch Institute in Berlin, and his Japanese colleague Shibasaburo Kitasato, announced that serum—the amber-colored, cell-free portion of blood—held the key to an animal's self-defense against infections. They immunized rabbits to tetanus by injecting into them weakened forms of the tetanus bacteria, similar to what Pasteur had done during his vaccination procedures. But Behring and Kitasato went further. They transferred serum from tetanus-resistant rabbits into mice, after which they gave the mice around three hundred times the lethal dose of tetanus poison. Shockingly, the mice who had gotten the rabbit serum did not fall ill and simply scampered around their cages. The mice already infected with tetanus, suffering with painful body spasms, should have been dead in hours. But after receiving the rabbit serum, they recovered entirely. Behring repeated the experiments with diphtheria, and the results were the same. "Blood is a very special juice," he wrote, concluding the paper with a line from Goethe's *Faust*.

Serum therapy hinted at a new explanation for the immunity that followed vaccination—one based on some sort of protective substance in the blood. And unlike vaccination, serum therapy could not only prevent but also cure disease. Behring referred to Metchnikoff's phagocyte theory as a "metaphysical speculation," one that "relied on the mysterious powers of the living cell."

Ultimately, despite the initial excitement and promise, serum therapy would prove to be ineffective against most other diseases. But Behring and Kitasato, unbeknownst to them, had discovered the modern-day antibody molecules, then called *antitoxins*. In doing so, they paved the way for a rival theory of immunity, one that championed curative body fluids, namely blood. It was referred to

as the *humoral* theory, given the ancient notion of the body's healing vital humors. The humoral theory would help to paint a more
exhaustive portrait of the immune system and its inflammatory
response.

The two immunity factions were split by location, along political
lines: Metchnikoff's wandering cells in France and Behring's serum
therapy in Germany. At first, the tide was strongly in favor of the
humoral theory. In 1901, Behring would be awarded the first Nobel
Prize in Physiology or Medicine for his work on serum therapy. The
battle that ensued between the immunity camps was vicious, with
a "degree of vilification unknown almost in present-day science,"
as Joseph Lister observed. Pathologists and microbiologists all over
Europe wrote Metchnikoff scathing letters in scientific journals.
"Were I small as a snail, I would hide myself in my shell," he would
lament after his papers were rejected.

Metchnikoff was a Russian, an outsider, but when he joined
Pasteur's group, he became enmeshed in a brawl between nations.
Reverberations from the Franco-Prussian War manifested not
on a blood-singed battleground but in echoes that hummed on
in scientific journals. It was no accident that Metchnikoff's most
vocal critics hailed from Prussia or that the immunity camps were
divided along the same lines as those that had fractured the microbiologists. In his book *Microbe Hunters*, Paul de Kruif suggests that
the immunology war may even have contributed to the start of
World War I.[†]

The drama took a toll on Metchnikoff's sleep and mental peace,
but he responded by working harder than ever, running fresh

---

* The historical "cellular" and "humoral" theories of immunology refer to early work
on phagocytes and antibodies, respectively. Today, the adaptive immune system
is typically divided into two branches: cell-mediated and humoral. In this context,
the cell-mediated branch includes T cells, while the humoral branch includes serum
antibodies.

† Microbiologist Paul de Kruif suggests in his book *Microbe Hunters* that scientific
struggles in immunology may have contributed to the start of World War I, but this
perhaps goes too far, as science historian Arthur Silverstein notes. Silverstein writes
that "it does seem probable that, in a minor way at least, it [the immunology struggles]
did represent one of the continuing reverberations of the Franco-Prussian War of 1870."

experiments. His assistants helped him brew all kinds of microbes in incubator ovens, scouring farms and woodlands for a variety of animals—frogs, salamanders, beetles, scorpions, flies, lizards—to infect. He studied phagocytes' roles in a variety of bacterial infections, including anthrax, erysipelas, typhus, and tuberculosis. He was armed with the support of the prestigious Pasteur Institute, where Louis Pasteur stood staunchly by his side, a liaison Metchnikoff would forever be deeply grateful for. He would run to Pasteur's deathbed in the autumn of 1895 with a deadened heart.

Remembering the simple life-forms that sparked his ideas would help keep Metchnikoff afloat. "When this theory was attacked from all sides and I asked myself if I hadn't, after all, gotten on the wrong track, all I had to do was to recall the fungal disease of the *Daphnia* in order to feel I was on firm ground," he wrote many years later.

As he struggled to keep his phagocytes from being devoured by the scientific community, Metchnikoff began to dissect their role in the body through the lens of an ancient phenomenon that had held fast in his mind since Messina days. While most pathologists and other scientists continued to argue that inflammation was wholly noxious, a threat to the organism, Metchnikoff loudly maintained that inflammation was meant to be helpful, creating a new notion of selfhood. Beneath it, he said, lay the heated force of immunity, the idea that the body could wage war to defend itself against intruders. He studied the evolution of the inflammatory process along a biological hierarchy, from single-celled beings to humans. In all organisms, phagocytes—macrophages and microphages— digested foreign matter, actively participating in inflammation. In more complex animals, blood vessels served as conduits through which phagocytes and other white blood cells could rush to embattled areas.

But evolution, he argued, had created an imperfect weapon, one that was molded over time in response to environmental challenges. Speed, not precision, was essential as ancient forms of life sought to ward off deadly attacks. Thus, inflammation, the healing force, was prone to causing collateral damage, with less adept control mechanisms than trigger functions. "The curative force of nature, the most important element of which is the inflammatory reaction, is not yet perfectly adapted to its object," Metchnikoff noted in an 1891 lecture at the Pasteur Institute.

While Metchnikoff remained enthralled with macrophages, a "brilliant eccentric" with a penchant for Havana cigars had been laboring to strengthen the humoral theory of immunity at a rival camp in Germany. Paul Ehrlich, who worked in Robert Koch's laboratory, did not have much interest in the arts, poetry, or popular music but consumed all kinds of medical journals and—for relaxation—Sherlock Holmes detective novels. As a young doctor, his love of chemistry had led him to experiment with myriad colorful cloth dyes that poured into his laboratory from the burgeoning German dye industry. His colleagues laughed at him as he walked around with rainbow-colored hands and occasionally a smudged face. On a whim, Ehrlich decided to use these dyes to stain animal cells, hoping they would make microscopy easier.

To his shock, they did much more. Some dyes were attracted to only certain cells or parts of cells and left other elements untouched, neatly highlighting structures and making them pop out from the background like lights on a Christmas tree. Ehrlich used his skills to create popular methods for staining various germs and cells, eventually discovering a way to visualize tuberculosis germs better than even Koch had. He was also able to identify different types of white blood cells—including lymphocytes, basophils, eosinophils, and neutrophils, some named according to their ability to take up basic, acidic, or neutral dyes—and helped found the field of hematology.

Ehrlich became obsessed with the core notion behind the staining experiments. The dyes were choosy, hunting for a specific match before they clung onto molecules, like keys fitting locks. If the biological world was filled with such precise relationships, he thought, could this be extrapolated to explain the mechanics of immunity? The idea was monumental. Ehrlich came up with a theory to explain *why* blood was such a special juice in Behring's experiments. He introduced the term *antikorper*, or "antibody." Antibodies, he said, were blood proteins that were produced by cells. They could target specific germs, toxins, or other foreign substances thanks to a lock-and-key fit. The secret was a tiny latch that allowed them to bind to these substances and disarm them, impeding their ability to survive in the body. In Ehrlich's vision, antibodies were branched, with multiple "receptors," or sites for binding foreign material. Technological limitations of the time prevented visualization of these interactions, but Ehrlich used his imagination to illustrate his ideas with vivid drawings that convinced a generation of scientists that they could actually "see" the Y-shaped antibody molecule in action. He compared his antibody receptors to the "grasping arms" of a sundew's tentacles, drawing pictures on whatever material was handy: the doors and walls of his office, a listener's cuff, and even formal dinner tablecloths.[*]

The discovery of the antibody tipped the scales in favor of the humoral theory of immunity, steering immunology away from Metchnikoff's wandering cells and toward chemistry, the interaction between molecules—the minuscule components that make up cells. Ehrlich introduced rigorous quantitation into immunity research, running clinical trials on diphtheria toxin and anti-diphtheria antibodies in order to standardize doses used in treatment. His work showed that the antibody was more than an abstract image: it was an entity that one could measure and manipulate in a test tube. For most immunologists, this made antibodies preferable to the elusive phagocyte. Suddenly, scientists young and old were working on antibodies. Over the course of the 1890s, new antibodies against different microorganisms were reported regularly, and Ehrlich became the new face of humoral immunity, which seemed to be winning a long-standing war.

Metchnikoff agonized over the fate of his macrophages, the

---

[*] Ehrlich had initially labeled antibody receptors "side chains." Although many of the particulars of the side-chain theory were eventually proven incorrect, the theory still had a significant influence on future generations of scientists. We now know that certain types of B cells are the antibody-producing cells of the immune system. Each B cell makes an antibody with a unique tip, which sticks to an antigen. The shape of each antibody's tip is made by random rearrangement of the genes that create the antibody. B cells that make an antibody that might cling to healthy cells are killed off or inactivated.

eater-cells that were being smashed into oblivion despite all of his efforts to give them a secure position in immunology. In 1896, commenting on the epic struggle between Metchnikoff's and Ehrlich's opposing schools of thought, the famed surgeon Joseph Lister remarked: "If there was ever a romantic chapter in pathology, it was surely that concerned with the theory of immunity." Metchnikoff himself was a romantic figure, akin to a Dostoyevsky character, sensitive and pessimistic and prone to depression. He reflected, perhaps, the greater pessimism that developed during the nineteenth century, a sentiment largely prompted by the dread of frequent disease and death against which humanity was mostly powerless.

But in 1908, in a surprisingly prescient move, a Stockholm committee awarded both Metchnikoff and Ehrlich the Nobel Prize in Physiology or Medicine, an honor they were to share "in recognition of their work in immunity." The Nobel Committee's reluctance to definitively side with one immunity camp over the other was a prudent decision that foreshadowed new developments in years to come. It was a move spurred in part by a modest renewal of interest in macrophages after British microbiologist Almroth Wright, drawing from both cellular and humoral theories of immunity, showed that certain blood proteins called opsonins bind to foreign substances and make them easier for macrophages to ingest. (Wright's theory was so popular that it was depicted in his close friend George Bernard Shaw's *The Doctor's Dilemma*.) Had the warring factions stopped to heed each other, they may have come to the correct conclusion that cellular and humoral immunity were two sides of the same coin—working together as their respective scientists failed to do.

———

The essential elements of the immune system and its inflammatory response first proposed by Metchnikoff and Ehrlich, molded by rich imagination and rudimentary experimentation, still hold fast today. We now know that the immune system is divided into two major branches: innate and adaptive. An inflammatory response can involve either branch—or both. The innate immune system* is our first defense against foreign threats. Its initial barriers are both physical and chemical, including layered sheaths of skin and body orifices opening into hollow tubes—like the airways, intestines, and genitalia— that are lined with sticky, protective liquids. The innate immune system also includes the hair on our bodies, brows, and nares and even the fine lashes on our eyelids. It defends with bodily secretions such as mucus, bile, and acid. Or spit, sweat, and tears. It is made up of our most ancient immune mechanisms, the ones we share with our primordial ancestors, and it is the main force behind acute inflammation.

Acute inflammation comes and leaves quickly, typically in a few days, fighting intruders while minimizing damage to healthy tissue. Phagocytes, including neutrophils and macrophages, rush to the site of tissue damage, ingesting germs or damaged cells and other foreign materials. Other types of white blood cells, like basophils and eosinophils, may join the fray. The four cardinal signs of inflammation noted by Celsus—redness, heat, swelling, and pain— usually accompany acute inflammation. In injured tissues, blood vessels widen and blood flow surges, causing redness and heat. The walls of inflamed vessels become more porous, allowing inflammatory cells, protein, and fluid to leak into tissues, creating swelling and putting painful pressure on nerve endings. Endothelial cells, which line blood vessels, become injured. Clotting systems are activated, making blood stickier by rushing platelets—small, colorless fragments that congregate in clumps—and other special substances into the area.

The adaptive immune system, which involves defenses that only exist in vertebrates, is more complex and slower to react, but also more targeted. The key players are lymphocytes—small, round white blood cells that can be divided into B cells and T cells.† B cells express antibodies on their cell surfaces that can bind to specific antigens,

---

* Innate immune cells include macrophages, neutrophils, dendritic cells, eosinophils, basophils, mast cells, and natural killer cells.

† Lymphocytes include not only the B and T cells of the adaptive immune system but also natural killer cells, which function as lymphocytes of the innate immune system.

molecules that are capable of stimulating an immune response. Antigens may be present on all kinds of foreign material, including germs, toxins, food components, tissues from other individuals (like transplanted organs), and even cancer cells. T cells take on various personalities, including "helper" T cells, which assist in activating other immune cells, and "killer" T cells, which focus on removing pathogens.

In many ways, the distinction between innate and adaptive branches is artificial—the two are inextricably linked. A dangerous germ, if it makes it into the body, will first face innate immune defenses: neutrophils, macrophages, and other cells rushing to contain the damage. But eventually, these innate immune cells might call on help from the more sophisticated adaptive arm. They scramble toward a nearby lymph node or the spleen, where they present fragments of the germ to other immune cells, prompting some of these cells to travel to the infected area. Lymph vessels, holding clear yellow fluid that runs alongside blood vessels throughout the body, carry the immune cells to the lymph nodes. Over time, B cells secrete specific antibodies into the blood that aim to neutralize the germ. These antibodies will stick around and recognize the germ if it ever reenters the body, a remarkable biological memory that largely explains the success of vaccinations.

If germs and foreign substances linger, if wounds fail to heal, or if autoimmunity and allergic reactions persist, inflammation can become chronic, lasting for months or even years, destroying tissues. While neutrophils and—to a lesser extent—macrophages are found in acute inflammation, macrophages and lymphocytes predominate in chronic inflammation.

Macrophages, like all blood cells, are made in the bone marrow, the spongy fat in the hollow of bones. They don unique names and features, depending on their location in the body: immature monocytes in the bone marrow and blood awaiting the call to tissues and organs; Kupffer cells in the liver and dust cells in the lungs; microglia in the brain and osteoclasts in the bone; Hofbauer cells in the placenta and red pulp cells in the spleen. A tattoo owes its existence to macrophages in the skin, which gobble up the ink.

When these macrophages die, they regurgitate the ink, which is then taken up by new macrophages, preserving the tattoo.

The journey into inflammation's inner machinery exposes a feverish history of the major cells and reactions behind the benevolent force of immunity. It is evident, however, that these same elements may deceive the body, fostering disability and even death. Autoimmunity was once viewed as a horror so unimaginable that scientists dared not consider it feasible in the natural world.

CHAPTER 2

Horror Autotoxicus

W hat Jay liked about Dr. Carter was his precise, laconic speech. No room for false hopes or glib declarations. Carter, a professor of medicine in rheumatology at the University of Chicago, was a renowned expert in inflammatory muscle diseases, with a research laboratory focused on improving treatments for patients with autoimmune conditions. He wore a dark suit and tie, and his white mustache was neatly trimmed. By this point, three months into his illness, Jay had already been evaluated by a neurologist and a rheumatologist, but his diagnosis remained unclear. Carter had never encountered an illness exactly like Jay's. He wondered, as he examined limp muscles behind a clunky brace, if this was yet another face of autoimmunity.

The immune system distinguishes self, the body it serves, from non-self, material that is foreign to it—benign or otherwise—through identification molecules called antigens. These are expressed not only on foreign material, allowing antibodies to bind them, but on the surfaces of all cells—an extension of Ehrlich's theory that came to light in later years. Germs and other non-self substances have foreign identification molecules on their surfaces, helping the immune system to recognize and destroy them. Autoimmune diseases can involve both an adaptive and an innate immune response. In a typical autoimmune disease, however, the immune system reacts against a self antigen, one that is normally present in the body of the host. Acute inflammation, a salutary defense when it tackles situations like infections or wounds, morphs into a chronic, destructive force in the setting of autoimmunity.

While the four cardinal signs of inflammation noted by Celsus— redness, heat, swelling, and pain—usually accompany acute inflammation, Virchow's *functio laesa* is the only universal sign common to most inflammatory processes. In many autoimmune diseases, inflammation is well hidden from the naked eye, but loss of function, whether detected on a physical exam or with the help of medical tests, is a glaring sign of underlying chaos. Autoimmune destruction of the insulin-producing cells in the pancreas causes insulin deficiency and high blood sugars, leading to type 1 diabetes.* In multiple sclerosis, inflammation damages nerves in the brain and spinal cord, causing neurological dysfunction. On the other hand, rheumatoid arthritis, a disease in which the immune system attacks the joints, can produce visible redness, heat, swelling, and pain.

Going over Jay's files, Carter was unable to find a pattern that pointed to any of the common autoimmune muscle diseases. Jay had lost critical muscle function in the back of his neck in what seemed like a sudden hit. More insidious inflammatory reactions against his diaphragm and the muscles of his throat were causing breathing and swallowing problems. Electromyography and nerve conduction studies, which evaluate how well muscle and nerve fibers are working, confirmed that the problem lay within Jay's muscles rather than his nerves. Laboratory testing revealed high levels of creatine kinase, an indicator of muscle injury seen in autoimmune muscle diseases. While strenuous exercise can also increase creatine kinase levels by creating micro-tears in the muscle fibers that the body regenerates as it builds muscle, these elevations are typically mild and

---

* While "diabetes" encompasses both type 1 and type 2 diabetes, the word is used throughout this book to refer predominantly to type 2 diabetes, the most common type of diabetes in the world. Type 1 and type 2 diabetes are discussed in further detail in Chapter 5.

brief—making strenuous exercise alone an unlikely explanation for Jay's condition given the severity and duration of his disability.

Despite the high creatine kinase levels, other clues pointing toward autoimmunity were missing. Jay's blood did not reveal elevated levels of specific autoantibodies, or antibodies that target one's own tissues. A biopsy taken from a muscle near Jay's neck showed that his muscle cells were dead or dying, but inflammation—with its telltale infiltration of immune cells—was minimal, possibly because he had been taking prednisone, a potent anti-inflammatory steroid medication, for quite some time. Still, the case for an autoimmune condition was weak. Perhaps Jay's muscles were simply wasting away due to a genetic disorder, as in patients with myotonic dystrophy, with no possible cure.

But Carter had a hunch that inflammation was at play, despite the equivocal evidence. He knew autoimmunity was a capricious, complex adversary, a perfect storm lit by both predisposing genetic factors and environmental exposures. Autoantibodies or obvious inflammation may not always be present or even caught by available medical tests. Sometimes the trigger is known, as with celiac disease, a serious autoimmune condition in which gluten ingestion results in intestinal damage. Other times it remains a mystery. The hour in the gym with the body angled anew, a walk in the summer heat, an unknown germ making its way across weakened barriers: these incidental scenarios—or dozens of others— could have coalesced to goad Jay's immune system into madness.

What was clear to Carter, as he examined his patient's neck and reviewed imaging scans, was that muscle had been irrevocably lost, enough to cause significant weakness. Worse, the damage was ongoing, since Jay's creatine kinase levels were still high. Carter knew he had to come up with a plan—one drawn from his decades of clinical experience treating complex, amorphous autoimmune diseases—and act fast. An uphill battle loomed. Even if the presumed inflammation could be controlled, there was no telling if Jay would ever regain enough muscle to discard the brace or if and when another episode would manifest. If triggered again, the immune system would attack the weakest spots in the body.

Inflammation would gravitate toward areas of trauma, pathology, or overuse, such as previously injured muscle or other diseased areas—or areas in which the immune system itself was vulnerable or inattentive. And there would be no way to anticipate or defend against its whims. It was the worst kind of enemy.

———————

Inflammation, our natural protection from harm in the context of immunity, extracts a price from us—one that evinces the cruel imperfections of the biological world, a reality brought on by the pressure of evolutionary selection in lieu of an infinitely intelligent designer. Metchnikoff recognized that inflammation, while on the whole a global benefit to the host organism, could also produce tissue damage. His army of macrophages were designed to gobble up invaders or swallow dead cells and debris, taking part in the turnover and regeneration of various tissues. But he also believed that macrophages played a role in the aging process, fostering wrinkled skin, gray hair, and a decline in the brain and other organs, ideas that modern science supports.

Ehrlich, on the other hand, refused to believe that an evolutionary downside existed for antibodies, which targeted foreign substances with a lock-and-key specificity. In 1900, he and his colleague Julius Morgenroth reported that the injection of goats with their own red blood cells failed to produce antibodies. Ehrlich wrestled with the question of how the immune system distinguished self from non-self, attacking invaders while tolerating the body's own constituents. He concluded that the body had an aversion to harming itself, writing:

> The organism possesses certain contrivances by means of which the immunity reaction, so easily produced by all kinds of cells, is prevented from acting against the organism's own elements and so giving rise to autotoxins, so that we might be justified in speaking of a "horror autotoxicus" of the organism.

Horror autotoxicus, Ehrlich's famous dictum, had instant appeal. Why would an immune system that had evolved to protect an organism try to destroy it? The idea of self-harm arising from physiological processes gone awry was not new. In 1887, the French pathologist Charles Bouchard had proposed a theory of "autointoxication," which held that toxic products formed in the intestine after poor digestion could lead to a variety of diseases. Hundreds of papers were written on autointoxication in the years prior to World War I. For example, autointoxication from colonic stasis, or a sluggish colon, was thought to be responsible for a wide range of ailments, from fatigue to seizures, and was treated with surgical removal of the colon. It was in this environment, at the height of interest in autointoxication, that Ehrlich considered autoimmunity.

He did not explicitly deny the existence of autoantibodies. Several scientists, including those at the Pasteur Institute in Paris, had shown that antibodies could be formed against a variety of normal cells in the body. But surprisingly, Ehrlich concluded that autoantibodies, even if present, would not harm the host. "An autotoxin... one that destroys the cells of the organism that formed it, does not exist," he wrote. The idea of an immunological autotoxicity was so horrific that he could not find a place for it in his logical mind.

Ehrlich's theory, especially given the popularity of his antibody-mediated immunity over Metchnikoff's macrophages, was widely influential in the early twentieth century. The notion of horror autotoxicus, which fiercely denied that a biological price might exist for the benefits that antibodies confer, made it challenging for scientists to accept the truth of autoimmunity for over fifty years.

Yet some scientists were willing to speculate on the existence of autoimmune diseases. In 1904, Vienna physicians Karl Landsteiner and Julius Donath studied a rare disease called paroxysmal cold hemoglobinuria, one of the first hematological syndromes, or disorders of the blood, to be recognized clinically. Minutes to hours after being exposed to cold temperatures, affected patients found their urine altered, from clear yellow to dark red or brown. They also often experienced fever, stomach issues, leg pain, and back aches, and the episodes returned with each cold exposure. In paroxysmal cold hemoglobinuria, red blood cells coursing through the vessels are destroyed. Hemoglobin, the red protein in blood cells responsible for transporting oxygen, leaks into the urine. In carefully controlled experiments, Landsteiner and Donath showed that a special autoantibody running through the blood of these patients was the destructive component. When exposed to cold, the antibody attached itself to specific antigens on red blood cells. Later, upon rewarming, the red blood cells would rupture.

Once Landsteiner and Donath published the results of their experiment, even Ehrlich had to concede that paroxysmal cold hemoglobinuria was an exception to the rule, that dysregulated immune cells had struck viciously against their host. It was the first discovery of an autoimmune disease in human beings, a clear depiction of a destructive autoantibody causing serious damage by turning on the body. The finding threw a jagged crack into Ehrlich's grand theory, small at first but ready to be wedged wide open by generations of scientists interested in the darker realities of immunology.

———

Allergic disease—another cost of inflammation—initially suffered similar ideological setbacks as it sought to claw its way through a scientific environment hostile to accepting immune reactivity against the self. An allergic reaction is one in which the immune system attacks substances that are typically harmless for most people, including food, environmental agents (like pollen and dust mites), and drugs, causing inflammation. A variety of inflammatory cells and proteins can be involved, including antibodies. Symptoms may be mild or deadly, ranging from issues like a stuffy nose and watery eyes to breathing problems, low blood pressure, and even death.

Interest in allergic disease grew during the "golden age" of immunological research in the decades before World War I, years during which the foundation was established for many future subspecialties in immunology. But the earliest observations were quickly dismissed. In the 1880s, Robert Koch, who noticed that inoculations of tuberculosis bacteria could cause an inflammatory skin reaction, attributed the finding to an excess of local bacterial toxins rather than to any component of the immune response. Emil von Behring described a "hypersensitivity" to diphtheria toxin in guinea pigs that had been immunized, but decided it was a "paradoxical reaction" solely due to toxins. It was no surprise that the pioneers in this field, those who first ventured to suggest that these reactions might be an essential part of the immune response, did not come from Ehrlich's retinue and the classical tradition of immunology. Paul Portier and Charles Richet, who in 1902 first described anaphylaxis—an acute, potentially life-threatening allergic reaction—were physiologists.

Despite Landsteiner and Donath's research on paroxysmal cold hemoglobinuria, enthusiasm for exploring autoantibodies and autoimmune diseases, the unsavory faces of inflammation, continued to decline after World War I. A focus on the antibody molecule itself, however, persisted. After the war, the field of immunology shifted gears. Its nascent thirty years had been spent under the wings of those interested in biology and medicine in the quest to unlock the mysteries of disease causation and prevention. As investigators tired of hunting for vaccines against elusive pathogens, and as Metchnikoff's theory of phagocytes was eclipsed by antibodies, biologists at the vanguard of immunological research were displaced by chemists. These scientists had a bull's-eye fixation on the antibody molecule, not a wide view of the organism in which it operated. They were more interested in how to manipulate the antibody and what it looked like—its size, shape, and structure—rather than its role in health and disease. And so immunology became a chemical science.

It was only after the Second World War that immunology was oriented back to its traditional biomedical origins, which would help to better elucidate the bodily price of inflammation. A fresh crop of scientists with few loyalties to old teachings tackled new problems. World War II inspired basic science research in many different areas, among them the search for a solution to improve skin and other tissue grafts for burn and wound victims. British biologist Peter Medawar and his colleagues found that the body rejected foreign skin grafts due to some of the same immune responses that underlay protection against germs. He helped to clarify the critical role of immunology in transplantation outcomes, establishing the field of transplant immunology, a subspecialty that would become crucial for the success of organ transplantation in humans.

At the University of Wisconsin, postdoctoral student Ray Owen came across a curious phenomenon while studying blood samples from fraternal cattle twins. They had shared the same circulatory system in utero, and their immune systems did not mount an attack against the antigens on one another's blood cells. The twins had become *chimeric*, each calf with both his own blood cells and those derived from the other twin. The finding excited Australian physician MacFarlane Burnet, who came up with a theoretical explanation known as immunological tolerance—the ability of the immune system to remain inert when confronted with foreign tissues that would normally elicit an adverse response. A few years later, Medawar and his colleagues confirmed Burnet's hypothesis in animal experiments. He and Burnet were awarded the 1960 Nobel Prize in Physiology or Medicine for their work on tissue grafts and the discovery of immunological tolerance.

All of these new concepts led to a radical shift in the direction of immunology, away from chemistry and once again toward biology. The intellectual block restraining many other biological areas, including autoimmunity, broke wide open, unleashing a flurry of activity. Autoimmune thyroid, adrenal, skin, eye, and testicular diseases were discovered, among many others, firmly establishing the potential of the immune system and its inflammatory response to cause disease by turning against the self.

Immunology research flourished in the second half of the twentieth century, but antibodies still took center stage, leaving phagocytes in the shadows. Scientists were enthralled by the questions posed by the adaptive immune system. In a world with infinite possibilities, how were antibodies able to counter the unlimited variety of foreign materials? The mystery ignited their work for decades. The Y-shaped antibody molecule was intensely dissected. Its branches came to be known as light and heavy chains, with each chain composed of folding, compact protein structures called domains. Even when the focus shifted to cells in the 1960s and 1970s, lymphocytes captured the most attention. T and B cells were easy to obtain and work with, and they were a part of nature's most sophisticated immune defense arsenal, an elite army. Phagocytes, on the other hand, shrouded in their humble origins at a quiet Italian seaside, charged with an aura of mysticism, suffocated in scientific wars, were thought unworthy of too much attention, a vestigial link to the most rudimentary organisms. But as Metchnikoff had begun to envision, their role in the biological costs of inflammation would become more central than ever imagined. And their absence, as scientists soon found, would prove to be deadly.

In 1950, a twelve-month-old boy arrived at the University of Minnesota Hospital with a strange constellation of symptoms. He had an enlarged liver, a lung infection, and a scaly rash around his eyes, nose, and mouth. Doctors could not figure out what was wrong with him, and the child eventually died. Similar cases followed. At the 1954 meeting of the American Pediatric Society, Boston physicians presented reports of babies and young children who had spent most of their short lives visiting hospitals, suffering an onslaught of recurrent infections that eventually killed them before they reached their tenth birthday. They called it an "immunological paradox," a "fatal granulomatous disease of childhood," so termed because of the pockets of inflammation made up of immune cells and other tissues, or granulomas, that are found in patients' bodies in response to infections.

The children were suffering from immune deficiency. If autoimmunity was disease caused by the immune system's inappropriate overreaction, immune deficiency sprang from the opposite state, a weakened immune system that left the body vulnerable to germs. Scientists had already described deficiencies in the adaptive immune response, such as a shortage of B lymphocytes. A dearth of these cells leaves the body bereft of important antibodies and patients get severe bacterial infections. But the children with fatal granulomatous disease had high levels of antibodies in their blood.

At the meeting, several doctors in the audience said they had come across similar rare cases. One physician dared to guess that damaged or missing phagocytes like neutrophils and macrophages—Metchnikoff's beloved eater-cells—were culpable, but no one paid much attention to him at the time. Years later, well into the 1960s, studies confirmed his assertion. Genetic defects can impede the ability of phagocytes to destroy certain microbes, leading to deadly infections. Scientists subsequently developed new treatments, lowering death rates for those afflicted, and the disease was renamed "chronic granulomatous disease." Phagocytes, once reviled, proved that their presence was essential to life.

Over time, phagocytes finally received their due revival in modern science. On December 7, 2011, one hundred years after Elie Metchnikoff had received a Nobel Prize, French biologist Jules Hoffmann delivered his Nobel Lecture at the Karolinska Institute in Stockholm, Sweden. Hoffmann, along with American immunologist Bruce Beutler, had been awarded one-half of the Nobel Prize in Physiology or Medicine for "discoveries concerning the activation of innate immunity." It was the first time since Metchnikoff and Ehrlich had shared the prize in 1908 that the award was given for advances in the innate immune system. All prizes in immunology over the last century had been bestowed on research relating to the adaptive immune system.

Hoffmann leaned toward the podium as he spoke softly into the microphone, both arms stretched out in front of him, his patterned maroon tie matching the color of the flowers in a scalloped gold vase behind him. He began the lecture by talking about his father, an entomologist, who had instilled in him a love for studying insects. Like Metchnikoff, Hoffmann had conducted his most important research on invertebrates. He showcased his experiments on *Drosophila*, fruit flies. As in all insects, phagocytosis was an important part of their defense against germs. Hoffmann and Beutler had identified proteins called toll-like receptors, which help phagocytes and other cells recognize invaders like bacteria, viruses, and fungi, triggering an immune response.

Early in the lecture, Hoffmann put up a stunning black-and-white electron micrograph image of a dead fruit fly, its magnified body recalling characters in a science fiction horror movie. The fly, which was deficient in toll-like receptors, had succumbed to an overwhelming fungal infection. Its large compound eyes gaped at the audience as its body lay lifeless under a hairy carpet of fungus, its legs mottled and twisted every which way. It was a victim in the same deadly battle that Metchnikoff had witnessed over a century earlier as phagocytes roamed in water fleas that were infested with fungus, ingesting the needle-shaped spores. Those images had given Metchnikoff the fortitude to continue with his immunity research. They were now reborn, splayed out on a wide screen and cosseted by the technological advances of twenty-first-century medicine that unveiled more of the mystery than he could have ever imagined. Hoffmann spoke of Metchnikoff's historical findings in his lecture and even named one of the antimicrobial fruit fly proteins "Metchnikowin."

Canadian physician Ralph Steinman, who was awarded the second half of the 2011 prize, had passed away from pancreatic cancer three days before the prize winners had been announced. When the Nobel Committee made their announcement, they had not yet heard the news of his death. As a postdoctoral fellow at Rockefeller University in the 1970s, Steinman had discovered a new type of innate immune cell with branch-like projections radiating from its body. He named it a dendritic cell. Like macrophages, dendritic cells are phagocytes that devour invaders. But they are even more adept than macrophages* in uniting the innate and adaptive arms of the immune system. Compared with other phagocytes, dendritic cells spend less of their time destroying germs and more on alerting adaptive immune cells, recruiting T and B lymphocytes to mount organized, targeted attacks against pathogens. Thus, over a century after the world's first theory of immunity was proposed, the 2011 Nobel Prize celebrated the essential, interwoven roles of both innate and adaptive immune responses.

---

* The first time a certain germ enters the body, dendritic cells are crucial to connecting the innate and adaptive immune responses. Other cells in the body, including macrophages, also do this, but typically when the body needs to reignite an immune response against germs that have previously been encountered.

———————

Macrophages, dubbed "Metchnikoff's policemen," had been pushed aside in the late-nineteenth-century frenzy over antibodies. Despite Metchnikoff's Nobel Prize, macrophages were ignored by scientists for over half a century. Pasteur had reassured Metchnikoff, "Generations of students will memorize your theory from their textbooks without any idea of what you have had to endure to establish it." And indeed, macrophages have persisted—through the arcs of time and space and through the tissues in our bodies that narrate their biological histories.

In 1996, when a thirty-three-year-old French explosives worker encountered a malfunctioning rocket that tore through his hands and forearms, it turned him into a bilateral amputee with stump levels 3 inches above the wrists. Four years later, in Lyon, the man underwent the world's first double hand transplant. The recovery was grueling, but nearly five years after the transplant, both hands and forearms grew hair and nails. The man could feel pain and heat and even light touch. The hands perspired, held a pen or glass, shaved his beard—previously unthinkable acts.

Peering through a microscope at minuscule punched-out skin biopsies of the donated hands, doctors noticed something strange. Specialized skin macrophages known as Langerhans cells dotted the samples, but the cells did not belong to the explosives worker. Genetic testing showed that they derived from the hand donor. These foreign macrophages could not claim immortality. Rather, they had simply maintained their population within the tissue.

Toward the end of the twentieth century, scientists began to realize that macrophages were often born outside of the bone marrow. Many originate from fetal matter seeded into tissues before birth and can self-renew, dividing to replenish themselves. This diminutive, physiological rebirth of macrophages within human tissue is but an echo of their dramatic renewal in modern science. Metchnikoff could not have foreseen the explosion of interest in the macrophage that would help redefine inflammation and disease in twenty-first-century medicine.

In the salient historical stories of immunology, a common theme that emerges is one of strife. The human struggle to unveil the science persists. On a microscopic level, innumerable battles rage between disparate species attempting to outwit each other in order to evade harm and death. The notion of an immune system conjures images of war, of victors and victims, of the body defending itself against intruders like germs. It calls to mind the dark, disabling costs of inflammation, as an intricate force meant to protect us paradoxically turns against us, leading to collateral tissue damage, autoimmune diseases, allergies, and more. But the familiar war metaphor does not capture the varied incarnations of inflammation that are relevant to modern disease. In the twenty-first century, frequent, fatal infections and traumas, routine calamities of the past, have mostly yielded to contemporary killers like heart disease and cancer. These illnesses are linked to a type of inflammation that is occult and insidious, a leaky hose that must be repaired. The innate immune system plays an important role in this type of inflammation. It is less a war and more a struggle for balance, an attempt to capture an inflammatory equilibrium. Inflammation, as scientists began to realize over the last decades, is more costly than ever imagined. It is involved not only in a few select disorders but in a wide range of ailments that are the most common causes of death in the world today.

CHAPTER 3

A Sense of Strangling

I woke around two in the morning, pager blaring, hospital
alarms ringing. A voice on the intercom called out the code
blue location, which was close to my call room. Still half asleep,
I ran down the halls of Menino Pavilion at Boston University Medical Center, where Jay and I were medical interns. The fated summer
day on which his illness first struck was still a few years away.

That night, it was my first cardiopulmonary arrest. I carried
with me, in the large pockets of my white coat, a copy of Sabatine's
Pocket Medicine, a printed email from our chiefs on "intern emergen-
cies," four pens, a few paper clips, a penlight, a knee hammer, a hair
tie, and some loose change. Also, fear.

When I walked into the man's room there was chaos and
motion, voices rising and falling, a hand working to turn the television off, the smells from a half-eaten food tray, warm and buttery
and rich, fresh in the air. I found myself climbing onto the side of
his bed, knees at his hips, putting the weight of my body into both
arms and pounding compressions deep into his chest. I felt the
crack of his ribs beneath my fingers. After a few minutes someone
relieved me, bringing new hands and energy to the work of reviving
his limp body.

Later, when we pronounced him dead, all the air and life from
the frenzy of our work evaporated from the room in an instant, in

a shock, as if it had never been there in the first place. The scene would repeat itself throughout my training and beyond, unfolding in hospitals across the country with a variety of actors, until I became inured to the emotion and drama, complicit in acknowledging that our efforts in these last moments of life are nearly always futile.

The next week I ate scrambled eggs and hash browns for breakfast while listening to a rheumatologist lecture on lupus, a disease in which the body attacks its own tissues. After eight years of structured learning in college and medical school, I had landed at one of the busiest city hospitals in the country, a tertiary care center in Boston's South End that was built in the nineteenth century to provide medical care for the destitute, a mission that continues to this day. Prisoners, alcoholics, new immigrants, battered women, and the homeless filled our wards, some with diseases I would never again encounter. That morning, as I tried to pay attention to the different body parts lupus can attack by causing inflammation—skin, joints, kidneys, stomach, lungs, heart, brain—I had been on call for over twenty-four hours straight. I was sleepy, but I was also bored. Rheumatology was not my favorite topic. The diseases felt abstract and disjointed, with no clear beginning or end, melanges of nebulous symptoms waxing and waning according to the moods of the body's immune system and its nonsensical capacity for self-harm.

Heart disease, on the other hand, seemed ostensibly simple, grounded in logic and reason. We had learned that a four-chambered muscle pumped blood through the body, large blood vessels branching off into smaller ones, a network that fed tissues and organs like the pipes that brought water and gas to a house. Too much cholesterol in the blood, from bad genes or poor food choices, could build up in the walls of arteries and slow or stop blood flow to the heart, leading to a heart attack. This was what had happened to the man from the code blue. Heart disease was a plumbing problem, and cardiologists used their tools to fix clogged pipes. The narrative stuck in our collective consciousness, passed on to each new generation of medical students. But as heart disease continued to claim a significant number of lives in developed nations in the late twentieth and early twenty-first centuries, sci-entists began to understand that cholesterol was not the only cul-prit: inflammation also played an essential part.

As a medical intern, obvious inflammation pervaded my daily work. Inflammation, the ancestral response that evolved to main-tain the body's integrity against a deluge of daily environmental attacks, could be set off by trauma, as in the burn victims rushed to the emergency room gasping for breath, their skin raw and irate and peeling. Or the kid who swallowed cleaning fluid in a suicide attempt, with cratered, bleeding ulcers in his stomach. Infections from all kinds of germs could evoke an inflammatory response, as could autoimmune and allergic diseases. In clinics, patients disfig-ured and crippled by rheumatoid arthritis, barely able to use their knobby, knotty hands, or those covered in the scaly, itchy patches of psoriasis struggled to tell their stories. Women with lupus, fingers and toes turning white on cold New England winter days, covered the burning rashes spreading across their nose and cheeks with hats and scarves. One long call night, supervised by my senior, I sliced through a patch of warm, tight skin on a man's back with a small scal-pel. The collection of pus and inflamed tissue that had walled off an infection finally opened to the world as yellow-green, foul-smelling fluid drained onto my gloves, spilling dead and dying immune cells, bacteria, and waste tissue.

In stark contrast, the inflammation in heart disease is typi-cally invisible to the naked eye. Brewing in the depths of modern medicine, just as I was getting my first grasp of it, was a new way of looking at heart disease and other common killers. Evidence to date reveals that cardiologists deal with a muted version of the force that plagues rheumatologists. This chronic, low-level, "hid-den" inflammation is also linked to other diseases of modernity, like obesity, diabetes, cancer, and autoimmune diseases. It even plays a role in aging, neurodegenerative diseases like Alzheimer's, and psychiatric illnesses. Inflammation is far more pervasive than I had ever imagined as a student. Metchnikoff's macrophages, which figure prominently in hidden inflammation, hang out in all kinds of places: the blood vessels traveling to the heart, body fat, the pancreas, cancer tissue, and the brain. Inflammation—one of the oldest diseases known to man—is perhaps a common thread running through nearly *all* disease.

Hidden inflammation had helped to cause the death of the man from my first code blue. But it would take decades for this notion to come to light, from its slow and subversive inception in basic science laboratories to affirmative trials in human beings as it battled for recognition alongside cholesterol. The roots of this idea reach far into history: the modern view of inflammation in heart disease is but a rebirth of the once-heretical teachings of Rudolf Virchow.

In 1969, on his first day at the brand new University of California in San Diego (UCSD) School of Medicine, Peter Libby listened to legendary cardiologist Eugene Braunwald lecture on rheumatic heart disease and fell in love with the field of cardiology. Libby's journey began in an era when medical knowledge had exploded. In the second half of the twentieth century, rigorous research methods yielded major advances in understanding the causes of illness, and the number of remedies for various ailments dramatically increased. Polio, the devastating infectious disease that left countless children paralyzed and crippled, resigned to living lives in iron lungs, could for the first time be prevented with Jonas Salk's miraculous 1952 vaccine. Death rates from heart attacks, deemed midcentury to be "bolts from the blue," fell dramatically in ensuing decades in the United States. A witness to these major advances that transformed untreatable diseases into treatable or curable ones could not help but lavish fervent belief in scientific progress, tempering the medical fatalism of centuries past. Libby, inspired by the successes that surrounded him, also had a sense of what underlay them. Science, he knew, was not a medley of eureka moments culminating in awards and absolute truths. It was discovery and rediscovery, with haphazard and incremental gains, filled with reversals and bolstered by cross-pollination between disciplines. It was, as in physics, brute force clawing through distance but lacking linear roads or concrete harbors. Truly novel ideas were scant, and wrenching secrets from nature was a laborious task. As he took his first steps forward in cardiology, Libby had yet to realize how far backward the journey would take him, how his incipient work would become subsumed with the past, rising from the buried history of one of medicine's most ancient branches.

After graduation, Libby became a medical intern at Brigham and Women's Hospital in Boston, a teaching hospital tied to Harvard Medical School. There, under Braunwald's direction, he worked on projects related to myocardial infarcts—dead, scarred heart tissue due to compromised blood flow—a field that eventually spawned the modern era of therapies to prevent or treat heart disease. Heart attacks were thought to be all-or-nothing phenomena at the time, lightning flashes that struck with little warning. Braunwald hoped to figure out how to decrease dead tissue, the damage from the storm's wreckage. Libby was assiduous in the laboratory, but he soon grew restless. His curiosity was sparked not by the final stages of heart disease, but its beginnings. What was the initial insult, the *cause* of heart disease? History had pointed to cholesterol as the offender.

More than two centuries ago, in 1768, William Heberden, the personal physician to the queen of England, had struggled with a question similar to Libby's when he first described heart disease to his colleagues as

a disorder of the breast marked with strong and peculiar symptoms, considerable for the kind of danger belonging to it. . . . The seat of it, and sense of strangling, and anxiety with which it is attended, may make it not improperly be called angina pectoris. They who are afflicted with it, are seized while they are walking, (more especially if it be up hill, and soon after eating) with a painful and most disagreeable sensation in the breast, which seems as if it would extinguish life, if it were to increase or to continue; but the moment they stand still, all this uneasiness vanishes.

Heberden's report was lauded not only for being clear and detailed, but also poetic—there was surprising cadence in his written words. He described chest pain due to heart disease, or angina pectoris, as we understand it today. But Heberden did not know what was causing his patients' symptoms. He supposed the cause was an ulcer, or cramps.

One of his students, Edward Jenner, who would later develop the world's first smallpox vaccine, suspected that clogged coronary arteries could be related to a patient's chest pain from heart disease. He carefully dissected a patient's heart during an autopsy one day, writing:

> After having examined the most remote parts of the heart, without finding any means for which I could account for his sudden death, or the symptoms preceding it, I was making a transverse section of the heart pretty near its base, when my knife struck against something so hard and gritty, as to notch it. I well remember looking up at the ceiling, which was old and crumbling, conceiving that some plaster had fallen down. But on further scrutiny the real cause appeared: the coronaries were becoming bony canals ... and the concretions were deposits from the coagulable lymph, or other fluids, which had oozed out on the internal surface of the artery.

He wrote to Heberden in 1778, "How much the heart must suffer from the coronary arteries not being able to perform their functions ... should it be admitted that this is the cause of the disease I fear the medical world may seek in vain for a remedy." But Jenner did not understand what the deposits were made of.

In 1829, French pathologist and surgeon Jean Lobstein* named what Jenner had observed, describing "a yellowish matter, comparable to a pea puree, interposed between the internal and medium layers" of blood vessels. Lobstein called this condition *arteriosclerosis* (now often used interchangeably with the word

---

* Lobstein, credited with being the first to use the term *arteriosclerosis*, also noted inflammation in the internal layer of the aorta. He attributed an earlier recognition of the disease to Aretaeus of Cappadocia, a Greek physician from the first century.

atherosclerosis). But the cause of the buildup remained a mystery. Most people believed that heart disease was due to old age, as inevitable as wrinkled skin or worn joints or death itself, an indelible mark of the passage of time. The physician William Osler deemed longevity "a vascular question, well expressed in the axiom that 'a man is as old as his arteries.'"

The first link in the chain that bound cholesterol to heart disease began nearly a century later with a young Russian doctor. Nikolai Anitschkow, his dark-rimmed glasses resting on high cheekbones, was obsessed with white rabbits. It was 1913, the year a Bolshevik activist with revolutionary ideas first published an article under the pseudonym Stalin. The man was eventually arrested by the Russian secret police and exiled to Siberia. In his laboratory at the Royal Military Academy in St. Petersburg, Anitschkow had ideas of his own that would spark a serendipitous discovery.

Atherosclerosis was well described in the literature by this time, but it was still cloaked in mystery and thought to be a natural—and untreatable—consequence of aging. Anitschkow was inspired by Alexander Ignatowski, a physician colleague who experimented on rabbits to find out if dietary protein was toxic and caused premature aging. The idea had been suggested some years earlier by Elie Metchnikoff. Ignatowski had attempted what no other scientist ever had: to induce atherosclerosis in an animal. He fed his rabbits beef brain, meat, milk, and eggs, and a few weeks later, he was excited to see plaques in the rabbits' aortic vessels—the same plaques one would find in human cases of atherosclerosis. But when he repeated the experiments with egg whites, the plaques did not form.

Anitschkow had a hypothesis of his own. He had noted that egg yolks and brain, foods rich in cholesterol, produced the strongest changes in Ignatowski's experiments. He also remembered German

---

* Arteriosclerosis is a general term used to describe the hardening and narrowing of arteries that lead to poor blood circulation. Atherosclerosis, the most common type of arteriosclerosis, refers to the buildup of fat, cholesterol, and other substances on the walls of the blood vessels.

chemist Adolf Windaus's 1910 paper that described much higher concentrations of cholesterol in atherosclerotic plaques than in the walls of normal arteries. With the help of a medical student, Anitschkow decided to repeat the rabbit experiments. He pumped pure cholesterol into his rabbits' stomachs with a feeding tube. After a couple of months, plaques started to form in the aortas. When he analyzed the plaques under the microscope in polarized light, glittering cholesterol esters sprung into view:

> The main thrust of our investigations is that . . . it becomes totally clear why only certain nutrients, for example egg yolks or brain, can evoke specific changes in the organism. Since the same changes can be observed by feeding pure cholesterol, there remains no doubt that it is precisely this substance that is laid down in the organism as liquid-crystal droplets and evokes extraordinarily damaging effects in various organs.

Anitschkow's ideas met with early skepticism. Rabbits were not humans, after all, and the dose of pure cholesterol he had pumped into their stomachs was enormous, much more than what a person would normally consume. But Anitschkow's experiments led to the development of the "lipid hypothesis"—the idea that high blood cholesterol is a major cause of atherosclerosis. And with time, effort, and further studies, the lipid hypothesis would become widely accepted.

In the mid-twentieth century, the lipid hypothesis gained traction with the help of the landmark Framingham Heart Study, which sought to identify risk factors for heart disease by observing residents in the small town of Framingham, Massachusetts. Before the Framingham study, little was known about the factors linked to heart disease. In fact, the term *risk factor* originated with this study. In the 1950s, clogged arteries, high blood cholesterol, and high blood pressure were still thought to be the inevitable consequences of aging, and no treatments were offered. But by the 1960s, the Framingham study found that cigarette smoking, high blood pressure, diabetes, obesity, and high blood cholesterol were correlated with an increased risk of heart disease, while exercise was linked to a decreased risk. Minnesota physiologist Ancel Keys had observed that high intake of saturated fats from animal products was tied to an increase in blood cholesterol levels and subsequent heart attacks. Genetic studies, too, revealed connections between inherited high-cholesterol syndromes and premature heart problems in childhood.

As high-quality observational trials continued to correlate high cholesterol levels with heart disease, a strong focus on studying the impact of lowering cholesterol levels ensued. A now-famous 1984 coronary primary prevention trial showed that treatment with a cholesterol-lowering drug decreased heart attacks in men with high cholesterol by 19 percent. After that, the National Institutes of Health (NIH) pushed for routine screening to check for high cholesterol and recommended aggressive treatment for those at increased risk. The National Cholesterol Education Program was created with the mission of preventing coronary heart disease, then at last a public health goal. Soon after, the discovery of powerful new drugs called statins, which inhibit cholesterol synthesis in the body and decrease blood cholesterol levels, ushered in a revolutionary era in preventive cardiology. Later, new drugs called PCSK9 inhibitors were also prescribed to lower blood cholesterol. Epidemiological evidence from Framingham and other studies as well as data from clinical intervention trials helped secure a lasting tenure for the lipid hypothesis in the scientific world.

At UCSD in the early 1970s, Libby had learned that elevated blood cholesterol was an important element in atherosclerosis. Physician Joseph Stokes, dean of the medical school and one of the early investigators of the Framingham study, would paint a vivid picture for the medical students. Excess cholesterol wandered into the arteries that fed the heart, clogging their walls and blocking blood flow, creating the fatty plaques of atherosclerosis, the yellowish pea puree Lobstein had described. It was largely a matter of poor plumbing, stoked by fat and grease and lethargy: a bland, mechanical process that was easy to envisage, an intuitively satisfying explanation for one of the deadliest maladies of the age.

But after medical school, as he progressed through residency training at Brigham and Women's Hospital, Libby felt the story was incomplete, though he could not put his finger on exactly why or how. His favorite textbooks, Stanley Robbins's *Pathologic Basis of Disease* and Howard Florey's *General Pathology*, approached disease from distinct angles. Robbins focused on what could go wrong with individual organ systems, an approach that paralleled the layout of physician specialty domains, and included chapters on the head, neck, heart, lungs, intestines, liver, kidney, skin, bones, and nerves. *General Pathology*, on the other hand, delved into the commonalities *between* diseases, the processes underlying mutations from health to sickness, such as the ignition of inflammation in injured tissues and the wound healing that followed, creating thick scars. Or cells succumbing to death by suicide, a routine as predictable and controlled as a nightly repose, the obverse of the haphazard, parasitic replication and growth of cancer.

Libby was mesmerized by general pathology, particularly by inflammation and the branch of medicine that hovered at its roots, buried in fertile soil: immunology. So central was the Florey textbook to his intellectual growth in science and medicine that for the rest of his life he would never store the large blue book in an office but always in his home, encased in a favorite wooden bookshelf.

In Seattle, the rudimentary caricature of the coronary artery as an inert tube clogged with cholesterol, as lifeless as the metal pipes under a kitchen sink that were apt to rust and corrode, was already being challenged. In August of 1976, pathologist Russell Ross and his colleague John Glomset, at the University of Washington in Seattle, published a paper in the prestigious *New England Journal of Medicine* urging scientists to look beyond lipids and consider the role of the arterial wall in atherosclerosis. Ross proposed a "response to injury" hypothesis. Atherosclerosis, he said, began with an injury to the simple, single-layered endothelial cells lining the inner walls of coronary arteries, the vessels supplying blood to the heart. This initial injury could be caused by excess blood cholesterol, but perhaps also stress on the walls from conditions like high blood pressure, or hypertension. The damage would lead to abnormal growth of under-lying smooth muscle cells and other tissues, comparable to a benign tumor, leading to even more cholesterol buildup in the area as the body attempted to heal the wound.

Ross called for more research into the mechanisms by which risk factors—including high cholesterol, hypertension, diabetes, smoking, and even genetic factors—were tied to heart disease, espe-cially their effects on a cellular and molecular level. He wondered if these effects could all be explained on the basis of endothelial cell injury. Ross's hypothesis pointed out current knowledge gaps and suggested directions for future research. But inflammation was not mentioned in the paper, and the obvious question to follow had not yet been asked: did injury to the arterial wall mean ensuing inflammation?

As he floated through his final year of residency, Libby knew he could continue successfully churning out papers on myocardial infarcts, settling into Harvard. But he had a stubborn, contrarian streak that would carry him away from worn paths, having grown up in Berkeley, California, steeped in the student activism of the 1960s. He had a hunch that understanding the biology of the arterial wall would unravel more of the mystery. But it would mean taking the dif-ficult step of leaving his mentor, Braunwald.

Libby considered joining Russel Ross's laboratory in Seat-tle, but his wife was rooted in Boston. So he decided he had to find another way. In June of 1976, he marched into Braunwald's office and started off simply. "I want to study something else," he declared, and described his ideas. Puzzled, Braunwald stared at Libby for a few moments. He then picked up the phone and proceeded to tell his con-tacts about a bright young doctor who sought basic science training in the emerging field of vascular biology.

Soon, Libby was immersing himself in immunology research alongside his cardiology training, following an instinctual spark, living what Braunwald had once called the "thrill of the chase." The two fields—cardiology and immunology—were as dichoto-mous as the pages that separated them in medical texts. But Libby was no stranger to chance intellectual interactions. He had enthu-siastically studied both biochemistry and French literature as an undergraduate in Berkeley. His wide-ranging interests ran from musings on Bach as a medical healer to devouring historical novels.

Libby had first encountered Rudolf Virchow's teachings during his undergraduate biology classes at Berkeley, with the popular notion *omnis cellula e cellula*—all cells arise only from preexisting cells. He now found himself paging through obscure papers Virchow had written on inflammation's role in atherosclerosis, fascinated. Virchow had hypothesized that inflammation played a key role in heart disease as early as 1858, speaking in a lecture of

> a state of irritation preceding the fatty metamorphosis, comparable to the stage of swelling, cloudiness, and enlargement which we see in the inflamed parts. I have therefore felt no hesitation in siding with the old view in this matter, and in admitting an inflammation of the inner arterial coat to be the starting point of the so-called atheromatous degeneration.

The "old view" referred to earlier physicians like Joseph Hodgson and Pierre Rayer, who had speculated on the link between inflammation and heart disease. But Virchow pushed our understanding of the process. Performing meticulous and elegant experiments in dogs, he demonstrated that mechanical and chemical stressors on blood vessels triggered robust inflammation along nearly the entire thickness of their walls. He coined the term *endarteritis deformans* to describe what was happening to the vessels. The plaque buildup, or atheroma, was the product of an inflammatory process in the inner vessel walls, a reaction to an irritative stimulus, a growth that took on the intensity of a cancer in an attempt to heal.

In Virchow's eyes, inflammation was an *active* participant in the drama, the guilty party rather than a mere bystander. He pointed out that while atherosclerosis starts as minor fatty swelling under the surface of the blood vessel wall, advanced stages involve "a depot which lies deep beneath the comparatively normal surface," which can burst into the lumen of the vessel and lead to "just as destructive results as we see in the course of other violent inflammatory processes."

But another leading pathologist of the day, Carl Rokitansky, had dismissed Virchow's ideas. He did not believe that atherosclerosis was an inflammatory process or that inflammation was central in any way. He thought that blood products formed sticky plaques in the coronary arteries because of a new type of dyscrasia, an imbalance in the four humors. Inflammation, he conceded, was present in the vessels, but it was probably secondary, a reaction to the plaque rather than a potential cause of disease. Their debate was fierce and would endure for the remainder of their professional lives.

After Virchow's death, the role of inflammation in atherosclerosis had largely been forgotten, disappearing from medical literature for much of the twentieth century, with few exceptions. The gap coincided with the emergence of "cardiovascular epidemiology," a field with a focus on nailing down risk factors for heart disease. Improved research methods, exemplified by the Framingham study, revolutionized the discipline, enhancing its success at predicting who would suffer from a heart attack and why. But the shift may have distracted from attempts to thoroughly understand, on a biological and cellular level, the forces that take part in initiating atherosclerosis, including inflammation. Virchow's uncanny intuition that there was more to clogged arteries than the passive buildup of physical matter—that there was a burning force beneath the skeletal story, perhaps triggered by an irritation—had become obscure.

The history of medicine, unlike classical history, is capricious and angular, bereft of the narrative framework provided by wars or shifts in political power. Movements rise and fall, sometimes in parallel iterations, and are eventually buried or resurrected across time and space. Awards are given for important discoveries—many unrecognized in their own time—but not necessarily for individual brilliance or even diligence. Virchow's ideas, submerged in silent snow, would not resurface for nearly a hundred years.

In his laboratory, Libby was able to see inflammation in ways that Virchow had not. As immunology research continued to flourish in the second half of the twentieth century, a more precise illustration of the inflammatory reaction emerged. New methods captured images of the smallest blood vessels, and mathematical and engineering approaches quantified changes. Chemical techniques revealed that immune cells—including macrophages, neutrophils, basophils, eosinophils, mast cells, and lymphocytes—were pumping out powerful inflammatory mediators, a collection of "messengers" that help to fuel or control the fire. Cytokines and chemokines, for example, are small protein messengers with big effects. When a germ makes its way into the body, cytokines and chemokines are among the first signals the immune system generates. They determine the quantity and quality of the inflammatory response, communicating with immune organs like the thymus, spleen, and lymph nodes to mobilize even more inflammatory cells into the bloodstream. They can inflame nearby blood vessels and tissues or affect the entire body with a fever or rapid heartbeat. Many cytokines, named for their ability to act between white blood cells, or leukocytes, are termed interleukins (ILs).

Meanwhile, as Libby continued his medical training, research on endothelial cells was exploding. The electron microscope detailed their fine structures, which lined the walls of all the arteries and veins in the body in a single layer, in direct contact with blood. They provided a tight barrier between the inside of vessels and surrounding tissue. Libby's Harvard colleagues, including pathologists Michael Gimbrone and Ramzi Cotran, figured out how to culture and grow endothelial cells and taught this skill to Libby. In cell culture experiments, Gimbrone and Cotran exposed endothelial cells to inflammatory cytokines and noticed something strange: goaded by cytokines, endothelial cells behaved differently. They recruited and conversed with other immune cells, favored the formation of blood clots, and relaxed their barrier defense, allowing seepage of fluids and cells into tissues.

Libby observed something even more surprising: endothelial cells pumped out their own inflammatory mediators. They were operating, in effect, like immune cells. The idea that endothelial cells could be both inflamed and inflammatory was heretical. Only a proper immune cell was supposed to produce inflammatory mediators. In his laboratory, Libby discovered that the cytokine interleukin 1 beta (IL-1β) had one of the strongest stimulatory effects on endothelial cells. It transformed them into inflammatory agents that could secrete more IL-1β as well as other cytokines like IL-6, attracting immune cells like macrophages. IL-1β also activated genes on the endothelium that initiated the first stages of atherosclerosis. Libby noted that cells in atherosclerotic plaques produced IL-1β when exposed to inflammatory stimuli. He was excited by the findings, which supported Virchow's early work. In 1986, he wrote that his results suggested that IL-1β promoted atherosclerosis, and raced to publish the data.

But cardiology journals—and cardiologists—were indifferent. Editors told him the findings were not highly relevant to the field, that readers would not be interested. Libby published the paper in a pathology journal, where it loomed quietly in the background, divorced from mainstream cardiology. His papers and grant proposals were shunned by his colleagues. Specialization, which allowed physicians and scientists to manage an increasing load of factual knowledge, also siloed them. The urge to traverse boundaries between disciplines, an important feature of modern science, grew in the 1960s. Immunology, as science historian Arthur Silverstein writes, was an important catalyst for this change.

Undeterred, Libby continued working. By the mid-1990s, Libby, along with other groups of scientists, had painted the mechanics of atherosclerosis anew, with inflammation taking part in every step. Far from resting in inanimate pipes, the living cells of blood vessels constantly communicated with each other and the environment. The convenient vision of atherosclerotic plaques growing large enough to obstruct blood flow in arteries, thereby causing heart attacks and strokes, accounted for only a minority of those disastrous events. In reality, most heart attacks and many strokes* happened after the fibrous cap of an inflamed atherosclerotic plaque burst open, unleashing a whirlwind of cholesterol deposits and inflammatory cells and molecules from the blood vessel wall into the

---

* The word "stroke" is used in this book to refer predominantly to ischemic stroke, the most common type of stroke. Ischemic strokes occur when blockages in blood vessels cut off the blood supply to the brain. Hemorrhagic strokes, on the other hand, are caused by bleeding in or around the brain.

lumen, ultimately creating a blood clot in the artery and an immediate heart attack or stroke. The rupture of the fibrous cap resulted
from the accumulation of cholesterol deposits *in*, not on, the vessel
walls as previously thought.

Delving further into the mechanisms of atherosclerosis, Libby
found that low-density lipoproteins—LDLs, or "bad" cholesterol
particles—made their way into the lining of coronary arteries,
where they sometimes injured the endothelium, leading to abnormal growth of underlying smooth muscle cells and other tissues,
just as Russell Ross had previously described. But they also incited
an inflammatory response. Just as a germ or traumatic injury can
inflame a body part with subsequent redness, heat, swelling, and
pain, LDL can inflame a coronary artery. LDL-mediated injury transforms endothelial cells much like cytokines do, disabling their original function and turning them into inflammatory powerhouses.
The cells weaken their tight, protective barrier, becoming leakier.
They fail to secrete enough nitric oxide, a critical molecule that
calms inflammation, widens blood vessels, and keeps blood flowing
smoothly, warding off blood clots. Endothelial cells in this setting
instead recruit immune cells and churn out inflammatory mediators. A dangerous, circular path ensues: inflammation begets blood
clots, and blood clots amplify inflammation.

As Libby and other scientists began to bare inflammation's critical role in atherosclerosis, they found macrophages at work in every
stage of the disease. Metchnikoff's "policemen," once ignored as
vestigial scavengers of the innate immune system, equipped to
fight ancestral killers like infections and wounds, landed at the center of humanity's deadliest modern illnesses.* Macrophages, which
make up most of the immune cells involved in atherosclerosis, vora-
ciously gobble up LDL particles. Ultimately, they become so packed
with fatty droplets that they appear foamy under a microscope,
earning them the name "foam cells," which have been recognized
as a hallmark of atherosclerosis since Virchow's day. Macrophages
are sophisticated warriors, able to assemble specialized platforms

---

* Regarding innate immune cells in atherosclerosis, much attention has been showered
on macrophages. Neutrophils, however, are beginning to receive some notice as well.

called inflammasomes that spew out dozens of inflammatory molecules. The NLRP3 inflammasome, for example, activates the inflammatory cytokines IL-1β and IL-18, both of which have been shown to play a role in heart disease. The T and B cells of the adaptive immune system also participate in the broil, albeit to a lesser extent. And when there are higher levels of inflammatory cells circulating in the blood, they leech onto fatty plaques in arteries, making these more likely to build up, rupture, and cause a heart attack or stroke. The inflammatory response, which evolved to protect and heal, operates in atherosclerosis as it does in autoimmunity by harming the body instead, perversely creating bigger plaques. The plaques most likely to rupture have a large lipid pool, a thin fibrous cap, and many macrophages. They are vulnerable, like a ticking bomb, because they are inflamed.

By the turn of the millennium, this new view of atherosclerosis as an inflammatory illness was expanding. In 1999, two months before his death, the Seattle pathologist Russell Ross—who had formulated the original "response to injury" hypothesis and called for more research into the mechanisms by which risk factors were tied to heart disease—published a paper in the *New England Journal of Medicine* declaring atherosclerosis "clearly an inflammatory disease" that "does not result simply from the accumulation of lipids." He went on to state that the earliest type of lesion, the "fatty streak," which was common in infants and kids, was a "pure inflammatory lesion, consisting only of macrophages and T cells." Ross guessed that endothelial inflammation and dysfunction leading to atherosclerosis could be caused not only by LDL but also other risk factors, like cigarette smoking, hypertension, diabetes, genetic changes, and even infections.

We now know that the risk factors for heart disease can indeed work together. Smokers, for example, have elevated levels of inflammatory markers in the blood.* Smoking causes oxidants to form (similar to the process that rusts pipes), making LDL cholesterol more inflammatory, fostering arterial inflammation even in people

---

* These markers include C-reactive protein (CRP), tumor necrosis factor alpha (TNF-α), and IL-6.

with average LDL levels. And inflammation is not just one mechanistic link between heart disease and its risk factors: it can be a cul-
prit in itself. Individuals with organ transplants may be chronically
inflamed, with a hyperactive immune system that attempts to reject
foreign organs. A kid who survives childhood leukemia, for example, may experience heart failure a few years later from the chemotherapy he once took and receive a heart transplant; though he has
no traditional risk factors for heart disease, he develops atherosclerosis in a few months, a complication driven purely by inflammation. Other illnesses that harbor chronic inflammation are also
linked to heart disease. In patients with rheumatoid arthritis, heart
disease occurs more often than in the general population and is
one of the most frequent causes of death. Common inflammatory
cytokines are involved in the development of both conditions, and
inflammation is an independent predictor of heart disease in these
individuals. Infections can generate a low-grade inflammation that
leaks into the blood and travels to distant sites, what scientists call
the "echo effect." Poor oral hygiene or habitual smoking, which can
lead to gingivitis, an infectious inflammation of the gums, can also
hasten the development of heart disease in this way.

Yet as the twenty-first century dawned, the idea of atherosclerosis as an inflammatory disease was still largely hidden from public view, including from physicians, patients, and medical trainees.
It had been developed by basic scientists working mostly with animals, laboring in laboratories. Studies in humans were lacking, and
the existing research was not free of controversy, with many arguing
that inflammation was invariably the result of heart disease, not a
potential root cause. Meanwhile, teaching and clinical practice clung
to old biology.

———

While Peter Libby was toiling in a basic science laboratory, Paul Ridker, another Harvard cardiologist, wanted answers from human
beings. As he worked in clinics and coronary care units, watching
patients suffer from heart disease, Ridker was haunted by many
unanswered questions: Why did half of all heart attacks and strokes

occur in people who did not have high cholesterol levels? Indeed, a quarter did not have any cardiovascular risk factors at all, including hypertension, diabetes, obesity, or a history of smoking. Was there an unanticipated aspect of the disease that studies like Framingham had missed? And many heart attacks seemed to strike out of the blue: the deadly plaques that ruptured were usually soft and shallow rather than hard and obstructive. Because these types of plaques did not impede blood flow until they ruptured, they might fail to cause chest pain or major abnormalities on imaging tests prior to becoming suddenly catastrophic. Moreover, traditional therapies that focused on relieving chest pain or shortness of breath caused by obstructive plaques—including balloon angioplasty, stent insertion, or outright surgical bypass grafts—did not address these unstable plaques and often failed to prevent future heart attacks.

Ridker had a hunch that the immune system played a crucial role, possibly by triggering an inflammatory reaction that caused vulnerable plaques to burst. He needed a simple blood test that could capture inflammation. But the type of inflammation he wanted to hunt down was neither acute nor chronic in the typical sense. It was both unseen and unnamed, a latent, low-level inflammation that seeped through the body in unassuming patients. He settled on C-reactive protein (CRP), a molecule produced by the liver in response to the cytokine IL-6, which is released by areas of inflammation. Measuring CRP is cheap and only requires a small amount of blood, and it acts like a thermometer, taking the "temperature" of a patient's inflammation. In acute illness, like a severe bacterial infection, arthritis flare, or trauma, blood levels of CRP shoot up. Levels also increase in patients suffering from all kinds of inflammatory ailments. Barring a serious inflammatory event like infection or trauma in the few weeks before a sample is taken, CRP levels tend to be stable in blood for decades. But Ridker was more interested in very minor elevations of CRP that swam in the blood of otherwise healthy people and could indicate chronic, low-level inflammation. The difference between normal and elevated is so minute in these cases that a special test called a high-sensitivity CRP test must be used.

Though CRP rises in response to any stressor that causes inflammation, it does not reveal *why* the inflammation is happening.

Levels are high in hospitalized patients who have had heart attacks, but inflammation can arise in response to injured and dying muscle. Ridker wanted to know if inflammation—the chronic, low-level variety—had been there long before the heart attack, portending calamity.

Ridker's interest in CRP had begun during his days as a Harvard medical student. Physician Charles Hennekens, who was one of Ridker's mentors as well as his tennis and squash partner, had published the landmark Physicians Health Study in the 1980s showing that taking a daily aspirin could lower the chance of getting a first heart attack. Ridker asked Hennekens if he had hung on to the blood samples used in the study. They were lying around in a freezer somewhere, Hennekens said.

Ridker then had access to baseline blood samples from around twenty thousand healthy, middle-aged physicians who had been assigned to aspirin or a placebo drug in order to prevent heart disease and were tracked over the next decade to see if they developed heart attacks or stroke. He started measuring CRP levels in some of the samples. The men had not yet had heart attacks, were not smokers, and did not always have other risk factors for heart disease. At the end of the experiment, a pattern emerged.

Ridker observed that otherwise healthy men in their forties with the highest CRP levels were three times as likely to suffer a heart attack and twice as likely to have a stroke over the next few years over men with little or no chronic inflammation. It was a chilling finding, one that pointed to chronic, low-level inflammation, roiling quietly in the body, as a marker of risk that preceded heart attacks and strokes by years. Ridker also noticed that the benefit of taking aspirin, as measured by the original study, was directly related to the underlying amount of inflammation. Aspirin yielded the highest gains in people with the highest CRP levels. It was prescribed as an antiplatelet drug thought to decrease the risk of heart attacks and strokes by helping to prevent blood clots, but it also had anti-inflammatory effects. The results suggested that targeting inflammation with drug therapy could be another way to mitigate risk.

Ridker's data did not prove that elevated CRP itself caused heart disease. Rather, CRP marked the presence of chronic, low-level inflammation. The study, performed in humans, uncovered a radically distinct biology from that being taught in lecture halls, supporting the work being done by basic scientists like Libby. It charged Ridker's career with direction, setting him on a path to chasing the inflammation trail for the next thirty years and beyond.

But Ridker's work, like Libby's, had initially been met with skepticism. For many doctors, the finding that measuring CRP could identify people at higher risk for future heart attacks and strokes was not enough to convince them to begin routinely checking patients' CRP levels. No specific anti-inflammatory treatments had been tested in this setting or been proven to effectively reduce risk. If they could not offer patients a way to alter the risk, what was the point of knowing about it in the first place?

Within a few years of his initial 1997 paper on CRP, Ridker published many additional studies that bolstered his early findings. Ominously, CRP levels in a sample from a decade ago could predict who had a heart attack yesterday. CRP is the downstream result of innate immune activation, produced in the liver when inflamed areas of the body release cytokines. Ridker was also able to link endothelial dysfunction—which Libby and other basic scientists had studied in the laboratory—to CRP levels. In patients with coronary artery disease, high CRP levels are associated with dysfunction of endothelial cells, and this dysfunction resolves when CRP levels normalize. Additionally, the higher the CRP levels, the lower the amount of protective nitric oxide produced by endothelial cells. Nitric oxide destroys foam cells, the lipid-laden macrophages that saturate atherosclerotic plaques most likely to rupture.

Other inflammatory markers also predict future heart attacks. IL-6 is increased decades in advance of heart disease. The chronic, low-level inflammation that markers like CRP and IL-6 pick up clearly come before, not after, disease and death strike. CRP is an independent risk factor for cardiovascular events, distinct from LDL cholesterol or other risk factors. It is at least as good as (or better than) LDL cholesterol at predicting risk. But since the tests identify different high-risk groups, relying on both is better than using either one alone.

Another breakthrough in understanding the importance of inflammation in heart disease came from the therapeutic angle. Cardiologists primarily thought that widely used statin drugs worked to prevent heart attacks by lowering cholesterol, but half of all heart attacks and strokes happen in people who do not have high cholesterol levels. Ridker believed the benefit of taking statins was far too large to be explained solely by their cholesterol-lowering effects. Some patients who were started on statins saw clinical improvement in symptoms like chest pain in just a couple of weeks, much too short a time frame for cholesterol levels to come down. Ridker knew that statins, like aspirin, were potent anti-inflammatory drugs as well. Statins had been shown to improve endothelial cell function, increasing their ability to pump out nitric oxide and widen blood vessels after just one month of treatment.

To test this hypothesis, in 2001 he launched JUPITER (short for Justification for the Use of Statins in Prevention: An Intervention Trial Evaluating Rosuvastatin), a trial in which nearly eighteen thousand patients with elevated CRP but normal cholesterol levels were given placebo or a statin medication. Shockingly, patients with high CRP and low cholesterol levels who had gotten statins lowered their risk of having a heart attack or stroke by 44 percent. They also had a 20 percent reduction in death from any cause. The drop in cardiovascular events was larger than that seen in any prior trial where statins had been prescribed to lower cholesterol. The trial suggested that elevated cholesterol was not the only offender in heart disease and that statins could treat the condition by acting like anti-inflammatory agents.

Again, however, the reaction from the cardiology community was mixed. If the trial had shown that statins could decrease heart attacks in people with supposedly normal cholesterol levels, perhaps the cholesterol cutoffs were just too lenient. Maybe Americans should be aiming for an even lower target. Hence, the trial altered preventive cardiology guidelines worldwide, urging doctors to further lower cholesterol levels in patients.

Ridker admitted that his results were only indirect suggestions about inflammation's role in heart disease. The study could not tease out how much of the benefit from being on a statin came from lowering cholesterol and how much came from lowering inflammation. It was not designed to answer those questions. It was an intermediate gain, a stepping stone to what would be the fundamental test of inflammation's role in heart disease: a large-scale clinical trial in humans looking at a targeted anti-inflammatory drug, one that did not lower cholesterol or any other risk factor except for inflammation. If this type of experiment succeeded, it would bring to light a biology both novel and veteran, drawing from a hypothesis that had been forsaken for a century. And it might—one day—play a part in radically altering the practice of medicine.

Ridker teamed up with Libby to discuss strategies for conducting such a trial. It would mean enlisting thousands of volunteers and spending tens of millions of dollars on a concept many cardiologists still considered embryonic, stuffed in journals but with little relevance to patient care. In the 1980s, Libby's paper on the cytokine IL-1β, which was ignored in its time, had noted that IL-1β could be an important part of the inflammatory pathways that led to atherosclerosis. IL-1β pushes endothelial and other cells to secrete cytokines like IL-6. IL-6, in turn, boosts the production of CRP by the liver. Ridker's work had shown that blood levels of CRP and IL-6 could predict the risk of heart attack and strokes. Interestingly, people with gene variants that blunt IL-6 activity and thereby systemic inflammation have a lower risk of heart disease. With the new experiment, Ridker and Libby aimed to target upstream cytokines like IL-1β, the molecules at the top of the food chain, rather than CRP. It would be an exacting, targeted attack against inflammation in heart disease. They focused on a drug called canakinumab, an antibody used to treat rare inflammatory diseases like juvenile arthritis that worked to block IL-1β.

In 2011, Ridker started recruiting patients for a massive randomized controlled trial called The Canakinumab Anti-Inflammatory Thrombosis Outcomes Study (CANTOS), funded by the pharmaceutical company Novartis. The study enrolled more than ten thousand patients with a history of heart attacks, all of whom were already taking high-dose statins. Despite this, they had high CRP levels and were defined as an "inflamed" group of patients. CANTOS looked at whether treatment with canakinumab could decrease the rate of repeat heart attacks and strokes in this population over a period of four years.

When the study began, many predicted the chance of success would be slim. Anti-inflammatory arthritis drugs had never been used in heart disease patients. The idea seemed intuitively outlandish, despite the supporting science that had accrued over the years. Colleagues warned Ridker that he would be putting his career on the line, but he was undeterred by the prospect of a failed hypothesis. Science, he thought, was largely about asking the essential questions, regardless of the outcome. Otherwise, what was the point of conducting research?

CANTOS, designed and run to rigorous standards, showed the world in 2017 that canakinumab, by decreasing IL-6 and CRP levels by about 40 percent (without any change in LDL cholesterol or other risk factors like diabetes and hypertension), lowered patients' risk of heart attacks, strokes, and cardiovascular death by 15 percent. It had also lowered the risk of unstable chest pain requiring urgent intervention. The results caused a stir, receiving national and international media coverage. For the first time, there was hard evidence that solely intervening on inflammation could improve outcomes in heart disease patients, that inflammation was a *cause* of heart disease. The trial also revealed a dose-dependent effect. Patients had received either a low, medium, or high dose of the drug. The low dose did not do much; the medium dose was more effective; and the high dose yielded the best results. It was a persuasive finding, indicating that the *magnitude* of inflammation inhibition drove the benefits in individual patients.

Libby, Ridker, and other scientists had slowly built a compelling narrative. It was unlikely that canakinumab would speedily make its way to physicians' prescription pads in cardiology clinics; the drug was unfamiliar to most, expensive, and, as with most biological medications suppressing the immune system, complicated by occasional infections. CANTOS's value, however, lay not in its immediate practical application but in that it exposed a new pathway for disease prevention and treatment, opening locks once soundly sealed. It established the role of inflammation in human atherosclerosis and brought laboratory work to life at the bedside, affirming not only three decades of painstaking scientific labor but also Rudolf Virchow's early experiments. Over 150 years after his initial lecture on inflammation in heart disease, the hard data Virchow had always championed finally manifested in support of his ideas.

Other trials followed suit. Colchicine, an anti-inflammatory drug made from the autumn crocus plant, used for centuries by ancient Greeks and Egyptians largely to suppress joint swelling, has a potential benefit in heart disease as well. Among its many anti-inflammatory mechanisms include an inhibition of the NLRP3 inflammasome and thus the production of IL-1β and other cytokines. It also affects abnormal movements of immune cells—commonly implicated in both gout and heart disease—in response to cytokines. While colchicine is used to treat gout, growing evidence shows that it also lowers the risk of adverse cardiac events in people with heart disease. In 2020, a large randomized controlled trial enlisting over five thousand patients found that the risk of cardiovascular events was significantly lower in those receiving low-dose colchicine rather than placebo.

Inflammation does not replace cholesterol as a risk factor. In fact, the two often work together to promote heart disease. But lingering inflammation, twice as common as excess cholesterol levels, has declared itself a novel enemy, taking part in every step of atherosclerosis and increasing the chances of plaque rupture leading to a heart attack.

CANTOS also revealed other benefits for patients taking canakinumab. They had a sizable reduction in their risk of developing inflammatory diseases like arthritis and gout. And there was yet another gain, one that astonished the researchers: patients' risk of dying from cancer went down by 50 percent, including a 75 percent decrease in lung cancer. CRP, a marker of inflammation that predicts the risk of heart attack and stroke, is also a marker of inflammation in the lungs and can predict the risk of lung cancer.

To enter the trial, patients had to be cancer-free. But in a study of middle-aged people, all kinds of early-stage cancers yet undetected were probably brewing. Maybe canakinumab prevented the progression, invasion, and metastases of these cancers. Did some element of the hidden inflammation that drove plaque formation in heart disease also spur cancer progression? Could calming this type of inflammation decrease cancer risk, too?

All of a sudden, cardiologists and oncologists looked at each other—and their patients—anew. Underlying the infamous top killers of modern humans, heart disease and cancer, was perhaps a common foe.

CHAPTER 4

Open Wounds

I n 1887, the German Crown Prince Frederick III fell ill. Known informally as "Fritz," the crown prince was a pacifist adored by liberal Germans, who anxiously awaited his ascent to the throne. But in March of 1887, Fritz complained of hoarseness. Using a hot wire, his doctor burned off a small growth he had found on his left vocal cord. In May, however, the growth returned. If it was malignant, Fritz would need a radical operation, one that removed his entire larynx. This might cost him his life, and would surely cost him his voice, making it impossible for him to function as emperor.

A distraught Fritz agreed to have a small piece of the affected vocal cord sent to Rudolf Virchow, who would determine if cancer cells were present. Virchow diagnosed a laryngeal wart, writing that he found inflammation but no cancer in the growth. By the next summer, Fritz was dead. Virchow, who performed the autopsy, found that "the larynx was completely destroyed through cancer. . . in its place was a hole as large as two fists." Fritz's missed diagnosis would forever haunt Virchow.

Today, cancer, like heart disease, is a major cause of death worldwide, stealing more collective years of human life than ever before. Its incidence is only increasing, making the question of *why* cancer arises more crucial than ever. In this, as with heart disease,

Virchow was visionary. He was one of the first physicians to point out that cancer and inflammation were intricately linked.

Two thousand years ago, the Greek physician Galen used the term *cancer* to describe inflammatory tumors of the breast in which superficial veins appeared swollen and radiated outward like the claws of a crab. Later, the name was extended to include all malignant growths. Galen thought cancer could evolve from inflammatory lesions. But it was "blacker in color than inflammation and not at all hot." In 1863, Virchow found white blood cells lodged in many cancers and guessed that repeated tissue damage and inflammation preceded tumor development. Cancer, he wrote, was a disease of chronic inflammation. But his theory was ignored at the time.

Over a century later, in the early 2000s, I was an undergraduate majoring in biology at the Massachusetts Institute of Technology. I would wake promptly to make it to a class called "7.012," an introductory biology course. One of my professors, Robert Weinberg, lectured on the laws that govern life on the planet—laws, he said, that could be used to solve big problems. Weinberg had thought deeply about some of these problems, including the answer to a challenging question: how did normal cells morph into cancer cells? Most cancers are caused by genetic changes that lead to uncontrolled cell growth and tumor formation. But cancer is a dauntingly complex disease. Hundreds of genetic mutations can reside in a single tumor, and every tumor is unique—even those from the same organ.

Weinberg, along with his biologist colleague Douglas Hanahan, published a seminal review paper in the journal *Cell* in the year 2000 titled "The Hallmarks of Cancer." In a graceful distillation, they describe six critical traits that all cancers share,* the

---

* Hanahan and Weinberg write that most, if not all, cancers acquire the same set of functional capabilities during their development, albeit through various mechanistic strategies. The hallmarks of cancer they describe in their 2000 paper as published in *Cell* are self-sufficiency in growth signals, evading apoptosis, insensitivity to antigrowth signals, sustained angiogenesis, tissue invasion and metastasis, and limitless replicative potential.

commonalities between varied, chaotic pathologies. These traits include, as cancer cells warp biology in their quest for immortality, the ability to reproduce endlessly, grow new blood vessels, evade death, invade tissues, and spread throughout the body. The article has garnered thousands of downloads and citations over the years, making it one of the most popular articles ever published in *Cell*. But inflammation had not been highlighted—not yet.

Scientists had long observed inflammation around cancer. But this inflammation was originally believed to be entirely salutary, attacking tumors the same way it attacked germs. In 1909, a year after he won the Nobel Prize with Elie Metchnikoff, Paul Ehrlich noticed that cancer cells, like germs, were monitored and destroyed by immune cells. He referred to this process as "immunologic surveillance," whereby the immune system keeps a close eye on the body and gets rid of most tumors before they turn into full-blown cancers.

However, as the twentieth century wore on, a darker reality was born, echoing Virchow's ideas. The immune system did indeed fight cancers, as Ehrlich had guessed, but it also betrayed the body by helping them to grow and spread. Cancers can outsmart the immune system, growing in ways that do not attract much attention, even directly subverting the antitumor response, using it for their own survival. This "tumor escape" is highlighted in advanced cancers, which always host plenty of immune cells but are rarely rejected by the immune system.

In the late 1990s, English scientist Frances Balkwill proposed that inflammation could help tumors to progress by propelling genetic changes. Balkwill studied tumor necrosis factor alpha (TNF-α), a classic inflammatory cytokine typically produced by macrophages and other immune cells. It was named for its ability to kill cancer cells when injected into tumors at high levels. But Balkwill found that TNF-α could behave differently when it lingered as a low-level force, promoting cancer instead. When she turned off the TNF-α gene in mice, eradicating even low levels of TNF-α activity, the mice did not develop tumors. Anyone working on TNF-α as an anticancer agent was horrified to see that the inflammatory molecule was instead working as a tumor promoter. Balkwill, along with

Italian physician Alberto Mantovani, continued to uncover many similarities in the microenvironments of chronic inflammatory conditions and tumors.

At Beth Israel Deaconess Medical Center in Boston, a teaching hospital of Harvard Medical School, pathologist Harold Dvorak came across the connection between cancer and a common chronic inflammatory condition—wounds. By drawing parallels between the two conditions, Dvorak drew attention to the ways in which inflammation could help to cause rather than curb cancer. Dvorak made a seminal discovery in 1983 when he found that cancer cells secreted an abundance of a protein called vascular endothelial growth factor (VEGF). VEGF makes blood vessels more "leaky" (allowing molecules or cells to move in and out of the vessel more easily) and stimulates angiogenesis, the growth of new blood vessels. Solid tumors cannot thrive without an adequate blood supply. Angiogenesis helps cancers grow as it does human embryos, allowing cells to obtain reams of nutrients and oxygen to support their wild expansion. Dvorak's findings, along with those of other scientists, helped pave the way for anti-angiogenesis treatments to halt or reverse tumor growth.

He then came across an unexpected link: macrophages in chronic inflammation expressed lots of VEGF as well. VGEF stimulates wound healing through various methods, including angiogenesis. Dvorak started to pay close attention to wounds, one of the historic scourges that inflammation had evolved to wrestle with. He noted that cancer and inflammation shared some basic developmental mechanisms. In 1986, he published an essay in the *New England Journal of Medicine* titled "Tumors: Wounds That Do Not Heal," highlighting the many similarities between solid tumors and wound healing, both of which deviated from the normal growth process.

It was a tongue-in-cheek proposal at the time, he admitted, but in the ensuing decades additional scientific research supported his

---

* In 1971, surgeon Judah Folkman proposed that solid tumors are dependent on angiogenesis. He and his team identified an angiogenic factor and inspired other labs to do so as well.

hypothesis, putting the idea that tumors act like wounds that do not heal on a firm molecular basis. The immune system senses that cancer, like a germ or wound, is a danger to the body's equilibrium that must be eliminated. If you slice your skin with a small knife, the bleeding soon stops, but the inflammation lingers for hours. There is redness, heat, swelling, and pain. Blood flow increases. Cells and fluid ooze out of leaky, inflamed vessels. The immune cells that rush to the area spur angiogenesis and secrete growth factors, pushing skin cells to regenerate and proliferate, healing the wound. Inflammation only subsides after the repair is completed.

In cancer, much of the same occurs. The mass of disturbed, multiplying cells in a tumor have undergone dramatic genetic changes that spark uncontrolled growth, but they also hijack the immune response to further their artful survival and spread. Just like wounds, tumors create stroma, a supportive network of connective tissue and blood vessels that mimics chronic inflammation. The immune cells that secrete growth factors to repair tissue in a wound serve to fuel a tumor's uncontrolled proliferation. Cancer, rather than being healed, is continuously fed.

Dvorak points out that he was not the first one to compare tumors with wounds. This idea, too, he says, went as far back as Virchow. The wound healing environment provides an opportunistic matrix for tumor growth. In the 1980s, scientists noted that chickens injected with Rous sarcoma virus, a virus that can cause cancer, developed tumors only at the injection sites or other wounded areas. Surgeons have long recognized the tendency of tumors to recur in healing resection margins—the sites of surgical extractions. "Cancer is not creative," Dvorak wrote in his eighties, in the last days before his retirement, as he looked forward to devoting more time to a lifelong passion for photography. "It is simply a nefarious parasite that takes advantage of host defense systems that were developed for other purposes—like inflammation."

———

How does a normal cell morph into a cancer cell? At the nexus of this transformation, or at least one critical catalyst, is inflammation, which can drive cancer cells to acquire all the essential hallmarks Weinberg and Hanahan described.

While the inflammation that precedes cancer can nudge cells toward malignancy or affect early growth, the inflammation elicited by cancer enables its continued growth and dissemination. Inflammation is not a requirement for genetic damage; cancers can arise in noninflamed tissues as well. All tumors give rise to an inflammatory environment, becoming surrounded by immune cells, even if the tumor does not develop from chronically inflamed tissues. These cells initially attempt to destroy the tumor. If they are unable to do so, the tumor corrupts the immune response for its own benefit.

A protein known as NF-κB has been dubbed the "first violin," an essential conductor that orchestrates an inflammatory response. NF-κB is a gene transcription factor, a regulatory protein that induces the expression of hundreds of genes involved in the inflammatory response, including ones that code for inflammatory enzymes and cytokines. It governs the behaviors of many different immune cells. NF-κB is activated in the immune cells infiltrating a tumor, which in turn induce NF-κB in cancer cells, attracting more immune cells into the tumor—a feed-forward loop culminating in a frenzy of unchecked growth.

Numerous immune cells are involved in this process, but the macrophage is one of the most crucial culprits. Macrophages, which typically chew up cancer cells as they would a germ, can turn into traitors. These corrupt macrophages, or tumor-associated macrophages, are found in most malignant tumors.* In some cases they comprise up to half the tumor mass. Patients with tumors hosting these types of macrophages tend to have a poorer prognosis. Since the 1980s, cancer researchers have linked macrophages to tumor growth and poor outcomes for cancer patients. In 2001, at the Albert Einstein College of Medicine, scientist Jeffrey Pollard showed that macrophages promoted the development of breast cancers in mice as well as their eventual spread to other sites of the body.

Macrophages help tumors acquire several of the classic hallmarks. They produce immunosuppressive molecules, protecting

---

* Tumor-associated neutrophils are also emerging as important components of the tumor microenvironment.

the tumor from being harmed by the immune system. They secrete growth factors and cytokines, like IL-6 and TNF-α, that promote a tumor's unchecked growth. They produce the angiogenesis-promoting VEGF, allowing tumors to drink from a robust blood supply. Macrophages degrade a tumor's extracellular matrix, its physical scaffolding, just like they swallow dead cells and debris in wounds, paving the way for cancer to spread to distant sites of the body.

In 2011, Weinberg and Hanahan added two additional hallmarks of cancer that could help cells achieve many of the other core hallmarks. One was genome instability and mutation. And the other was inflammation. Whether it shows up before or after the birth of a cancer, inflammation can affect all its life stages, from the initial genetic—or epigenetic*—influences that transform normal cells into malignant ones to the continued growth, spread, and immune evasion of the cancer as it thrives in an inflamed tissue atmosphere. Even internal signals in cancer cells can inflame. Oncogenes, or mutated genes that drive cancer cells to divide, promote the production of inflammatory molecules. In fact, many genes that play a role in cancer affect inflammation. They turn on inflammatory pathways within a cell or create an inflammatory milieu outside the cell. In cancer, inflammation is an integral part of the fatal conspiracy. It can at once be the match that lights the fire and the fuel that feeds the flames.

At least a quarter of all cancers originate from overt, chronic tissue inflammation. In my own specialty, inflammation can spring up anywhere in the digestive tract, from the mouth to the anus. Chronic, uncontrolled inflammation can cause mild to severe gastrointestinal issues and increase cancer risk. Bad heartburn bathes the esophagus, the hollow tube connecting the mouth to the stomach, in acidic fluid. The ensuing inflammation distorts the cells of the lower esophagus, leading to a precancerous condition called Barrett's esophagus.

---

* Epigenetic changes refer to modifications of DNA that affect gene expression by turning genes "on" or "off." These modifications do not change the DNA sequence but rather how cells "read" genes. Lifestyle and other environmental factors can prompt epigenetic changes to DNA.

Celiac disease patients who fail to strictly avoid gluten may have ongoing intestinal inflammation, increasing their risk for several types of intestinal cancers. A chronically inflamed colon in Crohn's disease or ulcerative colitis mandates frequent screenings with colonoscopies due to the increased risk of colon cancer. In healthy people, most colon cancers begin as polyps, fleshy protrusions in the inner colonic wall that can be removed during a colonoscopy. Certain types of polyps have the potential to turn into cancer one day, while others are harmless. Some polyps develop when colon cells try to wall off inflamed colonic tissue in an attempt to prevent chronic inflammation.

Chronic inflammation associated with infections can also lead to cancer. In the stomach, spiral-shaped *Helicobacter pylori* bacteria inflame gastric tissue, setting the stage for ulcers and stomach cancer. In the liver, infection with hepatitis B or C viruses or heavy drinking can inflame the tissue and increase cancer risk. Immunity protects us from cancer not only by destroying tumor cells but also by getting rid of germs that can cause chronic inflammation.

Other body systems are prone to becoming chronically inflamed as well. In the lungs, inflammation from injurious inhaled substances like cigarette smoke, air pollution, and asbestos predispose to lung cancer. Macrophages in the lungs, unable to digest asbestos fibers, mount a massive chronic inflammatory response in their attempts to defend and repair lung tissue. Where there is chronic, uncontrolled inflammation, cancer may follow, or have a better chance of following. But there are exceptions to the rule. Not all inflamed tissues have an equal rise in cancer risk. For example, even severe inflammation of the joints or brain carries little increased cancer risk, supporting the theory that some spaces with fewer resident germs might not be as prone to developing cancer related to inflammation.

Doctors are vigilant in dealing with inflamed tissues. The type of inflammation I routinely encounter can be seen with the naked eye during endoscopic testing, when flexible rods with cameras at their ends course through hollow intestines. Or it may be picked up when small pieces of tissue obtained during endoscopies are magnified and examined by a pathologist who looks for inflammatory cells. I also have a variety of imaging and blood tests at my disposal. When I diagnose inflammation in the digestive tract, I can attempt to treat it with medications, a change in diet and lifestyle, or both, decreasing cancer risk. Sometimes an inflammatory trigger needs to be removed: germs, chemicals, certain foods, or other environmental pollutants. Other times the inflammation itself needs to be suppressed, as in some autoimmune diseases with unknown triggers. In most cases of chronically inflamed tissues, the fear that a cancer is brewing looms large. I send biopsies to pathologists who rule out precancerous changes.

Often, however, the invisible drama unfolding in body tissues at the cellular level, the minute changes taking place as inflammation begets cancer—and vice versa—entirely escapes routine medical testing. In the summer of 2017, the results from CANTOS began to compel oncologists and other physicians to rethink inflammation and cancer. In my own practice, I treat patients with overtly inflamed tissues in order to prevent cancer. In contrast, the inflammation that Ridker and Libby had targeted hums in the background, low-level and silent, picked up only by testing for CRP. They guessed that cancer risk decreased in their middle-aged study population because suppressing hidden inflammation might have prevented early-stage, undetected tumors from progressing to full-blown cancers.

Most chronic inflammatory diseases carry an increased cancer risk, even in noninflamed body parts. But studies show inconsistencies. Some inflammatory diseases increase certain cancers and decrease others or barely affect the chance of getting cancer at all. The inflammation tied to cancer, whether hidden or overt, may be unique in every instance, a singular fingerprint drawing out select wings of the immune response with many different cells conversing through myriad signals. The possibilities are as infinite as the iterations of cancer itself.

Ruslan Medzhitov, an immunobiology professor at the Yale School of Medicine whose interest in the innate immune system began in the early 1990s during his student days at Moscow University, believes that hidden inflammation burns beneath many developing tumors. A healthy cell, he notes, depends on an adequate supply of oxygen, nutrients, and growth factors to remain robust. Deficiencies in these resources—or an overload of unwanted ones—cause a cell to become stressed as it adapts to adversity. If the stress is too much for the cell to handle, cell death ensues. Tissues can harbor a few or many dead or dying cells, malfunctioning to varying degrees.

Dead cells are typically swallowed up by macrophages, which constitute about 10 to 15 percent of most tissues. As Metchnikoff predicted centuries ago, macrophages play a central role not only in the body's defense but also in the maintenance of its tissues. For example, they help with bone turnover, tissue regeneration in the liver (they make up one-fifth of all liver cells), and the remodeling of nerve connections in the brain. Macrophages residing in tissues help cells stay healthy. They "taste" their surroundings, sensing tissue signals, and memorize the history of tissue health, adjusting their behaviors accordingly. But in the worst cases, when there are lots of dead cells and extreme tissue malfunction, many more macrophages and other cells are recruited to the area. A protective response becomes maladaptive as frank inflammation is detected.

Between obviously inflamed tissues and healthy tissues lies what Medzhitov calls *parainflammation,* a term he coined in 2008 to describe the bare whisper of inflammation triggered by stress in normal cells. Parainflammation, which can be as understated as an activation of innate immune genes, is yet another iteration of the hidden inflammation common to diseases of modernity. Parainflammation's initial intent is salutary, as it helps tissues adapt to stressful conditions and aims to restore function. But if tissue stress is ongoing, the inflammation can become maladaptive and chronic. Parainflammation is graded, mirroring the level of tissue malfunction, making the notion of overt and hidden inflammation less a dichotomy and more a continuum.

Several studies link hidden inflammation and different types of cancers, including lymphomas and cancers of the colon, lungs, prostate, pancreas, and ovaries. Studies also show that elevated markers of hidden inflammation, like CRP, are tied to a higher risk of some cancers and predict worse survival. Some anti-inflammatory drugs may work to prevent cancer by treating hidden inflammation. At the University of California in San Francisco, postdoctoral researcher Dvir Aran, working in physician-scientist Atul Butte's laboratory, wondered why aspirin and other nonsteroidal anti-inflammatory drugs reduced the risk of some cancers—including those of the colon—even when patients were not overtly inflamed. One reason for this, Aran thought, could be that parainflammation interacted with gene mutations to trigger cancer. In a 2017 study in mice, he looked at inflammation so minuscule it was only detected by analyzing proteins in the epithelial cells lining the intestines. Typical inflammatory cells that populate inflamed tissues were absent. Aran then identified a specific pattern of gene expression in these parainflamed tissues and looked for this unique signature—one that resembles macrophage gene expression—in thousands of human tumors and cancer cell lines. He found parainflammation in several cancer types. It was an important factor in tumor development, with a significant influence on prognosis, and cooperated with other genetic changes that sparked cancer. Aran also noticed that parainflammation could largely be shut down in human tissues by nonsteroidal anti-inflammatory drugs.

Targeting inflammation to prevent cancer is no simple task. Sight breeds vigilance: where inflammation is obvious, we can attempt to quench the flame. But often the story begins long before it finally reaches our eyes, in a small way, an isolated scene of lone cells suffering from environmental insults, perhaps in the air we breathe or the food we eat. Hidden inflammation, burning slowly, eventually strikes with ferocity. Maybe it floated in a relative of mine who was diagnosed, years after exposure to talc powder, with ovarian cancer. Or in the friend who had been breathing heavily polluted air throughout his childhood, becoming the father of a toddler before the fateful day on which a prickly collection of rapidly growing cells was found in his right lung. Of course, these are broad insinuations. It is impossible to view, in real time, the action as it initially unfolds or to get all of the minutiae on who was where and how and why one well-behaved cell—and then others—went rogue. But if we could indeed swing backward in time, hoping to find a perfect moment in which to stop cancer dead in its tracks, to prevent the wounds—both literal and figurative—from ever taking hold in the first place, hidden inflammation is one potential target. It may be just as important as overt inflammation in increasing cancer risk.

CHAPTER 5

Anatomical
Intimacies

C arrie, a forty-year-old doctor, had always struggled with her weight. She put in long days at work, rounding on her patients with meticulous care. Over the years, the flesh swaddling her midline grew and her blood pressure, cholesterol, and blood sugar levels ballooned. She knew that this left her at greater risk of being ambushed by a heart attack one day. Carrie soon went on a rigorous new diet and exercise regimen, never missing a day at the gym and counting every calorie. By the time I left our hospital to start a new job in a different city, she was barely recognizable. All the fat seemed to have melted off, and she was feeling better than ever. "But I'm always hungry," she would say.

A couple of years later, I met up with Carrie when I was in town. She had regained all the weight and more. Despite a masochistic level of discipline, she was heavier than she had ever been. Carrie asked me about bariatric surgery and about the inflatable balloons gastroenterologists insert into patients' stomachs to help them reduce weight. She knew she was fighting more than fat—she was fighting her own biology.

After we lose weight, our body is hardwired to regain the pounds.

We become hungrier and need to eat more to feel full. The brain rewards overeating and sends signals to our muscles so that they burn fewer calories. In short, the body defends its fat stores.

But there was another force at play in cases like Carrie's. Unlike the fat, it was unseen. It underlay not just obesity but also the related diseases like diabetes, warranting more sophisticated weapons to battle it. Carrie was terribly inflamed.

While Carrie's difficulty with weight is a common modern-day dilemma, malnutrition was the norm throughout most of human history (persisting well into the twentieth century, as exemplified by Herbert Hoover's 1928 campaign slogan of "a chicken in every pot" to feed the impoverished). Obesity rose steadily after the Second World War. No longer hunters and gatherers expending most of our energy to forage for plants and game, or mired in the early agrarian days of struggling with fickle farms, we humans had an increasingly endless surplus of foods available for effortless consumption. The number of overweight and obese people—the medical definitions of both conditions hinging on body mass index—climbed. And the likelihood of chronic diseases or premature death also rose in concert with body mass index.

Meanwhile, cultural opinion on fat started to shift. The fleshy female figures and depictions of portly industrial barons that dominated the art of the Middle Ages and Renaissance period waned. These older ideals of beauty, affluence, and power that had persisted well into the twentieth century were displaced by a frenzied public passion for slimming down. By the 1960s, English supermodel Lesly Hornby, popularly known as "Twiggy," exploded onto the fashion scene and became a teenage icon with her sticklike figure.

Physicians' outlook changed, too. While doctors had once encouraged carrying a few extra—even up to 50—pounds of flesh to provide a reserve of "vitality" that would shore up a person through illness, their attitudes started to change as the health consequences of obesity, which had been noted in the medical literature since the eighteenth century, became more apparent. In the early years of the twentieth century, excess weight was linked to a greater chance of death, and by the 1960s, obesity research began in earnest. Fat was associated with all kinds of maladies: diabetes, hypertension, heart disease, cancer, gallstones, heartburn, fatty liver, sleep apnea, auto-immune diseases, arthritis, neurological problems, kidney disease, infertility, depression, and more.

But *how* did it make people sick? The traditional view of fat, which focused on its role as a mere depot for energy storage as well as a convenient cushion and insulation for the body, yielded incomplete answers. As weight increased, adipocytes, or fat cells, crammed into a frame that bulged under pressure, creating an architectural defect, one that led to mechanical stress on various organ systems. This could explain issues like osteoarthritis and blood stagnating in veins, but it was not a satisfying answer for most of the other problems that came with excess fat.

In the early 1990s, while Libby, Ridker, and others were wrestling with the unpopular notion of inflammation in heart disease, graduate student Gokhan Hotamisligil, working in professor Bruce Spiegelman's laboratory at Harvard's Dana Farber Cancer Institute, stumbled into the relationship between inflammation and obesity. As he examined fat mice lined up alongside slim ones, Hotamisligil recalled his childhood. He had lived just a few miles from Pergamon, once a powerful ancient settlement near the Aegean coastal city of Izmir, home to the physician Galen. He remembered studying Galen's writings on the liver from 200 CE, which held that the architecture of an organ was essential to its function, to the flow of nutrients in and out, to maintaining a homeostasis, or balance, in the body—the same balance that prevented the round towers in Pergamon's great temples from crumbling to the ground. The fat mice, built with the same frame as the slim ones, were under an enormous structural pressure from their excess weight.

But Hotamisligil felt that secrets beyond structural stress abounded in their flesh. He turned his attention to the inflammatory cytokine TNF-α, the same molecule that seemed to promote tumors in Frances Balkwill's experiments. On a whim, he tested obese and lean mice for TNF-α expression. He found high levels in their spleens, an expected result for an organ rich in immune and inflammatory cells. But to his shock, he also found TNF-α in fat tissue. Fat tissue in the fat mice—but not in the lean ones—churned out high levels of TNF-α. A couple of years later, Hotamisligil found this to be true for humans as well. What, he wondered, was an inflammatory cytokine like TNF-α doing hanging around fat tissue?

By the time of Hotamisligil's experiments, body fat had shed its old image as a simple storage shed for excess energy. Molecules like leptin, an appetite-regulating protein made by fat cells, started to depict adipose tissue as a complex organ with its own hormones, receptors, and genetics. Fat is an organ involved in controlling *metabolic* processes, those that allow cells to convert protein, carbohydrates, and fats into the energy that builds tissues and fuels other activities that sustain life. Because these processes take place throughout the body, metabolic problems can affect multiple organs. Taking in too many unhealthy calories and moving too little means a higher risk of obesity and its related complications. The so-called metabolic syndrome, which Carrie had been diagnosed with, is defined as a group of risk factors associated with getting heart disease and diabetes: fat around the stomach, hypertension, high blood sugar, and unhealthy cholesterol levels.

Hotamisligil's experiment, announced in 1993, added a new twist to the picture. Traditionally, metabolism, or energy management, and immune reactions were viewed as separate entities with distinct functions, examined through specialized research tracks that seldom interacted. Metabolism governed life-sustaining processes like converting food to fuel and disposing of waste, while immune reactions were responsible for defending the body. But adipose tissue, beyond its known role in metabolism, was pumping out inflammatory cytokines—not just TNF-α, as later studies showed, but many others as well, like IL-6, IL-1β, IL-1, and interferon-gamma (IFN-γ). It was the beginning of a paradigm shift that connected fat and inflammation.

In 2003, two studies from scientists working at independent research centers came to the same groundbreaking conclusion, one that exposed the immune cells behind the inflammatory potential of fat. Scientists Anthony Ferrante, from Columbia Medical Center, and Hong Chen, then at Millennium Pharmaceuticals in

Cambridge, Massachusetts, discovered that macrophages lodged in the fat of mice and humans were responsible for most of the inflammatory cytokines coming from fat tissue. Metchnikoff's policemen had again landed in a most unexpected locale, rushing not toward an intruder but toward excess body flesh. In lean mice and humans, macrophages remained solitary and scattered among the fat cells. In contrast, macrophages clumped together in the flesh of obese animals, entirely surrounding fat cells in some cases and mimicking the layout of macrophages in chronic inflammatory diseases such as rheumatoid arthritis. Meanwhile, the number of macrophages in the fat of mice and humans grew in direct proportion to the size of fat cells and body mass indices. Their percentage in fat tissue ranged from under 10 percent in lean individuals to 40 percent in the obese and over 50 percent in the markedly obese. And greater numbers of macrophages even sailed through the blood of these markedly obese individuals.

Macrophages respond to stress. But the body has not evolved to manage overeating, which is one of the biggest stresses of modern life. In lean, healthy individuals, existing macrophages in fat tissue help to maintain an anti-inflammatory state through many pathways, including by secreting anti-inflammatory cytokines. But they alter their behavior in obesity, and an irate immune response ensues. Obesity stresses the body much like an infection does, activating some of the same intracellular stress pathways. Moreover, in obesity, a fat cell—which has been described by physician Stephen O'Rahilly as a fried egg atop a beach ball of fat—is at risk of rupture and death because of its architecture, a cell on edge. The bloated cell, stuffed with lipids, has little space for all the organelles that carry out the routine responsibilities of the cell. When overloaded fat cells start spilling their noxious contents, macrophages rush in to clean up, revving up inflammation. Stressed fat cells spit out inflammatory cytokines and produce less adiponectin, a critical protein that helps to control inflammation.

Ferrante, who devoted his laboratory to understanding how the immune and metabolic systems interact, estimated that more than half the cells making up fat tissue of obese individuals were actually immune cells. And while macrophages played a central role, data also implicated other innate immune cells as well as adaptive immune cells like T and B lymphocytes. T cells, for example, produce inflammatory cytokines and help to recruit macrophages to fat tissue. A wild idea started to take shape: beyond its role in architectural aesthetics and even metabolism, excess fat holds secrets more sinister than ever previously imagined. It was a bona fide immune organ.

Hotamisligil observed that the inflammation linked to obesity was not the flare of acute inflammation described since ancient times, a powerful fire—*calor, rubor, tumor, dolor*—quickly extinguished, or the loud inflammation of various autoimmune diseases. Rather, the new type of inflammation was low-level, chronic, and smoldering, escaping detection but as dangerous as spilled gasoline waiting for ignition, akin to the inflammation Ridker had been after with CRP testing or the parainflammation preceding cancer. And it needed a name. Hotamisligil dubbed it *metainflammation*.

Despite its low intensity and modest increase in blood levels of inflammatory molecules, evidence was mounting that metainflammation could exert profound effects on metabolic pathways, playing a role in diabetes, hypertension, high cholesterol, and heart disease. And metainflammation, emanating from fat, could perhaps also play a part in *making* a person fat. Among other things, chronic inflammation may prompt the body to store more calories as fat rather than burning them for energy. An original field of biomedical research called immunometabolism was emerging, with a focus on the tight connection between inflammation and metabolism.

———

From an evolutionary standpoint, the interdependence of the immune system and metabolism is not surprising. Metabolism, the most primordial process common to all species, emerged with life itself. The expensive immune response, younger than metabolism but older than many other systems, relies on enormous energy stores for host defense. Both energy efficiency and the ability to mount a potent attack against infections are crucial for survival. They coevolved, selecting for humans with a robust capacity to store nutrients and to wield strong—and sometimes overly sensitive—immune responses. In lower organisms like the *Drosophila* fruit fly, immune and metabolic responses are controlled by a common organ known as the fat body.

Macrophages and adipocytes derive from the same ancestral cell in our evolutionary past, one involved in both immune responses and metabolism, and share many functions. Both can secrete cytokines, and adipocytes have the capacity to turn into macrophage-like phagocytes, ingesting foreign substances. The receptors by which cytokines signal share some of the same pathways as the receptors that respond to insulin and other hormones.

Under a microscope, macrophages in fat tissue wrap themselves around the adipocytes, embracing them tightly, an anatomical intimacy that speaks to a complex functional relationship—one at the core of the link between immune reactions and metabolism. Many hormones in fat tissue, like leptin, have dual roles. Leptin controls food intake and regulates body weight, but it is also an inflammatory molecule. In obesity, as the body becomes resistant to leptin, fat cells secrete more and more of it, promoting hunger and inflammation.

The delicate balance between immune and metabolic responses, as they oversee each other like puppetmasters in tandem to maintain stability in the body, can be shattered by too little—or too much—food. On one end of the spectrum, starvation and malnutrition weaken the immune system, dampening its ability to fight intruders. On the other end, obesity leads to chronic, low-level inflammation, or metainflammation, a potential mechanistic link between fat and its complications.

Not all fat is created equal when it comes to inflammation. In the mid-twentieth century, French physician Jean Vague observed that body fat distribution was an important determinant of metabolic risks like diabetes and heart disease. His comments were ignored at the time, but scientists slowly started to realize that subcutaneous fat, or fat that lurks just beneath the skin, is largely harmless when it pads areas like the thighs, buttocks, or upper arms. In fact, it can act as a sink to protect other tissues from the toxic effects of too much nutrition. But subcutaneous fat around the stomach is dangerous. Excess belly fat indicates that a person also has visceral fat, the "deep" fat that wraps around the body's abdominal organs. This type of fat is highly inflammatory, hosting the most macrophages and spewing out an array of cytokines. It is tied to the complications of obesity, including diabetes, heart disease, and cancer. Visceral fat also portends a higher risk of early death from any cause. Excess fat around other organs, like the heart and its vessels, can also churn out inflammation, exacerbating the risk of chronic diseases.

Weight loss can decrease the number of macrophages in fat tissue, reversing its inflammatory potential and lowering the levels of inflammatory markers in the blood. But people with a normal body mass index who might not think of themselves as fat can have potent visceral fat hiding among their organs, leaking inflammatory molecules into the blood. This may arise from an unhealthy diet, a dearth of exercise, or both. This type of fat, just like its inflammation, is hidden.

Over the years, as her waistline failed to recede, Carrie was eventually diagnosed with type 2 diabetes, by far the most common type of adult diabetes in the world. Normally, after a meal, beta cells in the pancreas secrete insulin, a hormone that pushes glucose out of the bloodstream and into fat and muscle cells, where it can be used. In type 2 diabetes, the body becomes resistant to insulin, so that the pancreas must secrete higher amounts of the hormone to get the glucose into tissues. Eventually, the pancreas cannot compensate, nor is it as adept at producing insulin. Blood sugar levels rise, causing hunger, thirst, and frequent urination.

Although genetics can affect the risk of getting type 2 diabetes, more typically it is tied to critical lifestyle problems like poor diet, lack of physical activity, and obesity. Type 1 diabetes, on the other hand, is an autoimmune disease that usually starts in childhood. A violent inflammatory process destroys many of the pancreatic beta cells, rendering them incapable of producing enough insulin.

Genetic markers for type 1 diabetes are present at birth, and auto-antibodies against the pancreas show up in the blood long before laboratory tests indicate abnormal blood sugar levels.

While type 1 diabetes is a well-known autoimmune disease, type 2 diabetes had always been divorced from the immune system. But as the evidence mounted, type 2 diabetes started to veer from our traditional understanding of it, debuting as both a metabolic and an immunological disease. Like obesity and heart disease, it, too, was linked to chronic, low-level inflammation.

Insulin signaling, which affects how the body uses and stores sugar and fat, is one of the most crucial metabolic pathways in the human body. Insulin resistance, like inflammation, was shaped by protective evolutionary forces. Common to organisms as diverse as fruit flies and human beings, its essential functions survived millions of years of evolutionary divergence. In an age when humankind succumbed to an excess of stressors, including germs, predators, and starvation, insulin resistance helped to fight these ancestral killers, shunting glucose from muscle to inflammatory cells to fuel the immune response or preserving the brain's glucose supply during stressful situations. Cortisol, the major stress hormone in the fight-or-flight response, causes insulin resistance. In famines, insulin resistance helps humans store more calories as fat for later use. It also encourages reproduction, preserving nutrients for fetal development in pregnancy. Today, insulin resistance is observed in many cases of infections with an active inflammatory response, like hepatitis C, HIV, or sepsis, the body's life-threatening inflammatory response to an infection—typically a bacterial one.

In obesity, the initial aim of insulin resistance is salutary. An excess of calories can be toxic to fat and other cells. Without insulin resistance, which limits glucose uptake by fat cells, adipocytes would eventually swell and die, decreasing the storage pool for excess nutrients. A vicious cycle would ensue, with fewer and fewer cells available to shoulder the overwhelming burden of a body subsumed by too many calories. After the collapse of fat tissue, muscle and liver cells would follow a similar fate, with organisms literally eating themselves to death.

In patients like Carrie, however, excessive insulin resistance results in chronically uncontrolled high blood sugar levels, leading to a diagnosis of type 2 diabetes. Serious complications of this disease include heart disease, stroke, eye and nerve damage, and kidney disease. Diabetic patients take insulin-sensitizing medications or insulin injections to keep blood sugar levels within normal range.

The anti-inflammatory drug aspirin had once helped patients suffering from type 2 diabetes, hinting at a relationship between diabetes and inflammation. In the June 1876 edition of *Berlin Clinical Weekly*, physician Wilhelm Ebstein wrote: "It seems to be certain that in some cases of diabetes mellitus the use of sodium salicylate* has an influence on diabetic symptoms. These symptoms . . . can totally disappear. This is no small benefit for the patient." Twenty-five years after Ebstein's report was published, physician Richard Williamson of Manchester, England, chanced upon the same finding: by giving high doses of aspirin to his diabetic patients, he could decrease the glucose spilling out into their urine.

Their observations did not attract much attention until the second half of the twentieth century, when a British diabetic patient who was being treated with insulin walked into a hospital suffering with arthritis from rheumatic fever. After he was given high-dose aspirin to treat the arthritis, doctors were amazed to see that he no longer needed his daily insulin shots. Subsequent studies touted the use of high-dose aspirin for type 2 diabetes. But doctors did not know why or how aspirin helped their patients, and there was a steep price to pay, since side effects from high-dose aspirin were terrible, including a constant ringing in the ears, headaches, dizziness, and stomach ulcers. Physicians stuck with medications that replenished insulin, which were better tolerated and more in line with their understanding of how diabetes developed.

In the 1990s, when Hotamisligil found that fat tissue in obese mice behaved like an immune organ, pumping out inflammatory markers like TNF-α, he was struck by another phenomenon: the link

---

* Salicylic acid is the compound from which the active ingredient in aspirin was first derived. Sodium salicylate is a sodium salt of salicylic acid. Its characteristics are similar to those of aspirin, with a few exceptions.

between TNF-α and insulin resistance. Blocking TNF-α alleviated insulin resistance in the mice. Further, when Hotamisligil injected antibodies against TNF-α into obese, diabetic mice, their diabetes improved. TNF-α seemed to cause insulin resistance, and in 1996, Hotamisligil shed light on how this may happen: TNF-α disrupts insulin signaling after insulin binds to its receptors, preventing the insulin from doing its work.

In the ensuing years, Hotamisligil and other scientists continued to unveil a number of inflammatory molecules in fat tissue that played important roles in inducing insulin resistance. The cytokines released by fat cells, macrophages, and T cells, especially interleukin-1 (IL-1), were found to dampen insulin secretion from pancreatic beta cells and prompt their destruction. Surprisingly, studies showed that macrophages in type 2 diabetes patients populated not only fat tissue but also the pancreas, nestling alongside the cells they aimed to weaken and destroy. And the beta cells of the pancreas expressed their own inflammatory cytokines.

Importantly, taking fat out of the picture did not change the relationship between inflammation and insulin resistance. Slim animals injected with inflammatory molecules also developed insulin resistance, evidence that inflammation alone was an important piece of the puzzle. Inflamed cells tend to ignore insulin's directives, pushing the pancreas to secrete higher and higher doses of insulin in an attempt to nudge glucose into tissues.

Meanwhile, researchers conducting human observational studies reported on the close relationship between inflammatory markers in the blood and insulin resistance. In 1999, the Atherosclerosis Risk in Communities Study showed that inflammatory markers could predict the development of type 2 diabetes. Many additional studies have since confirmed this finding. For example, Ridker's team observed in 2001 that patients with high CRP or IL-6 levels in the blood were much more likely to be diagnosed with type 2 diabetes.

Still, some critics wonder if inflammation is merely a symptom of certain metabolic diseases, rather than a potential cause. After all, high blood sugar or abnormal cholesterol levels from overeating can also create inflammation that damages pancreatic beta cells. But Hotamisligil and other scientists remain firm, pointing to the role of evolutionary forces shaping obesity and insulin resistance. The complex interplay between inflammation, fat, and insulin, once adaptive, breeds toxicity in the modern age. Chronic inflammation, Hotamisligil notes, is damaging and absolutely causal to the process. He points to experiments showing that interfering with inflammation reverses disease, a biological response that dates back to ancestral times and is shared by different types of animals today. A diabetic *Drosophila* fruit fly, for example, could be cured of its diabetes by blocking inflammatory pathways, as can a diabetic mouse. But he cautions that in higher organisms, like humans, the pathways involved are more complex, warranting greater efforts to uncover the exact mechanisms to be manipulated in specific diseases.

A surplus of sugar in the blood or excess body fat is not always linked to chronic, hidden inflammation or detrimental to health. Obesity is rare in the wild, but some animals, like seals and polar bears, benefit from the mounds of flesh that equip them to survive in harsh environments. Their natural fat, however, differs from modern-day obesity in humans and is not tied to the same risks. Every autumn, hibernating animals like bears and bats eat themselves into what would be called morbid obesity in humans, with some species getting to a body fat level of 80 percent. For most hibernators, as in people, excess fat leads to other signs of the metabolic syndrome: high blood sugar, hypertension, and unhealthy cholesterol levels. But insulin resistance is helpful in these animals. Since insulin helps to shuttle glucose into cells, insulin resistance drives cells to use fat instead of glucose as a fuel, preparing the body for hibernation.

Hibernators do not have lasting consequences from their brief bout with obesity. They shed their extra fat during the winter season and fail to develop chronic problems like heart disease or diabetes. The constant, low-level inflammation that characterizes modern human obesity dares not take hold. Their transient state of insulin resistance, which flares up and down like acute inflammation, is put to good use before it fades away. This had not been the case with Carrie, in whom inflammation spilled out mostly from visceral fat tissue but also—to a lesser extent—from other organs, including her heart, brain, muscle, pancreas, and liver.

The liver, like fat, is a major metabolic organ, with roles in digestion, nutrient synthesis and storage, and detoxification. In obesity, fatty deposits form in the liver. Kupffer cells—the macrophages of the liver—become activated, turning the liver into a dominant secondary site of metabolic dysfunction and inflammation. Meta-inflammation itself can help to feed liver disease, from simple globs of fat in the liver—the most common type of liver disease in developed nations—to severe inflammation and end-stage scarring.

Obesity affects nearly every organ system, a breadth typically reserved for the worst cases of acute inflammation. Sepsis, for example, damages multiple organs, causing low blood pressure, a fast breathing rate, and mental status changes. For the septic patient, recovery or death loom in nearly equal measures, after which the state resolves. But for those with visceral fat, a low-level metainflammation may silently wreak havoc on their health, escaping detection and biding its time, lingering all hours of the day and with no resolution in sight.

———

Carrie had often struggled with getting restful sleep. For one thing, the buildup of fat in her neck and around her stomach increased pressure on her airways, causing breathing obstructions that led to sleep apnea. Intermittent episodes of low oxygen levels in her blood exacerbated hidden inflammation. But for a few dark, interminable nights one spring, as she awaited the results of a mammogram, her sleep troubles swelled. She had found a small lump in her left breast during a self-examination, and she knew this tiny mass of cells might curtail or irrevocably alter all her remaining days. She was haunted by the prospect of cancer, even though no one in her family had a history of it.

The strong link between obesity and cancer is bolstered by multiple mechanisms, including insulin resistance, high blood sugar, and high cholesterol—all evidence of disturbed metabolism. Inflammation confers an additional hefty risk. The cytokines produced by fat cells are the beginning of a dangerous symbiosis. Fat cells, when grown together with cancer cells, encourage the cancer cells to produce more cytokines as well. The cancer cells in turn coax fat cells into making even more cytokines, which help tumors to grow. If a cancer cell is able to metastasize to fatty tissue, it finds itself in a cesspit of cytokines that improve its ability to survive. In overweight and obese patients, fat cells tend to express an overload of inflammatory cytokines like IL-1β, IL-6, and TNF-α, which are linked to higher risks of developing cancer. It is thus no surprise that many cancers commonly develop right next to fat cells, as in breast cancer, or in places where they are close to fat cells, as in gastric, colon, and ovarian cancers. The environment that fat cells foster to help them survive in thick layers of fatty tissue is the same one that helps tumors to survive.

The lump in her breast, Carrie soon learned, was not cancerous. But she would need another mammogram in a year to make sure the lump stayed small and quiet. She was filled with relief. At a little over forty, the "old age of youth," as Victor Hugo once wrote, time still seemed ample. She resolved to try, yet again, to fight her fat and to resort to medical and surgical therapies as needed. But for Carrie, remaining healthy meant not just losing excess flesh, removing a physical encumbrance, but struggling against a power that hovered below the perception of pain or the glare of a mirror, altering every cell in her body. Hidden inflammation, warping insulin signaling and that of other metabolic regulators like leptin, fueled her hunger. Any diet she sought would need to be adept at fighting this low-level inflammation.

Silent inflammation may shed light on how obesity generates a slew of debilitating pathologies, helping to explain why different types of risk factors for both heart disease and diabetes—the metabolic syndrome—appear in the same person. What emerges is a complex, interconnected web with inflammation at its core, a "common soil" hypothesis that links our greatest modern health threats. Chronic diseases that tend to appear together include not only heart disease, stroke, cancer, diabetes, and obesity but also neurodegenerative conditions like Alzheimer's disease. An individual with one of these problems is more likely to get any of the others. It is no accident, in fact, that these metabolic diseases emerge during aging.

CHAPTER 6

Gray Matter

*T*he flesh around Elie Metchnikoff's belly broadened as he aged. Journalists commented on his "undisciplined hair, lying in heaps, tangled like wheat after a thunderstorm." The pockets of his worn suits that were overstuffed with newspapers and letters as he wandered about "much like one of those Bohemian book hunters or print collectors who haunt the quays of the Seine." He left umbrellas on trains, forgot to remove his galoshes indoors, and often misplaced his felt hat and glasses. But when sorting through stacks of research papers or laboratory equipment, he always immediately found what he was looking for. A lifelong atheist, Metchnikoff eschewed visions of an afterlife. Instead, he became obsessed with finding a "cure" for old age. He initiated the first methodical study of aging, inventing the term *gerontology* in 1903. Most centenarians, he noted, were people of humble means. Much of his scientific attention, as in his youth, centered on macrophages. But he now took a more nuanced view of his cherished eater-cells. Macrophages defended an organism against intruders, but they also maintained body tissues. Metchnikoff suspected that macrophages played a part in many of the debilitating diseases of old age. Perhaps, he mused, they were not only heroes, but villains as well, and all the shades of gray in between.

Metchnikoff's ideas on macrophages and aging, esoteric in their time, began to enjoy a revival in the late twentieth century. Scientists found that macrophages played a part not only in aging but also its accompanying ailments, including neurodegenerative conditions like Alzheimer's and even psychiatric illnesses like depression.

In the 1990s, Italian geriatrician Luigi Ferrucci, now the scientific director of the National Institute of Aging in Baltimore, Maryland, mulled over the mysterious mechanics of aging. He glimpsed some of his elderly patients struggling through the days in wheelchairs while others ran marathons. A person's chronological age seemed to be a fickle measure of health. Was there some way, Ferrucci wondered, to assess *biological* age? Perhaps a telling blood marker that could determine which individuals would succumb sooner than others to the passage of time?

Ferrucci was struck by the burgeoning research that linked chronic, low-level inflammation to heart disease, obesity, diabetes, cancer, dementia, and other conditions his aging patients commonly suffered from. He decided to focus on the cytokine IL-6, which had been shown to portend the risk of heart attacks, strokes, diabetes, and even death. It floated in the blood of obese people, churned out not only by immune cells but also by fat and muscle tissue. Ferrucci measured IL-6 levels in the blood of healthy and mobile middle-aged people, followed them for four years, and found that higher levels could predict who would eventually become disabled, needing assistance with daily activities like walking, eating, grooming, or using the toilet.

Astonished by the discovery, he sent his results to editors at prestigious journals. But he was met with apathy and even ridicule. Reviewers declared Ferrucci's findings absurd, insisting that the inflammation he had unmasked was merely a consequence of disease, possibly due to occult infections in the elderly. That inflammation could be central to—or even an important part of—a process as complex as aging seemed preposterous. Ferrucci published his data only after many rejections. Later, he realized that IL-6 was an imperfect harbinger: some people who did not have high levels began to decline regardless. But his enthusiasm persisted. IL-6, he imagined, was only a part of the story—not the whole story.

One tranquil summer evening at the turn of the century,

Ferrucci was enjoying dinner with his friend and colleague Claudio Franceschi in Modena, Italy. Franceschi was then the director of the National Institute of Aging in Italy, and as usual, the friends' talk turned to the links between inflammation and aging. Low-level inflammation, they knew, typically rose with age; the elderly had higher blood levels of inflammatory cytokines and expressed more genes involved in inflammation compared with their youthful counterparts. Studies had shown ties between markers like IL-6, CRP, TNF-α, IL-1β, and IL-18 and increased risk of chronic diseases, disability, and even death in old people. This type of inflammation needed to be taken seriously, but if it was going to be, Franceschi decided, it needed a proper name. He called it *inflammaging*: the hidden inflammation of old age.

———————

Beyond everything that can inflame young people, the old face additional challenges that contribute to inflammaging. Some are obvious to the naked eye. Body fat tends to travel from the periphery to the midsection with age, creating highly inflammatory visceral abdominal fat. The fall of sex hormones like estrogen and testosterone during menopause and andropause can also inflame. Sex hormones control macrophages and other immune cells, which become unruly and unregulated in their absence. Estrogen, for example, dials down the activity of macrophages in bone tissue. Lower estrogen after menopause contributes to bone breaking down faster than it is rebuilt. Chronic inflammation itself can lower sex hormones, sustaining a damaging cycle.

Inflammaging became increasingly important in the years since Franceschi coined the term, and in 2013, it was recognized as one of the "hallmarks of aging," several essential biological mechanisms that likely drive aging in humans. Genes, for example, tend to grow fragile and more prone to mutations and epigenetic alterations over time. Injured molecules and other stressors accumulate within aging bodies, which, paradoxically, become less adept at handling such stress. Stem cells, which replenish specialized cells throughout the body, begin to wane, as do telomeres. Telomeres are short

DNA sequences that act like buffers at the ends of chromosomes, the fine threads within our cells that hold most of our genetic information, protecting them much like plastic caps on shoelaces that keep the lace from fraying. Studies have shown that people with long telomeres live longer and healthier lives than people with short telomeres.

Inflammaging was both a hallmark and a common force uniting the others. The hallmarks are not independent entities but highly interconnected processes, converging on inflammaging, radiating through it like spokes on a wheel. Any single hallmark can fuel inflammaging, a low and slow-burning flame that pervades old age. At the core of this flame lies an ongoing goading of the innate immune system, in which, as Metchnikoff suspected, the macrophage has a central role.

As we grow old, our bodies accrue biological debris. Damaged or dead cells, misfolded proteins, and other shattered, mutated, or misplaced fragments of our molecular selves pile up in tissues and organs, stressing the system. Mitochondria, those omnipresent organelles within cells that produce the energy that is indispensable to human life, can become injured and inept. All of this detritus excites macrophages and other immune cells. They attempt to clear the "garbage," reacting to it as though faced with a rogue germ and contributing to inflammaging in the process. Our body's tendency to build up biological debris is not merely a function of time but also of our environment. In fact, any inflammatory or anti-inflammatory environmental influence—including food choices—can firmly impinge on one or more hallmarks of aging.

The inflammatory potential of damaged cells reaches far beyond simply striking up an immune response. Damaged cells can fix their faults or self-destruct, protecting the body from unintended consequences—like the tumors that arise from flawed genetic material. But a third option exists: some damaged cells simply surrender to time. No longer able to carry out their duties, they stop growing and dividing yet remain alive, entering a retirement phase known as *senescence*. Senescent cells initially secrete substances that help to repair damaged tissue. But over time, as their numbers increase, they disrupt the structures of organs and tissues. Senescence is not a flip of the switch but rather a gradual evolution, from a transient, reversible quality to a chronic, immutable one. And senescent cells, having divested themselves of their traditional responsibilities, are far from idle. They become potent inflammatory agents, churning out cytokines like IL-6 and IL-1β, altering the behaviors of normal cells nearby and those all around the body, including immune cells. As they accumulate with age, they can be found in all sorts of human tissues, including the skin, liver, lungs, brain, blood vessels, joints, and kidneys. They populate the insulin-producing cells of the pancreas and the muscles of the heart. They fan inflammation in atherosclerotic plaques—making them more likely to rupture—and in visceral fat, where they often take shelter. They promote a multitude of chronic inflammatory diseases. Senescent cells may be one underlying cause of many ailments we associate with aging.

In his fifties, Metchnikoff began to fear death. He spoke of it often, in lectures and books and papers. In his laboratory, he grasped for clarity on its haunting totality. His anxiety became especially acute when his kidneys, unable to filter toxins as they once had, started to flounder. He blamed macrophages for his graying beard, asserting that they devoured hair pigment. This is a partial truth, as we know today: macrophages do play a part in removing pigment during hair growth cycles, and animal studies suggest that sustained inflammation from innate immune activation may contribute to fading hair color. But the process is complex and partly rooted in genetics. In his early seventies, after suffering multiple heart attacks, Metchnikoff finally succumbed to heart failure, unaware that macrophages had saturated his death as they had his life. He donated his body to medical research, allowing it to become dissected and disfigured in the name of science.

Metchnikoff knew that the maladies of aging bodies tended to cluster together. Failing kidneys could herald other illnesses—maybe even ones that would turn fatal. Aging occurs slowly, making it impossible to diagnose the tipping point, the culprit snowflake in a messy avalanche. An organ grows sluggish. Bones and muscles slacken. One becomes frail and disabled and perhaps demented and depressed. Threading through these diseases, which have traditionally been viewed as distinct entities, is a silent, brooding inflammation. Research links inflammaging to many common conditions of old age, including heart disease, obesity, diabetes, cancer, kidney disease, arthritis, lung disease, muscle wasting, bone loss, frailty, dementia, and depression.

Hidden inflammation may increase the risk of developing kidney disease, and the obverse is even more firmly established. When the kidneys malfunction, the body builds up inflammatory triggers, including waste material. It bloats up with fluid, which inflames the gut and causes toxins like lipopolysaccharide to leak out. Lipopolysaccharide, found in the outer walls of certain types of bacteria, is an endotoxin, a bacterial toxin that triggers inflammation. It is one of the most potent activators of the innate immune system. It circulates in the blood and mimics a low-grade infection, provoking the release of inflammatory cytokines like IL-1β, IL-6, and TNF-α. Beyond this, sick kidneys are unable to adequately clear inflammatory cytokines, which linger in the blood and tissues. Inflammation worsens existing kidney disease and catalyzes other inflammatory complications, like heart disease, which is the most common cause of death in patients with kidney disease.

Inflammation also contributes to conditions like bone loss and arthritis. It interferes with bone repair and pushes osteoclasts—the macrophages of bone tissue—and other immune cells to actively break bone down, playing a part in excessive bone loss. Osteoarthritis, once thought to be a simple "wear and tear" joint disease, is now known to be, in part, an inflammatory one. The inflammation in osteoarthritis is more subtle than that seen in rheumatoid arthritis, and the innate immune response prevails.* Macrophages, responding to mechanical stress and tissue damage that occur over years of use or overuse, make their way into the joints of osteoarthritis patients, while the white blood cells of the adaptive immune system, T and B lymphocytes, tend to dominate in rheumatoid

---

* Macrophages make up around 60 percent of the inflammatory cells found in the synovial tissue of osteoarthritis patients, while T cells make up around 20 percent.

arthritis. The cells of a stressed joint also produce their own inflammatory factors that help destroy normal tissue. Many inflammatory diseases increase the risk of bone loss, including autoimmune diseases, obesity, and the dearth of sex hormones in old age.

Muscles, like bones, suffer the ravages of time. Most people begin to lose a little muscle mass and strength by early middle age, but the process markedly speeds up in some as the years go by. While this can happen due to issues like malnutrition and lack of exercise, scientists also point to the harmful effects of hidden inflammation upon muscles. And as with bones, macrophages play a major role.

The picture of old age that inflammaging illustrates is grim. Organs deteriorate. Walking slows and may even stall. Strength plummets, as even a handshake becomes laborious. A fall, a fractured bone, or any manner of inadvertent injury becomes nearly unavoidable. Hospitalizations, disability, and death loom.

These frail individuals are inevitably more susceptible to diseases of the brain and mind. Perhaps this was what Metchnikoff had feared most, beyond the loss of his body: a failing of his mind, and a deluge of unfinished work. And nearly a hundred years after his death, modern science began to reveal the intriguing role of inflammation in these dreaded illnesses.

In 1906, German psychiatrist and neuroanatomist Alois Alzheimer examined the microscopic brain elements of fifty-six-year-old Auguste Deter, who had recently passed away. For a few years, Deter had been one of his patients in an asylum near Frankfurt, where her husband brought her after she began to exhibit uncharacteristic paranoia, agitation, confusion, and memory problems. Alzheimer saw unusual fibers and clumps of stained matter in and around Deter's nerve cells, or neurons. "Peculiar material," he wrote of his findings, drawing what he saw by hand. He also sketched microglia, the macrophages of the brain, which nestled right next to the neurons. When he presented the case at a psychiatry conference in Tubingen later that year, it excited little interest.

Once rare, Alzheimer's disease is now the leading cause of

dementia in aging societies. Dementia is a broad term for defects in a person's ability to think, remember, or make decisions. The "peculiar material" Alzheimer had glimpsed consists of proteins called amyloid and tau. In Alzheimer's disease, the brain builds up misfolded amyloid plaques and tangles of tau, both considered hallmarks of the illness. Genes are involved in the development of Alzheimer's disease, but at least a third of cases worldwide are attributed to acquired risk factors like lack of exercise, smoking, air pollution, hypertension, obesity, diabetes, high cholesterol levels, and brain trauma.

But not everyone suffers Deter's fate. Autopsies show that some people build up brain plaques and tangles in old age but fail to get symptoms of Alzheimer's disease. Scientists working to explain this enigma point, in part, to the immune system and inflammation.

The idea, like many on inflammation and disease, goes as far back as Virchow, who had linked inflammation not only to ailments like heart disease, diabetes, and cancer but also to diseases of the brain. In fact, in his 1858 *Cellular Pathology*, Virchow's small, spherical sketches of nonneuronal supporting cells in the brain closely mirrored modern microglia. When studies like Ferrucci's began to turn up links between inflammation and dementia in the mid-1990s, most scientists assumed that the inflammation was a response to tissue damage. A neuroscientist at the University of Bonn named Michael Heneka guessed differently, however. Recalling Alois Alzheimer's diligent drawings of microglia, he had always suspected that inflammation might also be an active player, provoking illness. In May of 2010, he removed an inflammatory gene—one that coded for the NLRP3 inflammasome—from a dementia-prone strain of laboratory mice. Afterward, the animals aced memory tests and failed to develop any mental handicaps.

Heneka was incredulous. His experiment suggested that inflammation might play a part in the onset of dementia. Research has increasingly pointed to an important role for inflammation in neurodegenerative diseases like Alzheimer's, Parkinson's, and other dementias or even milder types of brain dysfunction. In recent years, scientists have begun to uncover more and more genes involved in Alzheimer's disease. Nearly all are related to the immune system, particularly the innate immune system, including variants that transform microglia from protective to pathological.

Microglia, the most numerous cells in our brains aside from neurons, are central to the link between inflammation and neurodegeneration. Like macrophages elsewhere in the body, they engage in tissue repair and attempt to protect the brain from infections, toxins, or anything else that can damage neurons. The crystal-like amyloid plaques that characterize Alzheimer's, with surface molecules reminiscent of some bacteria, are foreigners that elicit an innate immune response honed over millennia. Microglia attempt to eat and digest amyloid, much like their counterparts stuff themselves with cholesterol in atherosclerotic plaques. They release loads of cytokines in the process, like TNF-α, IL-1β, and IL-6, activating other microglia. This process, typically fleeting, becomes ongoing in Alzheimer's disease, where microglia are high-strung and hyperactive. The ensuing chronic inflammation maims neurons, feeding dementia. When nerve cells become inflamed, the connections between them are less adept at learning and storing information. Inflammation may also fuel the growth of plaques and tangles directly or by hindering the ability of microglia to clear them.

In 2013, Harvard neurologist Teresa Gomez-Isla autopsied brains filled with amyloid plaques and tau tangles. Strangely, only some of her subjects had shown signs of dementia when they were alive. Gomez came across a striking finding: the brains of the demented patients contained many more inflammatory cells than those who had been free of symptoms. Plaques and tangles in the brain, she thought, might need inflammation in order for disease to manifest, akin to the relationship between plaques and inflammation in heart disease. Perhaps inflammation, more so than the plaques and tangles themselves, drove the death of neurons, leading to memory loss and other cognitive issues.

For centuries, scientists believed the brain to be immune to the inflammatory whims of the body, protected by a rigid blood-brain barrier, a thicket of endothelial cells breached only in dire circumstances such as catastrophic brain injury. As early as a few decades ago, the brain was considered to be impermeable to immune cells and inflammatory proteins floating elsewhere in the body. We now know that macrophages and cytokines can cross the blood-brain barrier or relay inflammatory signals through the endothelial cells, spurring the growth of amyloid plaques and tau tangles, stimulating microglia, and impairing neurons. In Alzheimer's disease, as microglia become less adept at clearing amyloid plaques, macrophages throughout the body are recruited to join in their efforts, intensifying inflammation.

An inflamed body in middle age, in one's forties and fifties, maybe even earlier, is linked to mental decline in later years. High blood levels of inflammatory markers in middle age forecast smaller brain volumes, memory problems, and other cognitive issues or full-blown dementias like Alzheimer's. This holds true for those struggling with chronic inflammatory diseases, including heart disease, diabetes, obesity, hypertension, arthritis, and inflammatory bowel disease. Preventing or treating chronic inflammatory illnesses is an essential step toward maintaining mental agility in old age.

Chronic, common infections like cold sores or gum disease may increase the risk of future dementia, too. So can transient yet severe infections, like pneumonias requiring hospitalization. Patients with sepsis, in which an overwhelming immune response to infection leads to bodywide inflammation and malfunctioning of multiple organs, tend to experience rapid mental decline and delirium. In some cases, the inflammation might also cause a degree of permanent damage to the brain. And during milder infections like the common cold or flu, brain inflammation can contribute to the temporary "brain fog" that tends to accompany these illnesses.

A stroke or a traumatic brain injury can also increase the risk of dementia. Inflammation is both a cause and a consequence of many strokes. The same inflammatory processes that harm the blood vessels supplying the heart can damage those that feed the brain. People who suffer from traumatic brain injuries, like football players diagnosed with one or more concussions, may develop ongoing inflammation and a buildup of tau and amyloid proteins in their brains. Strokes and traumatic brain injuries can inflame the brain and lead to not only cognitive issues but also mood disorders like depression. Today, scientists link hidden inflammation to both neurodegenerative and psychiatric health.

———

In the 1980s, over a century after a melancholic Metchnikoff tried to inflame his body with an infectious illness, scientists began to point out parallels between illness and depression. Illness, which inflames the body, also alters the mind. People who are ill feel mental as well as physical fatigue. They may become gloomy and anxious, withdrawing from other people. They eat less and move slowly. Their sleep suffers and the pleasures of life fail to excite. Sickness behaviors likely evolved to help us focus on fighting sickness over lesser concerns—like searching for and digesting food—and to keep us from infecting others. These behaviors, which stem from an innate immune response, are eerily similar to the symptoms of depression.

The first time Metchnikoff tried to commit suicide, he was a young man in love with his wife, Ludmilla Feodorovitch, a sickly woman stricken with tuberculosis. She died only a few short years after their marriage, driving Metchnikoff to suffocate his sorrows with a large vial of morphine. He continued to battle bouts of depression while toiling in laboratories, until Olga, his second wife, came down with typhoid fever, and political unrest worsened working conditions at his university. This time, Metchnikoff attempted suicide by inoculating himself with the spiraled *Borrelia* bacteria, which causes relapsing fever. He survived, as did Olga, and eventually enjoyed brighter days at the Pasteur Institute.

Depression is traditionally thought to arise from low serotonin levels in the brain, and many antidepressant drugs are presumed to increase serotonin. One such drug called Prozac hit the market in the early 1990s, drawing millions and eventually billions in annual revenue. Around the same time, San Jose psychiatrist R. S. Smith quietly published—in an obscure medical journal—a scandalous paper titled "The Macrophage Theory of Depression." He suggested that inflammatory cytokines spewed out by macrophages could travel to the brain and contribute to depression.* Other largely over-looked papers linking the immune system to depression followed suit. Meanwhile, basic scientists found that injecting inflammatory stimuli such as infectious bacteria, lipopolysaccharide, or cytokines into laboratory mice made them act just as they did when subjected to psychological stressors.

That macrophages could affect mood was a wildly subversive notion, challenging not only biological dogma but a deep-seated dualist philosophy dating back to Descartes that staunchly sepa-rated mind and body, with mood disorders languishing far outside the realm of physical illness. The blood-brain barrier was a bodily manifestation of this belief, but it soon began to waver—as did its philosophical base.

Research to date reveals that inflammation may affect not only physical health but also mental health, including depression, suicide, anxiety, post-traumatic stress disorder, schizophrenia, bipolar dis-order, and autism. Genes do play a part in mental health (more so in some conditions, like schizophrenia and bipolar disorder, than oth-ers, like depression). Many of the genes tied to depression relate to the nervous and immune systems.

In the early twenty-first century, scientists began to find higher levels, on average, of inflammatory markers like CRP in depressed individuals or low-level inflammation lurking in the blood before some diagnoses of depression. Research also points to a linear rela-tionship: the more inflamed a depressed patient, the more challeng-ing the depressive symptoms. One large 2013 study, for example, evaluated over seventy thousand individuals for symptoms of psy-chological distress or depression. The researchers found that as subjects became progressively inflamed—as measured by CRP levels—their risk of psychological distress, depression, or even hos-pitalization for depression grew. Meanwhile, many studies involv-ing advanced brain imaging techniques show that inflammation of the body can directly affect the brain and mood. In both humans and animals, brain imaging reveals that body inflammation affects

---

* Research to date reveals that dysregulation of both innate and adaptive immune responses can be seen in depressed patients.

particular parts of the brain, including those involved in mood disorders like depression, altering the brain's functions.

Patients with chronic inflammatory illnesses, including those with heart disease, cancer, obesity, diabetes, dementia, stroke, rheumatoid arthritis, inflammatory bowel disease, lupus, asthma, and allergies, tend to develop depression more often than their healthy counterparts. The effect cannot wholly be attributed to the quality of life changes brought on by living with one or more of these diseases. Meanwhile, depression itself increases the risk for chronic inflammatory diseases and worsens their course. Further, the introduction of inflammatory stimuli seems to be able to induce depression in humans, as with early studies in animals, even in those without any prior mood disorders. Studies show that in patients who receive the inflammatory cytokine interferon-alpha (IFN-α) for cancer or chronic infections, around half go on to suffer from depression. Even a vaccine can elicit enough inflammation to provoke mild depressive symptoms, elevated blood cytokines, and altered brain activity in regions of the brain known to be implicated in depression.

Experiments in animals shed light on possible ways in which inflammation can cause depression. Microglia activated by body inflammation run amok, producing inflammatory cytokines that destroy neural connections in the brain that affect mood and behavior. Brain inflammation reduces the amount of serotonin flowing between neurons. Cell studies suggest that inflammation may also increase the risk of depression by suppressing the birth of new brain cells and quickening the death of existing ones.

We cannot yet claim that inflammation rests at the root of every single case of depression or dementia. But the evidence that continues to accumulate points to an integral role for inflammation in many neurodegenerative and psychiatric diseases, fueling ongoing explorations in this vast and largely uncharted territory. This evidence also aligns with the expanding body of knowledge on inflammation and modern chronic diseases, a remarkable narrative that emerges from consilience between disparate fields of medicine and threads hidden inflammation through the disabling and fatal diseases of modernity. Silent, simmering inflammation is not merely a consequence of these diseases but also a potential cause. And even when hidden inflammation does not directly contribute to disease, it may lower the body's thresholds for tolerating adversity.

The serendipitous and nearly simultaneous discovery of metainflammation and inflammaging in the late twentieth century illustrates the tight links between varied forms of hidden inflammation. While the notion of metainflammation stemmed from studies in obese mice and that of inflammaging was born of research in aging humans, both arrived at similar views on hidden inflammation and pointed to the critical role of chronic, low-level inflammation in metabolic diseases and aging, respectively. And although the inflammatory triggers for metainflammation and inflammaging may differ at times and at other times overlap, some of the molecular mechanisms that sustain them are shared.

The idea that hidden inflammation could be at the root of so many different chronic illnesses, that seemingly unconnected diseases of various body parts and systems share a deep biological link, was once an astonishing claim without much proof. Hidden inflammation lives and breathes wordlessly for long durations. Then one day it speaks up. A heart attack, seemingly out of the blue. The cancer diagnosis. Maybe even a routine day at the gym gone awry, culminating in devastating disability. Perhaps hidden inflammation had been wafting in Jay's blood years before his neck injury, playing some part, large or small, in what happened to him. Rheumatologists conceded it was probable. A silently inflamed individual is more prone to picking up an overt autoimmune disease.

In 1812, a few years before Virchow was born, the *New England Journal of Medicine* published its very first issue. It deemed infections among the most common causes of death, including cholera, consumption, smallpox, and fevers of unknown origin (it pointed to other fatal diseases of the day as well, like diarrhea, fainting spells, drinking cold water, and insanity). By the mid-twentieth century, as deaths from infections—immunology's classic struggle—spiraled into relative oblivion in many parts of the world with the advent of improved sanitation and antibiotics, new diseases began to bloom. A scourge of inflammatory autoimmune and allergic diseases spread through Western nations over the next decades.

Horror autotoxicus, the body's ability to attack itself, once an out-landish idea shunned by science, had now become a daily reality for many struggling with rare autoimmune diseases—like Jay—or common ones, including inflammatory bowel disease, celiac dis-ease, rheumatoid arthritis, multiple sclerosis, asthma, type 1 diabe-tes, and lupus. Alongside autoimmunity, other chronic conditions rose: heart disease, strokes, various cancers, obesity, diabetes, hypertension, neurodegenerative and psychiatric diseases, and more. Like autoimmunity, today these afflictions are considered—at least in part—inflammatory illnesses, a perception undergirded by scientific advances over the past couple of decades. Inflamma-tory illnesses are the most common cause of sickness and death in the world today. Hidden inflammation, which seethes quietly in the body before triggering—or intensifying—a wide variety of disor-ders, may lie at their core. Many of Virchow's ideas on inflammation and disease, ignored in his own time, have undergone a remarkable revival.

Metchnikoff's "big eaters," the wandering cells he watched in baby starfish larvae centuries ago and fought to keep alive as col-leagues attempted to crush them into extinction, also made a slow ascent to spectacular heights. Macrophages are both wolf and sheep, involved in nearly every disease, central to immunology and inflammation. They fight germs and tumors and promote tissue repair. They abound in most cases of chronic inflammation, includ-ing autoimmune diseases, working next to the cells of the adaptive immune system. And they play fundamental roles in the hidden inflammation that weaves through deadly diseases of modernity. Macrophages appear in every stage of heart disease, gorging on lipids stuffed in atherosclerotic plaques, pumping out dozens of inflammatory molecules, and portending plaques most likely to rupture and cause a heart attack. They lodge in body fat, devotedly wrapping themselves around adipocytes and spewing out most of the inflammation from excess flesh. They huddle in the pancreas, plotting to dismantle the cells that create insulin. They allow them-selves to become hijacked by tumors, irrevocably betraying the body. They partake in aging and its attendant ailments. In the brain, they help the mind to unravel.

In the early twentieth century, Portuguese philosopher Fernando Pessoa lived and worked in a small flat in Lisbon, where he filled his solitary life with prolific writing stored in volumes of notebooks. Pessoa often wrote under new identities with alternate names. He did not call them pseudonyms, a word that he claimed failed to capture the true intellectual depth of his divergent selves, but rather *heteronyms*. There was Alberto Caeiro, who lived in the country with his great aunt and received only a primary school education, or Alvaro de Campos, who studied mechanical engineering and wrote violent, wrathful poems. Or Ricardo Reis, who urged his readers to find pleasure in the moment rather than pursuing ambitions. And so on with nearly a hundred strikingly singular selves produced by one man. Pessoa churned out page after page for most of his life, living in obscurity, drawing no attention to himself or his work until the day he suffered stomach pain and fevers. "I know not what tomorrow will bring," he wrote on his deathbed at the Hospital Sao Luis, well before the posthumous explosion of interest in his work.

Inflammation, like Pessoa's heteronyms, inhabits multiple forms. Acute or chronic, hidden or overt, the pathways and players change depending on why, how, and where the body is inflamed. A sprained ankle yields Celsus's cardinal signs of *rubor*, *tumor*, *calor*, and *dolor*. The coughs and sneezes of common colds are a product of inflamed membranes that line the nose and throat. A flu virus inflames the entire body, causing muscles and joints to ache. Chronic, hidden inflammation is distinct from glaring autoimmunity, both in how it is seen and felt. And yet one writer, one imagination, wields the pen. Inflamed tissues typically encounter immune cell infiltration, the birth of new blood vessels, and an increase in cell numbers. If I walk the halls of the hospital I work in, the names of cells embroiled in inflammation differ depending on what floor I am on—endothelial cells in the heart, enterocytes in the gut—but the underlying inflammatory responses are similar, the cellular interactions constant though unique to each tissue or organ. It seems we are all staring, in some way, at the same—yet distinct—thing. The masterful ability of hidden inflammation to adopt an array of guises, to proceed covertly until chaos, poses a quandary: how to prevent, catch, or treat this silent fire, intervening before an inferno arises?

CHAPTER 7

Resolution

T he smell of melena—dark, sticky feces containing digested blood—is singular and unforgettable. As I walked into New York Presbyterian's intensive care unit one fall morning, the stench hit me hard. Old congealed blood with copper and iron, burnt and mixed with stool, something sweet and sour and rotting, not quite what you would find in a slaughterhouse but just as sinister.

The patient had been vomiting up fistfuls of what looked like coarse ground coffee, the vomit filled with blood darkened from exposure to acids in her stomach. When I turned her body sideways to do a rectal exam, watery black clots poured out of her anus and onto my gloves. I put an endoscope down her throat, a snake-like tube about the thickness of my index finger. Once I reached her stomach, I could see a large, cratered white patch with bright blood spurting out from one vessel. I injected epinephrine, a medication that constricts blood vessels and helps to stop bleeding, around the ulcer. I passed a clip, a device with a pincerlike claw at its end that can grasp and compress tissue, down the endoscope's biopsy channel. After I placed two clips on the vessel, the bleeding finally stopped. The patient had been taking aspirin—lots of it— for back pain.

Anti-inflammatory drugs like aspirin were among the first

reported drugs used by humans. The *Ebers Papyrus*, an ancient Egyptian medical text, noted that the bark from a willow tree could be used as an anti-inflammatory medication to relieve aches and pains. Thousands of years later, when Celsus first described the redness, heat, swelling, and pain of inflammation, he used willow bark to mitigate these symptoms. In the late eighteenth century, Reverend Edward Stone, a country parson in England, accidentally tasted the willow bark he had been using to treat fevers. Its bitterness reminded him of the quinine-containing cinchona bark used to combat malaria.

In 1828, Johann Buchner, a professor of pharmacy at Munich University in Germany, succeeded in extracting the key ingredient responsible for willow's healing properties: bitter yellow crystals he named salicin. Then, in 1897, a bright young chemist named Felix Hoffman,* working for the pharmaceutical company Bayer, created a chemically pure and stable form of salicin that he used to relieve his father's rheumatism. In 1899, Bayer launched the substance under the trade name aspirin, distributing the powder in glass bottles. The drug became a wondrous hit, making the Bayer name world-famous.

Aspirin was initially used to treat pain, fever, and inflammation. It lowers the activity of NF-κB and many inflammatory molecules, including CRP, IL-6, and TNF-α. It is also an antiplatelet drug that helps to prevent blood clots. Today it is routinely prescribed to ward off recurrent heart attacks, strokes, and some cancers. With ibuprofen, naproxen, and other common medications, it falls in a class of drugs known as nonsteroidal anti-inflammatory drugs, or NSAIDs.

NSAIDs can be obtained over the counter and are used all over the world to alleviate the aches and pains of modern life, from muscle or joint pains to headaches, colds, flus, fevers, and more. They work by hindering the function of cyclooxygenase enzymes, which create hormones, known as prostaglandins, that induce inflammatory symptoms like pain and fever as part of a healing process. But dampening the inflammation comes with a price. Inhibiting

---

* Hoffman was possibly working under the direction of his colleague, Arthur Eichengrün. Hoffman would later go on to synthesize heroin.

prostaglandins, which also serve to protect the inner layer of the stomach from acid damage and promote blood clotting by clumping platelets, can lead to stomach and intestinal ulcers. Every year, thousands of Americans die from bleeding caused by NSAIDs. These drugs can also cause liver or kidney problems. Some NSAIDs limit the severity of gastrointestinal side effects by targeting only one kind of cyclooxygenase enzyme, but these varieties are not in common use because they paradoxically promote blood clotting and increase the risk of getting a heart attack or stroke.

Throughout history, anti-inflammatory drug development has focused on suppressing or blocking inflammation. From willow bark and NSAIDs to an enormous range of modern agents with new therapeutic targets, anti-inflammatory drugs aim to throw water on a burning fire. However, a growing body of research reveals a radical new strategy for combating inflammation. Scientists today are studying not only how to suppress inflammation but also how to *reverse* it.

On that summer day in 2012 when Jay first became sick, he took ibuprofen to help with the muscle pain in his neck. When it did not help, and when his condition worsened—progressing to a complete head drop and problems with breathing and swallowing—doctors prescribed prednisone, a steroid drug.

Steroids are powerful synthetic drugs that resemble cortisol, a hormone the body produces naturally. Unlike NSAIDs, they do not just calm inflammation. They also suppress the immune system in a broad, nonspecific way by inhibiting many different types of inflammatory cells and chemicals. Nearly a century ago, they were hailed as miracle drugs. When they were given in the late 1940s to a bedbound young woman at the Mayo Clinic with rheumatoid arthritis, her severe joint swelling and pain disappeared. She was able to walk up and out of the bed she had been confined to for years (and, as the story goes, downtown to Rochester to go shopping), stunning the world. The French painter Raoul Dufy, so disabled by rheumatoid arthritis he had to affix his brush to his hand with tape, improved remarkably with steroids, as did his colorful but clumsy paintings.

But once again, there is a cost. While NSAIDs are ostensibly portrayed as benign and are in frequent, cavalier use, steroids are well known for their horrific side effects (an issue that initially prompted the creation of NSAIDs in the first place). Over the months he was on steroids, Jay became restless and could not sleep well. He gained weight and his face swelled. His adrenal glands started to shut down and he was more fatigued than ever, with little appetite and weak muscles. He bruised easily. The longer he used steroids, the more he risked additional problems like serious infections, eroding bones, and diabetes. Once prescribed liberally in the 1950s, the side effects of steroids made doctors cautious about their use, especially for long durations. In my own patients I deploy them sparingly, only for the worst cases of inflammation, and mostly to buy time while newer, safer medications take effect.

Still, when there is an inflammatory crisis, steroids can save organs—and lives. As the days drummed on, I watched and waited for steroids to halt Jay's inflammation, to produce a miraculous recovery, or at least a slow, incremental restitution. But even the highest doses failed to help.

Many of our weapons against inflammation are frustratingly vague. NSAIDs and steroids, the former with the potency of a slingshot and the latter with that of a sledgehammer, are undiscerning when they seek to deter inflammation and suppress the immune system. They take broad aim, and varying degrees of collateral damage inevitably ensue. Central to the quest for new drugs that suppress the immune system is the idea of specificity, or the ability to inhibit the immune response as selectively as possible, to shut off a problem pathway while preserving the body's ability to defend itself. This harkens back to Paul Ehrlich's staining experiments and his vision of precise, lock-and-key relationships between antibodies and their targets. In the 1900s, Ehrlich himself applied the concept of antibody receptors to the search for drugs, envisioning that a drug could be tailored to kill a specific germ just like a bullet could be fired from a gun to hit a unique target. He named the hypothetical drug *Zauberkugel,* or "magic bullet," and eventually developed Salvarsan, the first effective drug for syphilis.

Today, efforts to create magic bullets to treat inflammation are ongoing. The challenge lies in that the gun is aimed inward, toward the body, targeting a mechanism molded by evolutionary forces, one that is integral to our survival. The early history of attempts to manipulate the immune system was dominated by efforts to elicit or enhance the immune response in order to resist germs, as with the development of vaccines. The notion of rigorously suppressing the immune system is a relatively new idea, born of scientists' increasing awareness that an overreactive immune drive demands restraint, whether the pathology stems from autoimmune disease and allergies or medical interventions like organ transplantation.

In the second half of the twentieth century, an arsenal of immune-modulating drugs was developed. These drugs, if visualized as a staircase, generally become more potent or specific—or both—with each step up. They alleviate autoimmune diseases, gifting patients with long-lost function. They allow foreign organs, once routinely cast out and rejected by angry immune systems, to be set in new bodies without discord. Hearts, lungs, livers, kidneys, and more make their way from brain-dead donors to new, hopeful homes. A clinician tasked with the art of treating inflammation decides not only which medications to use but also whether to start at the bottom or the top of the staircase, to start small and escalate therapy or to begin with the biggest guns and scale back as needed. The method chosen is dictated by the disease process, how acutely ill the patient is, how much disability or death is at stake.

Jay's rheumatologist at the University of Chicago, Dr. Carter, knew he would have to start at the top of the staircase, with the toughest tools at his disposal. Even on the steroids other doctors had prescribed, Jay was fading fast and risked severe, permanent impairment—or worse. What was once neck strain mandating the use of ibuprofen had morphed into something caustic, stealing more muscle function with each minute it persisted unassailed. Jay lumbered through the days in an awkward brace, a lifeline that allowed him to move through the world intact, since the muscles in his neck and back could no longer keep him upright. He feared that modern medicine was no match for his immune system, which might eventually finish what it had begun. He wondered if the brace would become a permanent extension of his body.

Carter's problem was grand. He could not entirely grasp the shape or form of the inflammation he was chasing. It had revealed its catastrophic wrath, its ability to engender extreme loss of function, but the underlying process remained shrouded in secrecy. The cardinal signs of inflammation were not only invisible to the naked eye but also largely undetected by modern medical testing. Carter could not predict how Jay would respond to therapy. Each individual has a singular response to an immune-modulating drug, depending in part on their genetic makeup. And each drug targets a unique aspect of the myriad inflammatory pathways in the body.

The challenges in targeting inflammation, a primordial mechanism that protects us from the environment—including the organisms with whom we share the planet—are vast. The immune response, critical to sustaining life, evolved redundancies. It is orchestrated by many molecules, so targeting one or a few may not be enough. The process is finely tuned, with inherent sensors and feedback pathways. Inhibiting one critical component of inflammation may simply trigger a compensatory response involving another pathway. Even when these challenges are overcome, the risks as compared to the benefits from taking a medication that suppresses the immune system may be unacceptable, such as fatal infections.

Carter decided to take a swing in the dark by attempting a forceful, urgent shutdown of Jay's immune system. He would use some of the most potent weapons in his arsenal, all wielded in concert. Jay was daunted by the idea, but he was also desperate enough to proceed with the gamble. Carter continued Jay's steroids. In addition, he prescribed tacrolimus, a drug commonly used after organ transplants to prevent the body from prompting an inflammatory attack against foreign tissue. Tacrolimus suppresses T lymphocytes, which help to protect the body from infections. Serious risks include kidney damage, cancer, and severe infections. Jay would need to avoid close contact with people who had infections, keep cuts and scratches clean, and stay away from certain foods like raw oysters or other shellfish. Carter also added azathioprine to Jay's regimen. Azathioprine is a chemical that suppresses the immune system by inhibiting DNA synthesis and decreasing the production of immune cells. Its side effects include gastrointestinal issues, certain cancers, and bone marrow suppression, which can lead to anemia.

There was something else Carter tried, too. I made a trip to the hospital's infusion ward that summer, where Jay sat in a reclining peach-colored chair. An intravenous bag with clear fluid hung from a cart next to him. The bag had the look of an innocent saline solution, but it held pooled immunoglobulins, or antibodies, from the blood of around a thousand donors. The mixture is called intravenous immunoglobulin. When used in autoimmune disease, it attempts to distract the immune system from whatever it is bent on destroying, like redirecting a crying child with a shiny new toy, displacing or neutralizing the antibodies and cytokines that have turned against the body.

But these are general theories—no one knows or has proven exactly how intravenous immunoglobulin works to help inflamed patients. The week before, Jay had anxiously read up on what could go wrong with intravenous immunoglobulin treatments, including kidney failure, blood clots to the heart or lungs, and brain inflammation. Yet it seemed fitting for a mysterious case of debilitating inflammation to be treated by a process just as inscrutable. He sat back in the chair for eight hours, eyes closed, taking slow breaths, while the catheter that punctured his vein drew the fluid into his body. It was the first of many identical infusions to come.

———

Carter's urgent treatment of Jay's illness was fundamentally traditional, aimed at suppressing inflammation. Around two decades before his daring attempts, in a laboratory nearly a thousand miles away, another physician embarked on an entirely new way of tackling inflammation. Charles Serhan sought to understand not only how to suppress the inflammatory response but also how to reverse—or *resolve*—it. On his return to Boston from a trip to Japan in 1989, where he had spent glorious days roaming the streets of Tokyo and Kyoto, lecturing at meetings and snacking on all kinds of delectable noodles and small fish, Serhan fell suddenly, violently ill. At first, he blamed the take-out meal he had eaten the night prior, having had no time to cook for himself since his arrival. Sharp pains shot through his stomach, which was oddly firm and exquisitely tender to touch. When his symptoms failed to abate and he developed a fever, Serhan was rushed to Brigham and Women's Hospital, where he had recently begun his career as a faculty research scientist in the Department of Medicine. Doctors diagnosed a bowel perforation, a small hole in the intestines. Surgeons operated urgently to close the gap.

No one knew why it had happened. Serhan, a young man in his thirties, had never been sick before. He did not have any intestinal issues like diverticulitis or inflammatory bowel disease. As he lay in a hospital bed with a newly created ostomy—an opening through the skin on the abdominal wall through which the intestines dump stool—he wondered what exactly was happening to his body. He asked a postdoctoral researcher in his laboratory to draw repeated blood samples from his vein, which revealed high levels of inflammatory markers. This was expected for someone with inflamed intestines who had just undergone surgery. But Serhan, in a dreamy haze from the effects of painkillers, was more curious than ever about the melodrama with minute actors taking place in his body. His doctors, he felt, did not understand inflammation.

Back in his laboratory, after a second surgery to reconnect his intestines, Serhan decided to investigate further. He recreated in mice a version of the inflammation that had ravaged his intestines. In the bodies of his mice, as in his own, acute inflammation, the first line of defense against injury, was defined by a congregation of various cells, including neutrophils and macrophages. These cells pumped out powerful inflammatory mediators to fuel the fire, like cytokines and chemokines. Blood vessels dilated, blood flow increased, and fluid and protein leaked out into tissues. Once the stimulus abated—be it a germ or injury—the inflammatory cells made their way out of the scene and the inflammation resolved.

Serhan, unlike most other scientists of his day, was most inter- ested in this latter phenomenon: the dwindling of the fire, the dead cells and debris in the battle's wake. This process of resolution had been well documented in the literature for as long as there had been microscopes, but it was thought to be passive, meaning that immune cells and the chemical mediators they secrete naturally decrease with time, diluting their effects. In his mice, Serhan took a closer look at the events that transpired during this phase. He noted the moments in which inflammatory cells first began their retreat and those in which they would vanish completely, detailing exactly how fast inflammatory responses died down. He wondered if there was more to the picture, if there were hidden signals guid- ing the fate of these cells. It could be, he thought, that the world had it backward. Maybe the overarching dilemma in inflammatory dis- eases involved not the ignition, the on switch, but a broken resolu- tion, the off switch.

Serhan and his team continued working throughout the 1990s, and by the turn of the twenty-first century, they revealed that the resolution of inflammation is indeed an active process. Inflamed tis- sue does not return to its unadulterated state as a matter of course. The movement to clean up the mess and repair systems relies on specific anti-inflammatory cytokines, growth factors, and other molecules. Macrophages and neutrophils, cells that prompt acute inflammation, switch gears when it comes time for inflammation to die out, releasing new chemicals, the invisible mediators Serhan had been looking for. He named these molecules *resolvins*.

Working with scientists around the world, Serhan and his team continued to discover more and more of these molecules in the ensuing decades, culminating in a superfamily of "specialized pro-resolving mediators"[*] that includes resolvins, lipoxins, pro-

---

[*] Macrophages and neutrophils produce most of the specialized pro-resolving mediators identified during the resolution of inflammation. Research indicates that muscle and fat tissue can also make and release certain specialized pro-resolving mediators. The origin of specialized pro-resolving mediators identified in other tissues (for example, the placenta, human breast milk) are not yet known. Specialized pro-resolving mediators are referred to as "pro-resolving mediators" throughout this text.

tectins, and maresins. Pro-resolving mediators are unique immune-signaling molecules. Most are derived from lipids, not proteins. They help turn off inflammation, ridding the body of any residual inflammatory cytokines and debris. They slow the infiltration of immune cells and push macrophages to ingest dead cells—one of the signals that drives macrophages to switch to an anti-inflammatory state. In many animal studies and some human ones, these small molecules have been shown to reverse inflammation in disease and amplify the healing response, prompting tissue regeneration and wound repair. Pro-resolving mediators reduce tumor growth, enhance cancer therapy, and lower inflamed body fat. They protect against ischemic stroke and Alzheimer's disease. Serhan's experiments have shown that pro-resolving mediators resolve the inflammation that occurs when blood flow returns to oxygen-deprived tissues during surgery. They are also especially adept at turning off one of the most aggravating symptoms of inflammation: pain.

Pro-resolving mediators stimulate special white blood cells known as regulatory T cells (Tregs). Tregs are critical for maintaining inflammatory homeostasis and subduing excessive inflammation. With a mere touch, they can dial down the activity of another immune cell. They manage and mollify all kinds of innate and adaptive immune cells, including macrophages, dendritic cells, B cells, and certain highly inflammatory types of T cells linked to chronic diseases and organ rejection. Tregs produce anti-inflammatory cytokines like IL-10 that are integral to protecting the body against unwanted immune reactions and resolving inflammation. They tell the body to tolerate its own antigens, preventing horrific, lethal autoimmunity. They help transplant patients accept foreign body parts, averting organ rejection. Tregs are useful in many diseases linked to hidden inflammation, including heart disease, obesity, diabetes, inflammatory bowel disease, rheumatoid arthritis, and lupus. They abound not only in lymph tissues like the thymus, bone marrow, lymph nodes, and spleen, but also in the skin, hair, lung, liver, fat, brain, and placenta. Obese individuals tend to build up inflammatory T cells in their adipose tissue at the expense of Tregs.

Resolving inflammation is not equivalent to just dampening it. When inflammatory pathways are turned off, as with traditional anti-inflammatory drugs, the risk of unwanted casualties looms large. Pro-resolving mediators, which have both anti-inflammatory and pro-resolving properties, aim to solve the underlying problem, not mask its effects. They strengthen rather than impede signaling pathways that have evolved over millennia, commandeering nature's own anti-inflammatory mechanisms, with little to no risk of immune suppression. In fact, they actively assist the body in killing and eliminating germs. Resolution pathways, like inflammatory ones, are distinct. Different people with the same disease, as well as different tissues in the same person, may have singular resolution pathways. It is, as one scientist said, like walking into a barroom brawl, unaware of who started it or how it is going to end. Still, resolution strategies can intrinsically afford greater scope than conventional approaches.

Most modern drugs do nothing to foster resolution—and some actively disrupt it. NSAIDs, for example, lower the amplitude of inflammation, decreasing redness, heat, swelling, and pain, but they also delay resolution. The quieter inflammation may linger longer in the body, hidden from view. The exception is aspirin,* which blocks inflammatory mediators but also triggers the production of some pro-resolving mediators. Aspirin, with its roots in ancient medicine, is one of the few modern drugs with the ability to both dampen and reverse inflammation. Pro-resolving drug development is still in its infancy. A mouth rinse deemed safe in periodontal disease, eye drops filled with the same pro-resolving mediators found in tears, and pro-resolving drugs for the prevention or treatment of neurodegeneration as well as autoimmune conditions like inflammatory bowel disease are underway.

The dilemma of treating hidden inflammation rests in part on how to test for it—how to capture a presence we cannot see or feel, yet one that may, in some years or decades, culminate in disaster. Here, too, resolvins are poised to play a role. Chronic, low-level inflammation can hide in tissues, organs, and blood vessels. Most

---

* Aspirin, rather than inactivating the enzyme cyclooxygenase, modifies its function so that it stops producing pro-inflammatory prostaglandins and starts to produce pro-resolving mediators.

testing approaches to date have focused on measuring inflammatory particles in the blood, the cost and utility of which can vary.
CRP, a molecule made by the liver in response to the cytokine IL-6,
rises in patients with all kinds of inflammatory triggers or illnesses.
High-sensitivity CRP, which Ridker had chased in his early experiments, detects the smallest elevations in CRP. It can unveil hidden
inflammation and is used to predict the risk of heart disease. More
expensive tests include those that search for important cytokines
like TNF-α, IL-1β, and IL-6, all of which play a part in inflammatory conditions, including heart disease, obesity, diabetes, and certain autoimmune diseases. While existing blood markers are tied to
various inflammatory diseases and even the risk of death, problems
can arise with using such markers to routinely catch hidden inflammation. They provide a polaroid snapshot of the blood, declaring
that inflammation is present, but do not reveal why it persists or
how long it has been around for. Acute inflammation from a cut or a
cold in a healthy person, for example, can generate some of the same
inflammatory markers as the more malevolent hidden inflammation tied to chronic disease.

As research accrues, specific patterns and clusters of many additional markers—inflammatory "signatures"—may help to not only
better define the state of being silently inflamed but also shed light
on its source in the body. We might also measure, in addition to simple concentrations of markers in a given individual, their relative rise
and fall in response to inflammatory challenges. Radiographic testing, like MRI or computed tomography scans, can add valuable quan-
titative information about inflammation. For example, these studies
can mark and measure the inflammation that surrounds blood vessels or point to inflammatory findings on atherosclerotic plaques
that portend rupture.

A few routine blood tests can also be useful surrogates for reveal-
ing inflammation. These include tests of fasting insulin and hemoglobin A1c levels. Hemoglobin A1c is a form of hemoglobin—a blood
pigment carrying oxygen—that is bound to glucose. High levels of
fasting insulin or hemoglobin A1c are indicative of excess inflammation tied to diabetes or other diseases. Hemoglobin A1c is also used
to monitor how well diabetes is controlled. Homocysteine, an amino

acid that travels in the blood, is a risk factor for heart disease and can be altered by diet and other lifestyle choices. A surplus of homocysteine is linked to markers of inflammation and to chronic inflammatory diseases.

But Serhan believes that quantifying resolvins can also help to diagnose hidden inflammation. He developed a method to measure circulating levels of pro-resolving mediators and noticed that individuals with chronic inflammatory diseases tended to have lower levels. People with diabetes, for example, had not only too many cytokines swimming in their blood but also too few resolvins. Inflammation's off switch, he speculated, probably played as much of a role as its on switch in a wide array of pathologies, including chronic wounds, typical autoimmune diseases like rheumatoid arthritis, and other modern afflictions tied to hidden inflammation—heart disease, cancer, obesity, diabetes, neurodegenerative diseases, and more.

In the twenty-first century, the burgeoning approach to testing for and treating hidden inflammation relies on tactics similar to those used for typical chronic inflammatory diseases. But the state of being silently inflamed is defined by unique regulatory circuits, and approaching it with the heavy weapons tailored for the classic state may be less than ideal. For a smoldering inflammation that does not know how to stop, a low-risk solution that allows it to fade gently into oblivion, minimizing ancillary damage, may be fitting.

No magic bullet existed for Jay's condition. But tacrolimus, azathioprine, and many infusions of intravenous immunoglobulin finally halted his ongoing inflammation, preventing fresh injuries. After a couple of years of grueling physical therapy, Jay regained around half of the normal strength in the muscles of his neck, some of which had hypertrophied to partially compensate for those destroyed. He was able to discard the brace, ride his bike, and hike. Running remained a challenge. Due to residual weakness in his muscles, it was difficult for him to keep his core stable and to maintain an upright posture during a run. A heaviness in his neck would return at the end of each day, recalling the terrifying loss of function in the past as well as the haunting possibility of a future recurrence.

Carter presented Jay's case at an international rheumatology meeting. Jay's disease was likely an atypical *necrotizing autoimmune myopathy*, an embodiment of the irate moments in which his immune system had raged against the muscles in his neck and other parts of his body. Much like the artificial boundaries between aging and age-related diseases, self and non-self, and even innate and adaptive immunity, the iterations of inflammation are shape-shifting entities that rest on a continuum. Traditionally, we have jumped from identifying inflammation to dampening it with drugs. Drugs can be potent antidotes to a variety of disabling or even fatal inflammatory processes, as evidenced by Jay and many other patients.

But hidden inflammation presents a unique dilemma. The idea that it may be a shared biological mechanism between diseases as distinct as obesity and aging, or depression and heart disease, fosters a new understanding of human health. It pushes us to consider preventing or treating these diseases in concert rather than in parts, taking into account the totality of a patient's body and mind, including the germs that live within and on top of us. The biomedical framework of the nineteenth century—divided by organ systems and an understanding that a specific cause results in a specific disease—is no longer effective for most health conditions that plague us today.

Effectively combating hidden inflammation begins with delving deeply into its root causes. Human genetics, which have remained relatively constant, or longer life spans cannot alone explain the steep rise in chronic inflammatory diseases over the last decades. Our destinies are largely shaped by lifestyle. Hidden inflammation is one important mechanistic link between many environmental triggers and the ailments they are associated with. For example, most cases of cancer and heart disease, two of the twenty-first century's top killers, are a product of modern life. Genes play a much smaller role. Environmental carcinogens and lifestyle habits—including smoking, alcohol use, pollution, diet, and chronic infections—can damage DNA and inflame tissues, cultivating an ideal context for cancer to take hold and grow in. They can also—along with high blood sugar and hypertension—injure and inflame the simple, single-layered endothelial cells lining the inner wall of coronary arteries supplying blood to the heart, the life support of the body's vasculature.

Hidden inflammation, rather than protecting us from our greatest modern hazards, helps to promote them. This mute heat is largely fueled by our modern environment, with food emerging as one of the deadliest factors.

CHAPTER 8

Quiet Conversations

O livia was in her mid-thirties, with straight, carefully cropped auburn hair. She put in long, arduous
hours at work, not for the love of labor or money,
but to ward off the anxiety that consumed her when she felt she
was not being productive. She wore a lace-backed coral summer
dress and tapped her painted nails on her purse throughout the
visit, keeping up with the rain's rhythmic drumming on the win-
dowpane. We were in an eighth-floor gastroenterology clinic at
Columbia's Herbert Irving Pavilion. The sky was dark and the
room felt small.

Months ago, Olivia had developed bad stomach pains and saw
blood in her feces. During a colonoscopy, her gastroenterologist
looked for gleaming, pale pink tissue, with saccular segments of the
colon neatly lined up like the folds of an accordion, findings typical
for normal colons. Instead, cratered ulcers interspersed with wild
red streaks colored Olivia's mucosa, the membrane lining her intestines. It was swollen and disfigured, with patches of disease transforming smooth folds into a misshapen mosaic. She had Crohn's
disease, a type of autoimmune inflammatory bowel disease in
which the immune system attacks the lining of the gastrointestinal
tract, causing inflammation and bleeding. Symptoms typically wax

and wane. Patients often have stomach pain, bloody diarrhea, and nutritional deficiencies, but they can also suffer problems related to inflammation outside of the gastrointestinal tract, like mouth sores, skin rashes, joint pains, and red eyes.

Olivia's doctor initially gave her steroids, which quickly calmed her inflamed intestines but could not be used for long duration because of their toxic side effects. Instead, she was maintained on azathioprine, one of the drugs Dr. Carter had prescribed for Jay. Every few weeks, Olivia also received infusions of a biologic drug, that is, a drug derived from living organisms. This drug, called infliximab, is a genetically engineered antibody that targets the inflammatory cytokine TNF-α, which is produced by macrophages and other immune cells. Blocking TNF-α has downstream effects, since TNF-α triggers a cascade of additional inflammatory molecules that play a role in various diseases, including IL-1β (the cytokine targeted by canakinumab, which is also classified as a biologic drug) and IL-6. Sometimes, for the sickest inflammatory bowel disease patients, a combination of drugs works best. Since each drug uses a different tactic to treat an inflamed gut, the drugs can work synergistically.

Olivia gained weight and felt better on these therapies. But they made her nervous. While the drugs were safer for long-term use than steroids, with a lower risk of serious infections, hazards still lurked. Patients on infliximab may succumb to certain infections that TNF-α protects against, like tuberculosis. They also suffer a small but real risk of lymphomas and other cancers. And over time, many inflammatory bowel disease drugs can lose their effectiveness.

Medications can improve symptoms and help prevent catastrophic complications in Crohn's disease, like intestinal fistulas and narrowings or even cancer, but there is generally no cure. Consequently, Olivia was desperate to do anything she could to prevent or treat her inflammation. In particular, she sought advice on nutrition. The Internet was awash with stories of inflammatory bowel disease patients altering their diets and remaking embattled, inflamed intestines—and their lives. A plethora of books and blogs claimed the anti-inflammatory diet could cure everything from autoimmune conditions to heart disease and cancer. But there didn't seem to be a consensus on what this diet should consist of. Were grains, particularly those containing gluten, acceptable? What about dairy or beans?

Olivia is not alone. Many patients walk into my clinic wanting to know what they should be eating to prevent or treat inflammation, from those with wheat sensitivity and food intolerances to others with serious autoimmune issues like inflammatory bowel disease. Even healthy people, increasingly aware of the link between inflammation and disease, hunt for concrete answers. Inflammation has become a buzzword in mainstream culture, and the anti-inflammatory diet is one of the most coveted yet confusing topics in nutrition. As the threat of epidemics and pandemics persists in the twenty-first century, people also long to eat and live in ways that can boost immunity.

Much of the immune system lives in the gut, which is heavily exposed to the external world. There are three major channels* through which inflammatory triggers enter our system: the skin, the lungs, and—an especially susceptible point of entry—the gastrointestinal tract. With a surface area much larger than that of the skin, the hollow tubes of the gastrointestinal tract exist, essentially, outside the body. The gastrointestinal tract starts at the mouth, where digestion begins and saliva helps to break down food while we chew. After we swallow, the muscular esophagus, a hollow pipe, pushes food into the stomach, where it is held and mixed. The small intestine, a long, coiled tube measuring around 20 feet when taut, uses digestive enzymes from the pancreas and bile from the liver to pulverize the broken bits and is also responsible for absorbing nutrients into the bloodstream. Waste made up of food debris and bacteria then passes through the colon, which absorbs water before stool is emptied into the rectum and flushed out of the body. If the gastrointestinal tract was sliced lengthwise and spread flat, the mucosal part—the barrier

---

* Other channels through which inflammatory triggers can enter our bodies include the eyes, ears, and urogenital tract.

side that comes into contact with outside materials—would have the surface area of a small studio apartment.

The innate immune response, an ancient mechanism shaped by evolutionary forces over hundreds of millions of years, is called on by the body to counter germs, poisons, and trauma. It also responds to an ostensibly harmless, common intruder in modern times: food. This most primitive aspect of our immune system is a central player in the interactions between food, the germs in our gut, and other elements of the environment that fuel hidden inflammation tied to modern chronic diseases.

The food we eat can directly spark or inhibit an inflammatory response by the immune system. Innate immune cells—including macrophages, neutrophils, and dendritic cells—and the epithelial cells lining the intestines rely on primordial pattern recognition receptors to closely examine germs and other matter.* These receptors, which are embedded in cell membranes, recognize alien material and trigger a powerful inflammatory or anti-inflammatory response, activating genes and generating a cascade of signals conserved across the phylogenetic ladder, from insects to plants to humans. They can even respond to distress signals spewed out by stressed cells all over the body. In short, the immune system is prepared to fight food as it would a germ.

Food can also influence inflammation through the microbes in our bodies. The comprehensive, nuanced answers to Olivia's questions rest in an exploration of the intricate web of relationships between food, germs, and inflammation. The true story of the anti-inflammatory diet hinges on decades of research on diet and disease as well as a burgeoning understanding of how germs, particularly those that live in the gut, affect human health. The origins of this story lead us back to macrophages, the eater-cells at the heart of modern disease, and their conversations with microbes.

---

* Pattern recognition receptors, which include toll-like receptors, are mainly expressed by phagocytes, but they are also expressed in other immune cells, including cells of the adaptive immune system. Indeed, almost all cells in the body have some types of pattern recognition receptors. Pattern recognition receptors can be stimulated not only by germs but also by noninfectious matter that may induce "sterile" inflammation.

In the early 1890s, the fifth and last devastating cholera pandemic swept throughout the world, claiming hundreds of thousands of lives. Elie Metchnikoff, like many scientists in Europe, raced to understand the disease. His initial experiments were risky: using himself as a guinea pig, he drank flask after flask of water filled with the comma-shaped *Vibrio cholerae* bacteria—from the Seine River, a fountain in Versailles, even the feces of the infected. Metchnikoff survived the cholera cocktails, but his nineteen-year-old laboratory assistant did not. Wracked with guilt, he vowed to abstain from reckless human experimentation.

But Metchnikoff was left to wonder *why* cholera chose to kill certain individuals rather than others within the same community. Hunched over petri dishes in his laboratory, he found that some microbes incited cholera growth while others impeded it. He wondered if analogous events transpired in the human gut. Maybe microbes dictated whether an infected person actually became sick. "The flora of the human stomach has hardly been investigated, and that of the gut—even less so," he wrote in an 1894 paper covering cholera.

Metchnikoff experimented with intestinal germs not only in humans but also in tadpoles, rats, rabbits, guinea pigs, and macaques. He also ordered some of the largest bats in the world, known as "flying foxes," from India. By 1901, late in his career, Metchnikoff pointed out that gut germs could be innocent or deadly. "The idea is to define precisely the two categories, and to engage the beneficial bacteria in fighting the harmful ones," he wrote. He suggested that some germs, particularly those in the large intestine, or colon, produced poisons that seeped through the intestinal wall and into the bloodstream, leading to arteriosclerosis, a hardening of the arteries, and injuries to other organs. "The intestinal flora," he wrote, "is the principal cause of the too short duration of our life, which flickers out before having reached its goal." He was taken with the idea that microbes might interact with the immune system, releasing toxins that stimulated the body's macrophages, obliquely contributing to aging and disease. This seemingly simple conjecture—one that bound microbes and macrophages—would have implications beyond his wildest imagination.

———————

Microbes have inhabited the earth for billions of years, long before complex animal life, helping plants to extract vital nutrients from soil. They live on, around, and inside our bodies. Bacteria, viruses, protozoa, and fungi teem on our skin, lungs, intestines, mouth, genitalia, and eyes. The intestines house more microbes—and more microbial species—than any other part of the body. Microbial density is greatest at body orifices like the mouth, drops off in the acidic stomach, rises in the small intestine, and explodes in the colon, where it comprises a "gut microbiome" that has captured the majority of scientific attention showered on the microbes that live alongside us. The gut microbiome has more cells and genetic information than the whole human host by severalfold. It functions like a vital organ, with a metabolic capacity that surpasses the liver's. It can become diseased, with harmful shifts in microbial communities; and like a diseased organ, it can be replaced by a transplant.

In an ancient world of scarcity, as humans foraged for food, a symbiosis between human and microbes evolved, a relationship that benefited both parties. Gut microbes ferment what we cannot digest, harvesting energy and producing vitamins, minerals, and other helpful compounds. They degrade toxic substances, including carcinogens. And they defend us against deadly germs. Germs create their own antibiotics to battle each other, ones that cause little collateral damage when compared with their human-made counterparts. Early microbiome research uncovered these and other key functions. Eventually, scientists realized that gut germs also play a central role in immunology and inflammation.

Pioneering scientists in the mid- to late twentieth century studied animal microbiomes and disease, but intense scrutiny of the gut microbiome began in the first decade of the twenty-first century. As researchers started to link hidden inflammation to common chronic diseases, studies on microbes flourished in concert. In the ensuing years, scientists eavesdropped on conversations between microbes and macrophages and other immune cells. Intestinal germs shape—and are shaped by—the immune system, playing an important role in immune reactions. The intimacy between microbes and immune cells affects seemingly fated incidents, such as succumbing to a deadly infection, being burdened by seasonal allergies, or failing to build immunity after a vaccine. It affects the risk of becoming silently inflamed and developing chronic inflammatory diseases.

One critical method by which the conversations between microbes and immune cells help prevent disease is by training our bodies to distinguish harmless food and germs from their toxic counterparts. A great deal of this dialogue takes place within the intestines. The intestines contain the largest reservoir of macrophages in the body. Their lives are often hard and brief, constantly yielding to younger replacements circulating in the blood. Not only do they heal wounds and fight germs, as in other tissues, but they also learn to live alongside a multitude of gut microbes, functioning within complex layers of the intestinal immune system.

At the inner lining of the intestines, the interface between "self" and the exterior world, rectangular epithelial cells shove tightly against each other like bricks, limiting the entry of harmful substances. These cells soak up nutrients and secrete a protective layer of slippery, watery mucus that coats the digestive system and contains an antibody called immunoglobulin A (IgA), which bars the entry of toxins and bad germs. The mucosal immune system exists not only in the intestines but also in other body cavities exposed to the environment, including the nares, lungs, eyes, mouth, and genitalia. Burrow deeper into the wall of the intestines and the immune system's vernacular bares a cohesiveness that unites historically fractured segments of immunology. Beyond the fence of epithelial cells lies the lamina propria, loosely packed tissue that holds most intestinal immune cells, the blood supply, and lymph vessels. Here, innate and adaptive immune cells, including macrophages, dendritic cells, and B and T lymphocytes, mingle. Within the lamina propria sits the gut-associated lymphoid tissue, the largest lymphatic organ in the body, consisting of special patches of lymphoid tissue found throughout the intestines and lymph nodes all over the abdominal

Tregs, which help to manage inflammation, are abundant in the gut and essential to its tolerance of germs. They also inhibit Th17 cells. Honda hunted for microbes that coaxed Tregs to flourish. He found not one but a group of anti-inflammatory bacteria dubbed "clostridial clusters," distant relatives of a noxious bacterium known as *Clostridium difficile* (*C. difficile*), but with the opposite effects in the body. Unlike *C. difficile*, clostridial clusters induce Tregs and calm inflammation. And in the first human study of its kind, biologist Joao Xavier and his colleagues at Memorial Sloan Kettering Cancer Center linked changes in the concentrations of different types of immune cells in the blood to alterations in gut microbial species, supporting the idea that microbes may affect the production of immune cells in the bone marrow and their subsequent proliferation throughout the body.

A complex picture of the immune system began to manifest. The presence of an animal's genetic code does not in itself suffice to create a mature, healthy immune system. Microbes must assist in this nuanced task, as revealed by the isolated anguish of germ-free mice. In species as diverse as humans, flies, and zebra fish, microbes are essential for immune system development. They partake in the creation of immune cells and the organs in which they are stored. They encounter immune cells in many parts of the body, like the airways, skin, and genitalia, but it is largely in the intestines that they engage in some of their most important conversations, ones that shape and define immune behavior. In the intestines, immune and epithelial cells sense a wide variety of bacteria, viruses, fungi, and parasites. Microbes can influence immune cells in many ways: physical contact akin to a human embrace, chemical signals from microbial molecules, or even by altering gene expression. An everlasting dance begins at birth as microbes manipulate not only innate immune cells but adaptive ones as well, including B and T lymphocytes. Sculpting the immune system is one of their most important undertakings.

An immune cell living in the gut can choose to depart for a new home elsewhere in the body. It may resurface in largely sterile areas like the heart, liver, or spinal fluid, sharing with its new community the lessons learned from gut germs, warning tissues of imminent danger. Thus, microbes calibrate the immune system not only in the intestines but throughout the body, with the ancient cells of the innate immune system making close inaugural contact.

The therapeutic potential of fecal transplants is yet another illustration of the conversations that take place between microbes and immune cells. Transplanting gut microbes—as one would transplant an organ—can alter the recipient's microbiome and calm intestinal inflammation. In the early days of modern fecal transplants, I had encountered Oscar, a wizened old man in his eighties who brought his wife's feces to his appointment in a plastic food container. While he lay on the bed awaiting a colonoscopy, a nurse and I added some water to the excrement and transferred it to a blender. The stench was overpowering, filling the endoscopy room and our senses as never before.

Over the last year, Oscar had been having agonizing bouts of severe intestinal inflammation, coming to the hospital with stomach pains and diarrhea brought on by *C. difficile*. The last episode had dropped his blood pressure and harmed his kidneys, nearly killing him. I had given him various antibiotics, including an intermittent dosing regimen that lasted for months, but to no avail—the infection persisted. And so, rather than annihilate one recalcitrant germ, we attempted the opposite. A fecal transplant would fill his gut with many germs and hopefully displace *C. difficile*. It was the dirtiest of cures. As I maneuvered the scope through Oscar's colon, I sprayed liquid feces along its length. After the procedure, he dozed in the recovery room. In a day, his diarrhea disappeared. In the months and years to come, *C. difficile* never again reared its head.

Fecal transplants are not new. Ancient Ayurvedic texts recommended eating cow dung for stomach issues, and fourth-century Chinese doctors used "yellow soup" filled with fresh or dry stool to cure patients with severe diarrhea. Many animals do not share humans' revulsion toward feces and routinely swallow each other's excrement to acquire microbes. In Western medicine, the first fecal transplant was performed in 1958 by Colorado surgeon Ben Eiseman, curing a critically ill patient with *C. difficile*. Soon after, the antibiotic vancomycin was introduced to combat *C. difficile* and fecal transplants receded into the background. But the early decades of the twenty-first century brought renewed interest in gut microbes and a resurgence in fecal transplants, including, eventually, those from volunteer stool donors that could be taken in pill form. Research increasingly supported the use of fecal transplants for certain cases of relentless intestinal inflammation, with a 90 percent cure rate in patients with recurrent *C. difficile* for whom antibiotics had proved futile.

Fecal transplants seemed to work by displacing deadly intestinal germs with harmless or helpful new ones. Curiously, however, they were effective for *C. difficile* infections even when sterile fecal filtrate—that devoid of microbes—was used. Fecal transplants had shed light on the stunning therapeutic potential of altering gut microbes, but they also highlighted the need to better understand the therapy. What explained the utility of germ-free feces? Moreover, what distinguished a sick microbiome from a healthy one?

In 2004, physician-scientist Jeffrey Gordon and his team at Washington University in Saint Louis helped answer this question by performing fecal transplants in mice. Gordon transferred gut microbes from fat mice into skinny, germ-free ones and watched in awe as the germ-free mice grew obese. A few years later, Gordon took microbes from fat and thin human twins and infused them into skinny, germ-free mice. He saw the same phenomenon: the mice that recieved the fat twin's microbes became fat. Obesity, it seemed, could be caught like an infectious disease from a sick microbiome.

Gordon, who spends most of his time in a laboratory and shies away from media attention, was the first to explore how microbes could potentially cause obesity in animals, drawing a glaring spotlight onto gut germs. Scientists began to link an imbalance in these microbes, or *dysbiosis*, to all kinds of illnesses, including obesity, heart disease, diabetes, autoimmune conditions, liver disease, cancer, and neurodegenerative and psychiatric ailments. Dysbiosis, a word often confined to describing microbiomes, formally translates to "living in distress." A dysbiotic microbiome is an ecosystem disrupted, with changes in the variety of microbial species, the chemicals they make, and the genes they stimulate. Dysbiosis seemed to be not only a potential cause of obesity and other chronic inflammatory diseases—with stronger data for some conditions than others—but also a consequence, part of a sinister loop that continually reinforced itself.

A dysbiotic microbiome is very often an *inflammatory* microbiome, one that is tied to hidden or obvious inflammation that may begin in the gut and weave throughout the body, contributing to illness. In France, at the University of Clermont-Ferrand, Gordon's early experiments inspired the young Benoit Chassaing to obtain a PhD in microbiology. While Chassaing was fascinated by the idea that gut germs could cause obesity, he wondered *how* this happened. The mechanisms were murky. Dysbiotic microbes could certainly affect energy harvest, forcing a person to absorb more calories from food and favoring the buildup of fat.* But Chassaing believed that the picture was incomplete. He thought of Gokhan Hotamisligil's experiments, which had shown that chronic, low-level inflammation, or metainflammation, may drive obesity, diabetes, heart disease, and other metabolic complications. Maybe the dysbiotic microbiome Gordon had transferred from obese to lean mice prompted hidden inflammation.

Gut microbes shape immune responses during health and disease, fine-tuning the magnitude and duration of inflammation like puppeteers. They help the immune system to trigger—and suppress—inflammation as needed, reacting to threats without overreacting, preventing deadly infections and chronic inflammatory diseases. As microbiome research progressed, scientists found—in both human and animal studies—that varying volumes of inflammation in the gut could be transmitted by transplanting gut microbes, as with obesity. Chassaing's research, along with the work of various colleagues around the world, suggests that one important mechanism by which a dysbiotic microbiome negatively impacts health is by creating chronic inflammation.

---

* Research continues to point out that the microbiome plays an important role in how we extract, store, and expend the calories obtained from food.

The tenor of an inflammatory microbiome is grimly altered, its mass of germs irritable and destructive, forcing macrophages and other immune cells to behave erratically without cause. But this state eludes rigid definitions. Certain microbes have emerged as being "anti-inflammatory," as with Honda's clostridial clusters, or "inflammatory." Some bacteria that possess slender, whiplike appendages called flagella and cell walls with specific toxins are more likely to be inflammatory. Still, it is difficult to label most germs wholly virtuous or evil. Like characters in a Dickens novel—and most humans—their personalities are layered and evolve over time with environmental inputs.

Many species play dual roles and are able to foster both health and disease, depending on their influences and surroundings. The bacterium *Helicobacter pylori* (*H. pylori*), for example, causes stomach ulcers and increases the risk of stomach cancer in a small number of people, which led to an aggressive campaign to eradicate it ("The only good *H. pylori* is a dead *H. pylori*," a 1997 *Lancet* article proclaimed). But *H. pylori*, present in mummies from northern Mexico before Columbus arrived in the New World, has populated humans for over fifty thousand years and helps to regulate the immune system. It sparks the production of Tregs and reduces the risk of heartburn, allergies, asthma, and other inflammatory diseases. It even appears to protect against some types of esophageal cancers. *H. pylori* is not passed on to offspring after parents are treated with antibiotics against this germ. With each generation, it has been increasingly absent from the guts of most children living in Western nations.

The symbiosis between humans and microbes, honed by evolution, is imperfect and fraught with potential conflict if not properly managed. An intent to harm the host lurks within both individual species and the consortium at large. *C. difficile* typically resides peacefully in the intestines of healthy humans, kept in check by competing bacteria that make use of food and space in the gut. When antibiotics destroy these bacteria, *C. difficile* proliferates and becomes toxic. In the wrong company, harmless—or even helpful—bacteria can alter their behavior and turn deadly.

Oscar's wife had cured a relentless disease by donating her

feces. She assumed her gut microbiome to be robust and healthy. One of the challenges in evaluating the health of a microbiome, however, is the lack of a reliable reference point. Microbiomes vary immensely both within and between bodies. A human hand harbors over a hundred different bacterial species, only a few of which are common to both hands or shared among different people. Although scientists have identified bacterial "cores," or frequently occurring species in certain populations, the number and range of microbes in an individual's intestines is as distinct as a fingerprint. The difference between a "fat" and "thin" microbiome is clear within any single study, but pinpointing steady disparities across studies is challenging. And gut germs are in constant flux across time and space, evolving along with the sun's rise and set, a recent meal, or chats with host cells. In truth, multiple iterations of "anti-inflammatory" and "inflammatory" microbiomes likely exist. But studies do reveal a vague yet defining characteristic of healthy microbiomes: a richness, or diversity, of species. A diverse microbiome is more likely to consist of germs that will counter—rather than propagate—inflammation. It is no surprise to find that diversity breeds strength in the human gut as it does in biological ecosystems around the world. In its absence, inflammatory diseases take hold and infectious germs find fertile ground for invasion.

Where microbes live and what they are doing may be more important than who they are. Studies show that the location and function of gut germs are central to igniting inflammation. A microbe can be helpful in the gut but deadly if it ends up in the blood. At the inner lining of the intestines, epithelial cells squeeze against each other to bar the entry of germs and toxins. They secrete a slippery, double-layered film of sticky mucus. Microbes anchor themselves to the loose outer layer, devouring nourishing carbohydrates found within it. But most hesitate to encroach upon the dense inner layer, which is filled with deadly antimicrobial molecules. Mucus provides personal space for both immune cells and microbes, a safe conduit through which conversations can flourish. Mucosal microbes may influence the immune system more so than those found in the lumen of the gut. Chassaing has found that when microbes breach etiquette by flooding the inner mucosal layer, a low-level inflammation ensues.

Microbes produce hundreds of thousands of chemical messages, or metabolites, through which they communicate with the body and contribute to health or disease. These metabolites can be inflammatory—driving macrophages to produce cytokines like TNF-α, IL-1β, and IL-6—or anti-inflammatory. They can mimic human antigens, encouraging autoimmunity, or even exit the body, wafting through the air to convey far-flung signals. While microbes are usually shunned from the blood, their metabolites are often free to make their way through the mucus and across the epithelial cell barrier, interacting not only with immune cells but with other types of cells as well, including neurons. They travel in the bloodstream to distant organs, such as the brain, riling up immune cells all around the body and impacting levels of inflammation. Metabolites are the lifeblood of the dialogue between microbes and immune cells, proof of *what* germs are doing. Sterile fecal filtrate, or fecal transplants that are missing microbes, may be filled with metabolites that can reprogram the immune system or perform other functions.

Unlike the linear, cause-and-effect dichotomy depicted by the traditional view of infectious germs and distinct infections, the relationship between inflammation, gut germs, and disease tends to be circular. At UCLouvain, an international university in Belgium, Professor Patrice Cani studies the interactions between food, gut germs, and the lingering low-level inflammation that leads to chronic diseases, hanging a sign with the motto "in gut we trust" on his office door. In 2007, he and his team put a group of mice on a high-fat diet of mostly lard and corn oil while giving a control group regular chow. A month later, the mice on the high-fat diet were considerably heavier and developed low-level inflammation, insulin resistance, and fatty liver.

Blood levels of the microbial molecule lipopolysaccharide, an endotoxin that activates the innate immune system and triggers inflammation, had risen between two and three times the normal values. When Cani injected a new group of mice with pure lipopolysaccharide, increasing their blood levels of lipopolysaccharide to match those of the mice fed corn oil and lard, he found, strikingly, that they developed the same health issues, gaining just as much weight as their predecessors and developing silent inflammation, insulin resistance, and fatty liver. Bacterial toxins could pass through the gut barrier, make their way into the blood, and inflame the body, potentially contributing to metabolic disease. Cani called this phenomenon "metabolic endotoxemia."

In Cani's initial experiment, inflammatory intestinal germs had bloomed in response to food. The gut microbiome is a major source of endotoxins in the body, and scientists have found that identical microbiomes can produce drastically different amounts of lipopolysaccharide and other inflammatory molecules. This is an important twist: the lard and corn oil probably affected not only the types of microbial species populating the gut but also their core behaviors, leading them to express their distaste for certain foods through the language of the immune system. In 2018, a similar experiment conducted in humans supported Cani's findings. In a six-month randomized controlled trial, researchers at Qingdao University in China fed over two hundred young adults mostly white rice and wheat flour with varying concentrations of fat—at 20, 30, or 40 percent of their total daily calories—in the form of soybean oil, the most widely used edible oil in Asia. As the amount of soybean oil in the participants' diets increased, their microbiomes became progressively inflamed, with less diversity, more lipopolysaccharide and other inflammatory metabolites, and fewer anti-inflammatory ones. Blood levels of the inflammatory marker CRP rose along with the amount of fat in the diet, suggesting bodywide inflammation.

Cani and other scientists paint an intricate, circular picture of inflammation, germs, and disease. One or more environmental insults—like food—may shift the balance and behavior of gut germs, which then intrude upon the intestinal mucous barrier and express inflammatory molecules like lipopolysaccharide, giving rise to hidden inflammation in the gut and—as more lipopolysaccharide is absorbed into the bloodstream—throughout the body. Inflamed intestinal epithelial cells slacken, loosening their barricades and creating a "leaky gut" that allows even more substances—like food antigens, bacterial toxins, or other unwanted elements—into deeper layers of the intestinal wall or through the bloodstream. Microbes can also directly alter the expression of genes and proteins that affect leaky gut. Inflammation can be both a cause and a consequence of leaky gut, which is tied to various chronic inflammatory disorders and can manifest not only in the intestines but also in other parts of the gastrointestinal tract, like the esophagus. A healthy gut, however, inevitably experiences small fits of leaks—including during exercise or other environmental stressors—that allow the immune system to encounter and learn from foreign matter.

The inflammatory microbiome propagates both hidden and overt inflammation, contributing to metabolic syndrome, obesity, heart disease, diabetes, liver disease, and other illnesses. A snapshot of people with one or more of these conditions reveals low bacterial gene counts due to a lack of microbial diversity and high levels of lipopolysaccharide and inflammatory cytokines in the blood. Lipopolysaccharide infiltrates fat and liver tissue, activating macrophages that incite inflammatory genes and proteins. It can even reach the brain, affecting mood and behavior. Cytokines alter the workings of receptors that respond to insulin, leptin, and other molecules controlling body fat, or they make their way into clogged vessels to rupture plaques. Disease reinforces dysbiosis, creating parallel currents. In obesity, fat tissue churns out additional inflammation, worsening leaky gut and the absorption of inflammatory microbial metabolites. An inflammatory microbiome, caught in a cycle of continual violence, can badly harm the host.

But even a healthy microbiome inevitably yields some level of inflammatory molecules like lipopolysaccharide. The mass of germs in a human intestine, an immune organ, cannot help but draw attention to itself. It elicits a whisper of inflammation, a low, constant hum—evidence of a profitable partnership between microbes and immune cells. Gut microbes attract macrophages from the blood, pulling them into the intestines. The pathologically inflammatory microbiome, then, is a relative rather than an absolute concept. The tipping point at which hidden inflammation from gut microbes tends to harm may be obscure, but the diseases that ensue from its rampage throughout the body are not—demanding efforts to prevent their ascent.

———————

For Olivia, an anti-inflammatory diet may begin with simple, somber meals. Clinical trials to date reveal that two types of dietary interventions are optimal for quelling inflammation and inducing remission in inflammatory bowel disease patients: elemental diets and polymeric diets. Elemental diets are made up of amino acids, simple sugars and fatty acids, the building blocks of protein, carbohydrates, and fat distilled into their most basic elements, as well as vitamins and minerals. Polymeric diets are slightly more sophisticated, with intact proteins and more complex carbohydrates and fats. These liquid mixtures are not palatable. For this reason they are often given in hospital settings, through a tube that travels from the nose to the stomach.

But despite their drawbacks, elemental and polymeric diets work. Deprived of the diverse load of food antigens a typical meal presents to the immune system, they are able to exert a powerful anti-inflammatory effect in inflammatory bowel disease, comparable to using steroid medications in some cases but without any of the harmful side effects. In fact, they are routinely used all over the world in children with inflammatory bowel disease flares, to shield them from problems with bone development, growth, and other unwanted side effects of steroids.

The evidence that food, albeit stripped to its most essential elements and its core definition—a substance one can eat or drink that nourishes the body and helps to sustain life—can exert a powerful anti-inflammatory effect, enough to put a serious autoimmune disease into remission, is a testament to the remarkable potential of "food as medicine." But elemental and polymeric diets are a sliver of the pie, a narrow window into the world of healing victuals.

Certain diets are not only enjoyable but can also help to both dampen *and* resolve inflammation. While most drugs are missing resolvins, some foods are full of them. A true anti-inflammatory diet is also pro-resolving and can be as powerful as a drug—even more so. Food and other lifestyle factors, independent of their effects on inflammation, can alter immunity as well. Vital nutrients feed immune cells, enhancing their ability to defend the body, while malnutrition or an unhealthy diet and lifestyle can inhibit immunity.

As the story of the anti-inflammatory diet advances, unearthing the fundamental elements of this fare, it sheds light on how food and germs conspire to influence inflammation, fostering health or illness. The earliest large-scale scientific endeavor that began to address these ideas, an iconic chapter in the history of nutrition science, was initiated nearly a century ago to explore the connection between diet and heart disease.

CHAPTER 9

Fat Wars

At the University of Minnesota, physiologist Ancel Keys spent his days running a research institution so large that it was housed in the football stadium. Initially, the stadium had been the only space available in which to study the athletes in the physical education program who were the subjects of Keys's early studies. While the Laboratory of Physiological Hygiene had started small, it gradually expanded over the decades to 20,000 square feet at Gate 27 of the stadium. From these make-shift quarters, Keys's work on food and health flourished, setting the stage for our modern understanding of nutrition science.

Born in 1904 to two Colorado Springs teenagers without college degrees, Keys was a subject in Lewis Terman's longitudinal study of 1500 "child geniuses." A restless, bohemian spirit carried him through brief and varied stints in his youth. He worked as a lumber-jack, as a clerk at Woolworth's, and as an oiler aboard the S.S. President Wilson. For a few long, hot months, he shoveled stacks of bat guano into gunnysacks in an Arizona desert. Eventually, Keys made his way through college, earned his PhD in biology from Berkeley, and secured research positions at Harvard and the Mayo Clinic before settling in Minneapolis.

In the late 1940s, Keys turned his attention to a select group of Minneapolis men who appeared in his laboratory for periodic physical checkups. On average, these men were slightly overweight and arrived in suits. Around a quarter were the presidents or vice presidents of notable businesses. Their wallets and appetites overflowed, as did their tendency to die prematurely from heart disease. Once a relative rarity in the 1920s, heart disease was felling middle-aged men at a frightening rate by mid-century, becoming the leading cause of death. The country's president, Dwight Eisenhower, would also suffer several heart attacks.

Keys drew blood from the Minnesotan businessmen and found high cholesterol levels, which he suspected was the culprit. As the lipid hypothesis began to gain acceptance, scientists generally agreed that the higher the blood cholesterol levels, the greater the chance of having a heart attack. But which foods, Keys wondered, might cause blood cholesterol to rise?

At the time, the incipient science of nutrition was largely known for the discovery of vitamins, but a few doctors were starting to formally experiment with diet and disease. Keys was struck by the work of Walter Kempner, a physician at Duke University who prescribed mostly white rice and fruit for patients with hypertension and heart disease. Kempner wrote that most saw improvement in their conditions, as well as a decrease in blood cholesterol. But Keys's historic questions on diet and disease encompassed more than cholesterol levels. His work, unbeknownst to him at the time, was the first rigorous attempt to understand how food might affect *inflammatory* diseases.

In 1951, Keys was on sabbatical in Oxford, England. He was accompanied by his biochemist wife, Margaret, who was also his research partner. At an international conference in Rome, Italian physiologist Gino Bergami mentioned to Keys that in Naples, where he practiced, "heart disease was no problem." Keys was eager to find out if this was true. He loaded some laboratory equipment into a little car, and in 1952, he and Margaret traveled to Naples as guests of the University of Naples. They drove to Switzerland in a bitter snowstorm, eventually hopping onto a train that traversed the 12-mile tunnel to Italy, where, taken with the balmy breezes and singing birds, they enjoyed their first espresso in Domodossola. Keys later wrote of that day, "We felt warm all over, not only from the strong sun but also from a sense of the warmth of the people, a feeling we were later to experience in all of [the] Mediterranean, that great stretch of land from the Strait of Gibraltar to where Europe ends."

In Naples, Keys measured blood cholesterol in firemen and other city employees. He found that the levels were much lower than those of the Minnesota businessmen. And as his Italian colleague had promised, heart disease was rare in public hospitals, a peculiarity. Southern Italy, in fact, had the highest concentration of centenarians in the world.

For most of his trip, Keys sampled the simple fare of common Neapolitans: homemade minestrone soup, endless varieties of freshly cooked pasta served with tomato sauce (and, occasionally, a sprinkle of cheese or bits of meat), hearty bean dishes, bare bread "never more than a few hours from the oven," vast quantities of fresh vegetables, and small portions of fish or other meats perhaps once or twice a week. Dessert was always fresh fruit—other sweet treats were reserved for special occasions. Compared with Americans, their diets contained minimal meat, dairy, and eggs. The meats they did consume, Keys noted, were intrinsically distinct from the American types. Codfish was popular, and the locals took a dim view of any fish that was not exceptionally fresh. The chickens were aged and scrawny, their skins yellow-tinged with carotene from the natural forage that made up a good part of their feed. The large, marbled beefsteaks of Minnesota were replaced with thin, lean cuts—veal, naturally leaner than the flesh of older animals, was favored. The demand for fatty, striped bacon yielded to a preference for prosciutto, slim and raw and cured, the most prized ham in the land.

Keys found some exceptions to his observations. The wealthy men of the Rotary Club in Naples, for example—unlike the firemen, city clerks, longshoremen, and steelworkers—enjoyed more meat and dairy. Their blood cholesterol levels were high, aligned with the amounts found in Minnesota businessmen rather than other Neapolitans. They succumbed to heart disease in comfortable private clinics. Keys was occasionally feted by these men at rich multicourse dinners. Sitting by a windowsill at the University of Naples, a nineteenth-century building with a courtyard and palm trees and sculpted gardens, he drew plans for an ambitious and grand study,* the very first of its kind, to test his growing hunch that diet and disease were intricately interwoven.

Keys's now-famous Seven Countries Study launched in 1958, a mere year after the Framingham study investigators announced their first major findings on the risk factors for heart disease. Like Framingham, the Seven Countries Study was known as an observational study. Keys aimed to record the diets of healthy men and women around the world and to collect additional data like blood pressure and cholesterol levels. He would then observe the subjects for many years to determine the rate of heart disease—or death—in each group and figure out if a certain type of diet (or any other baseline measurement) was correlated with an increased risk of heart disease.

In an age bereft of commercial jet planes, computers, and the Internet, Keys assembled an international team of collaborators, a group of renowned scientists who were critical to the study's success and a testament to Keys's political and networking skills. He enrolled approximately twelve thousand middle-aged men from the United States, Italy, Greece, Finland, the Netherlands, Japan, and Yugoslavia. All subjects came from rural areas or small communities. His focus on these locations was dictated not only by the obvious differences in dietary patterns between these countries but also by logistics and budget. Keys elicited support in places where he had established contacts. He sought countries that could provide the funding and infrastructure required to complete the study, with adequate interinstitutional and governmental cooperation that would yield well-organized medical systems and reliable census data. He was careful with certain areas that had recently been occupied by the Nazis, since, as he noted in a later paper, "the prolonged

---

* The entire period of 1952–1956, in which Keys and Margaret traveled extensively (usually accompanied by Boston cardiologist Paul Dudley White), led to planning the Seven Countries Study. They initially performed pilot studies in Madrid, Sardinia, South Africa, Finland, and Japan, which convinced Keys to embark on the formal systematic comparisons that made up the Seven Countries Study.

influence of the war and its aftermath on the diets of many countries cannot be ignored." Some countries, like France, Sweden, and Spain, declined to participate due to a lack of interest, funding, or both. Keys found the "purest" version of the Mediterranean diet in Portugal. But the dictator of Portugal did not want his country's name attached to what some started to call a "diet of the poor."

A credo of Keys's laboratory was that the data had to be "of the highest caliber, valid, reliable and most relevant to the scientific question." Frustrated that countries tended to describe the same diseases in different ways, he developed a standard code for categorizing health diagnoses in the Seven Countries Study and insisted that all electrocardiograms be sent to the University of Minnesota for analysis. He took great pains, even by modern standards, to confirm the authenticity of dietary data. Nutrition studies typically rely on participants' memories of foods consumed. Keys, in addition, had dietitians weigh the food and drink ingested by sample subjects in each group and send these foods, freeze-dried, to the University of Minnesota for chemical testing.

During his world travels, Keys took careful notes.* In Japan, as in Italy, heart disease was rare. Noboru Kimura, a young Japanese cardiologist who had previously worked in Keys's lab, collected data on ten thousand autopsies from his medical school at the University of Fukuoka and revealed that the hearts and arteries in his country were in far better shape than what Keys was used to seeing in America. The Japanese migrants to California, however, were no different from Minnesotans. Vegetables, rice, fish, and soy foods were central to Japanese cuisine. But, notably, Japan had the highest per capita intake of salt, and the incidence of hypertension and stroke was immense. Most salt came from the table, but also from soy sauce,

---

* Data gathering for the Seven Countries Study, which took place in systematic surveys at zero, five, and ten years and systematic follow-up for fifty years, has been ongoing. Ancel Keys provided an account, near the end of his long career, of the Seven Countries Study design, conduct, and findings, with his discussion on and interpretation of their importance. "Epidemiologists and historians alike," as Henry Blackburn writes, "look back to the 1986 article by Keys et al. because of its accessibility and its substantive content and summative conclusions, as well as because of the influence that the Seven Countries Study has had on epidemiology and public health."

which families bought by the quart. Keys, recalling that rats fed ample amounts of salt were likely to develop hypertension, advised against copious salt intake.

In Finland, Keys enjoyed saunas with local loggers in Karelia. These men were lean, strong, and fit, unlike the Minnesota businessmen. But despite their enviable physiques, they had the highest rates of heart disease and the shortest life spans in all of Europe. Keys watched them devour slabs of cheese the size of bread slices, which they smeared with a thick layer of butter and washed down with beer. By the end of his stay, he longed for leafy salads, vegetables, and fruits.

Keys observed that fat preferences in Finland, the Netherlands, and America contrasted sharply with those in areas like Italy and Greece. "The fat in the land," Keys wrote of the Mediterranean, "came from olives. Butter was almost unknown, milk was something out of a small can for scant use in cooking. But olive oil! It was the only cooking fat." On the Greek island of Crete, olive culture boasted a four-thousand-year history. Archaeological excavations uncovered the enormous pottery jars of Minoan civilization, up to 4 feet wide and 6 feet deep, for storing olive oil. The first pressing of the olives, which created the "virgin" oil, began soon after the olives were gathered. In sealed containers protected from light and heat, the oil kept for years and was often used raw to preserve its delicate taste. A fraction of the olives made their way whole to kitchen tables. Adults ate around six small olives a day, including those used in cooking. In most parts of the Mediterranean, 15 to 20 percent of a person's daily calories came from olives and olive oil.

The Cretes, however, consumed what Keys knew to be a "high-fat" diet, with around a third of their calories coming largely from olive oil. But they, like their Italian counterparts, seldom developed heart disease. "The important peculiarity of the American diet compared with the Mediterranean one," Keys mused, "is the large amount of invisible fat of the saturated type in meats, milk and other dairy products." Saturated fats include solid animal fats like lard, tallow, and butter. Fatty acid chains in saturated fats have mostly single bonds and are "saturated" with hydrogen. Many vegetable oils, including olive oil, are mono- or polyunsaturated, meaning they are missing one or more hydrogen atoms in their structure (coconut and palm oil, however, are saturated fats).

As the first five- and ten-year follow-up data from the Seven Countries Study emerged, Keys's data showed—consistent with findings from other studies—that a person's age, blood cholesterol, blood pressure, and smoking status were correlated with the risk of developing heart disease. But he was the first to find, on a large scale, that food was also tied to this risk. His global gustatory experiences, ones that had elicited all his senses, were muted onto black-and-white print graphs that emphasized a sole dietary element, saturated fat, that rose along with both blood cholesterol levels and the risk of developing heart disease. Diet was why, Keys thought, American men had twice as much heart disease as the Italians and four times as much as the Greeks, Japanese, and Yugoslavians. He stressed that the *type* of fat consumed, rather than the percentage of total fat intake, was a key factor. Americans needed to eat less meat and dairy foods, which were high in saturated fats, and more plant foods as well as the unsaturated fats found in olives, nuts, seeds, and avocados.

Keys noticed that saturated fat seemed to raise blood cholesterol levels much more than actual cholesterol in the diet. In Cagliari, on the Italian island of Sardinia, many families kept chickens and often ate eggs, one of the richest sources of dietary cholesterol, an "important and remarkable substance quite apart from its unfortunate tendency to be deposited in the walls of arteries," Keys wrote. Cholesterol is necessary to sustain life and is particularly important for brain and nerve cells. The body manufactures all it needs, even if our diet is entirely devoid of cholesterol-laden foods: nearly every human tissue can make cholesterol. Keys found that the men who ate an egg a day did not have much higher blood cholesterol levels than those who ate only one or two a week He wrote in a later paper that "for the purposes of controlling the serum level, dietary cholesterol should not be completely ignored but attention to this factor alone accomplishes little." The Mediterranean diet was typically low in both cholesterol and saturated fat, which tended to appear together in the same foods.

Like any scientific study, the Seven Countries Study faced limitations, including those presented by its design. For example, it intended to explore the relationships between diet and disease across a wide range of dietary patterns rather than among cohorts chosen at random. And projecting its findings onto populations outside of the study in time and space was not a seamless task. The mid-twentieth-century diets of all cohorts, for example, had yet to accrue the large volume of ultra-processed foods that are well known around the world today.

But the Seven Countries Study developed important tools for standardizing nutrition research. Keys's colleagues continue to report on over fifty years of follow-up data from its subjects. When interpreted in the context of the broad spectrum of nutrition studies conducted to date, including other large-scale observational studies, randomized, controlled dietary intervention trials, and the growing data on the relationships between food, germs, and inflammation, the Seven Countries Study appropriately raised an alarm on dietary patterns with a surplus of saturated fat. This type of fat, as we know today, is linked not only to high blood cholesterol levels but also to inflammation, both of which can promote heart disease.

Emerging evidence, including experiments in petri dishes and animals as well as observations and interventions in humans, points to how fat type can dictate the behavior of macrophages and other immune cells. Saturated fats activate NF-κB, stimulating inflammatory molecules like IL-6, CRP, and TNF-α. They drive macrophages to inflame and injure, pushing them to assemble the NLRP3 inflammasome, which pumps out dozens of inflammatory molecules, including the cytokine IL-1β. IL-1β has long been targeted by scientists focused on chronic inflammatory conditions like heart disease, diabetes, cancer, neurodegenerative diseases, and arthritis. Diets high in palmitic acid—the most common saturated fat, found in foods like butter, cheese, milk, meat, and palm oil—increase IL-1β levels in humans. Saturated fats are more likely than unsaturated fats to make their way into fat tissue, where they increase macrophage numbers and enhance the inflammatory potential of body fat. They also alter HDL, an anti-inflammatory "good" cholesterol, rendering it dysfunctional and inflammatory.

Fat is largely digested and absorbed in the small intestine, but a portion makes its way to the colon, where it feeds gut germs. An excess of saturated fat decreases microbial diversity and breeds inflammatory bacteria—including some species linked to inflammatory bowel disease and other types of intestinal inflammation—and their metabolites. Germs flooded with saturated fat inch closer to the intestinal mucosa and produce potent endotoxins like lipopolysaccharide, which circulate in the blood and mimic a low-grade bacterial infection, giving rise to hidden inflammation and leaky gut. With the help of saturated fat, lipopolysaccharide is more easily ferried across the intestinal barrier and into the bloodstream.

It is difficult to reconcile the thought that saturated fat is intimately entwined with human evolution yet tends to irritate the immune system in excess. After all, saturated fat flows through many foods, including—in small amounts—plant fats from olives, nuts, seeds, and avocados. Notably, not all saturated fats are created equal and may vary in their ability to inflame. The type that predominates in dark chocolate, for example, is less incendiary than that in a steak. While the dose of saturated fat affects inflammation, so does the context in which it is consumed, not only overall dietary patterns but also the particulars of individual foods. In most plant foods, saturated fat is overwhelmed by unsaturated fats, fiber, vitamins, minerals, and beneficial nutrients known as polyphenols. In breast milk, where saturated fat provides easily absorbed energy for infants, it travels alongside resolvins. In most modern animal foods, however, the company it keeps is often inflammatory.

———————

Not all fats affect the immune system—and hence health—in the same manner. Decades after Keys initiated the Seven Countries Study, another scientist helped to confirm this notion. As a boy, Walter Willett had discovered a knack for growing all things green, bringing home blue ribbons from his local 4-H club in Hart, Michigan. At thirteen, he lost his father—a man who had always encouraged Willett to aspire to academic success—to brain cancer. A few years later, when Willett matriculated at Michigan State University, he paid most of his tuition by growing sweet corn, tomatoes, and a few other vegetables on a neighbor's farm. Willett would go on to become one of the most esteemed nutrition scientists in the world, earning degrees in medicine and epidemiology, eventually joining the faculty in the Department of Nutrition at the Harvard School of Public Health.

In 1980, Willett began collecting data on the food habits of around one hundred thousand nurses who were part of the Nurses' Health Study, initiating one of the biggest epidemiological undertakings in the history of nutrition. He followed this up with the Nurses' Health Study II and III. His research yielded thousands of papers that elucidated the pervasive influence of diet on disease. He shed light on the nuanced effects of fats and carbohydrates, pointing out that the sources of these macronutrients mattered, drawing ire from mainstream nutritionists. At the time, the message to shy away from saturated fat had grown into a mantra to avoid *all* fat. And people replaced animal foods high in saturated fats with refined carbohydrates and sugar. Willett's data revealed that this was an unhealthy, lateral move. Importantly, his evidence supported Keys's historic observations, showing that monounsaturated and polyunsaturated fats—found largely in plant foods like olives, nuts, seeds, and avocados—could help ward off what we now recognize as chronic inflammatory diseases, including top killers like heart disease and cancer.

Amid the plethora of plant oils, olive oil—the principal source of fat in the traditional Mediterranean diet along with whole olives—captures the most attention in medical literature. When substituted for saturated fat from animal foods, it may prevent heart disease by lowering both LDL cholesterol levels and inflammation. Olive oil is largely made up of monounsaturated fats and contains anti-inflammatory polyphenols like oleocanthal. Like NSAIDs, oleocanthal inhibits cyclooxygenase enzymes.

Willett's favorite fats come from a variety of nuts, which he stores in his lunchbox—along with salad and fruits—during his frequent travels. Nuts are high in unsaturated fats. A wealth of data—including observational studies and randomized controlled trials—show that they protect against many chronic inflammatory diseases, including heart disease, stroke, cancer, and diabetes, as well as untimely deaths from any cause. They lower cholesterol levels and inflammation, especially when they replace meat, dairy, eggs, and refined carbohydrates, decreasing inflammatory biomarkers like CRP, IL-6, and TNF-α. They are rich in fiber, polyphenols, vitamins, and minerals. Eating a small handful of nuts each day can lower the risk of chronic inflammatory diseases and lengthen life.

In Boston, in the 1990s, physician Charles Serhan found an unexpected link between certain unsaturated fats and resolvins. As he began identifying resolvins in laboratory mice, he wondered how the body created these precious substances. To his shock, the raw material his mice used to make resolvins and most other pro-resolving mediators came from a special type of polyunsaturated fat called omega-3. The laboratory chow the mice feasted on, drab pellets that resembled excrement, was fortified with omega-3s.

Omega-3s are essential fatty acids. They are required by the body, but unlike cholesterol, which the body makes, omega-3s are obtained only from the diet. They are originally made in plants, which produce omega-3s during photosynthesis. They flood the cell membranes of chloroplasts and help them to collect light. They abound in many plant foods, such as dark leafy greens (particularly wild types), walnuts, flaxseeds, hempseeds, and chia seeds. Omega-3s flourish in algae, including seaweed, and weave their way through the aquatic food chain, building up in seafoods such as oysters, sardines, and salmon. A growing body of research emphasizes the importance of omega-3s in an array of chronic inflammatory diseases.

The human brain, one of the fattiest organs in the body, craves omega-3s. Some scientists hypothesize that a lack of omega-3s may contribute to problems like autism and attention deficit disorder. Omega-3s may thin out the blood, helping to prevent blood clots. In patients with atherosclerosis, they have been shown to prevent heart attacks, strokes, and even death. They shrink atherosclerotic plaques, as revealed by imaging studies, and help to stabilize them, decreasing the risk of plaque rupture. Population studies indicate that higher omega-3 levels in the diet are correlated with a lower risk of death from any cause.

Omega-3s have potent effects on the immune system. They inhibit inflammatory gene regulators like NF-κB and instead activate anti-inflammatory gene regulators, helping to prevent macrophage migration into fat tissue in both animals and humans, blunting the inflammation it spews out. They lower inflammatory cytokines and blood levels of biomarkers like CRP, IL-6, and TNF-α. Their by-products both reduce and resolve inflammation. For example, docosahexaenoic acid and eicosapentaenoic acid, two omega-3 fatty acids, are essential for making resolvins and other pro-resolving mediators.

Over the years, omega-3 fats have been fleeing the food supply. Since they grow rancid more readily than other types of fats, plant breeders often select for crops with fewer omega-3s. Food industries replace omega-3s with a more stable essential polyunsaturated fatty acid known as omega-6. In broad terms, omega-3 fats give rise to the most powerful anti-inflammatory compounds, while omega-6 fats tend to produce inflammatory ones and encourage blood clots. But despite this nominal dichotomy, both types of fats have important roles in the body, many of which researchers are still working to understand. In fact, omega-6 fats also travel along anti-inflammatory pathways at times. Nuts and seeds contain varying amounts of omega-6 and omega-3 fats working synergistically to promote health.

The true dilemma relates to the balance of omega-3s and omega-6s in the body. Because they compete for the same enzymes and space in cell membranes, they engage in a zero-sum game: an excess of omega-6s in the diet hampers the body's ability to process omega-3s and yield an abundance of compounds that rein in and resolve inflammation. Unlike olive oil, many—but not all—plant oils are exceedingly high in omega-6s. Methods employed to extract plant oils can concentrate omega-6 fats, which have become central components of processed foods and most restaurant industries; aside from table sugar, they are the cheapest source of calories. Our ancestors consumed around four times as many omega-6s as omega-3s, but modern diets unleash fifteen to twenty times as many omega-6s, leading to a profound dearth of omega-3s. In general, supplanting saturated fats with unsaturated fats is healthful, but dietary imbalances can inflame.

A balanced intake of unsaturated fats can favorably affect immune cells. Unsaturated fats, particularly omega-3s, elicit macrophages' latent, genial qualities involved in tissue repair and resolution of inflammation. When omega-3 fats influence macrophages, anti-inflammatory effects ensue, including a decrease in IL-1β, IL-6, and TNF-α and a boost in the anti-inflammatory cytokine IL-10. Unsaturated fats, especially those from intact plant foods like nuts, seeds, and avocados, act as *prebiotics*, or foods that promote human health by nourishing gut microbes. These fats fuel anti-inflammatory microbial species and behaviors. Omega-3s, in particular, promote microbial diversity and spur the growth of bacteria that produce short-chain fatty acids, metabolites that benefit human health in many ways. They can even help to counter the harmful effects of saturated fats upon gut germs.

———————

Well before Keys's Seven Countries Study, a new chemical process used to manufacture vegetable oils contributed to the shortage of omega-3s in the food supply and gave rise to perhaps the most insidious fat to date. In the early 1900s, pepole fried up hunks of meat in solid animal fats, usually lard from pigs. They also cooked with soft, creamy butter or the harder tallow from sheep and cattle. Vegetable oils, considered inedible, had no place in the kitchen. But when German chemist Wilhelm Normann figured out how to get liquid vegetable oils to mimic solid fats by adding hydrogen, a process called hydrogenation, American entrepreneurs William Procter and James Gamble seized on the opportunity. Hydrogenated cotton seed oil—agricultural waste from cotton production—looked and cooked just like lard, but it was dirt cheap, almost free. Crisco, the first cooking fat to be made entirely from vegetable oil, was born in 1911. Its name, meant to evoke freshness and cleanliness, was adapted from the

phrase "crystallized cottonseed oil." Crisco did not have a particular smell or taste. It was as unassuming as the individuals upon
whom it was unleashed.

In the wake of Upton Sinclair's 1906 *The Jungle*, which had
somewhat reduced the American appetite for meat, Crisco was
marketed as a "pure" alternative to animal fats. Sinclair, who was
born into an old Virginia family whose wealth had been wiped
out by the Civil War, was a "déclassé Southern aristocrat with a
boyish charm." He was horrified by the appalling conditions he
found inside slaughterhouses for both people and animals. In
large meatpacking districts, manure-filled stockyards crammed
with cattle stunk of rotten eggs and decaying flesh, odors that
mingled with the aroma of beef and potatoes cooking in nearby
hotel kitchens.

Sinclair described the immigrant men, women, and children
who crammed into tenement apartments next to city dumps and
did dangerous work in dark, stuffy rooms. They worked in assembly lines as "killing gangs," including "knockers," "rippers," "leg
breakers," and "gutters." Cattle covered in boils and suffering with
tuberculosis were still fed into the assembly lines. Ground-up rats
infested rotten hams and sausages, and "there were things that
went into the sausage in comparison with which a poisoned rat
was a tidbit." Animal scraps like organs, bones, and fat made their
way into lard and fertilizer. A can of "potted chicken" might contain pork fat and beef scraps. Workers routinely lost their own body
parts—or lives. Sinclair famously described the men who fell into
steaming lard vats, "overlooked for days, till all but the bones of
them had gone out to the world as Durham's Pure Leaf Lard."

"It seemed to me I was confronting a veritable fortress of
oppression," Sinclair wrote of the meatpacking industry, one of
the most powerful and successful businesses in the country. "How
to breach those walls, or to scale them, was a military problem."
Sinclair's work of realistic fiction stirred the country. Writer Jack
London called *The Jungle* the *Uncle Tom's Cabin* of wage slavery,
noting, "It is alive and warm. It is brutal with life. It is written
of sweat and blood, and groans with tears." The uproar over *The
Jungle* prompted president Theodore Roosevelt to sign the Pure

Food and Drug Act, which banned misleading labeling of foods and drugs, sparking the formation of the federal Food and Drug Administration (FDA).

Meanwhile, through one of the most brilliant marketing campaigns the young century had ever seen, a masterpiece in the subtle art of persuasion, Procter & Gamble convinced Americans to let a laboratory vegetable oil into their pots and pans, displacing familiar animal fats like butter and lard. The feathery, pearly white product was packed in tin cans and wrapped in white paper, emphasizing its pristine state.

Cheerful ads channeled a ring of progressivism in a country hungry for a cleaner, stylish century. For decades, America had been turning away from farms and toward scientists working in laboratories. Savants in suits could surely solve any problem. As historian Susan Strasser writes, Americans were using bizarre new things, including toothpaste and cornflakes. In this light, Crisco emerged as an "artifact of a culture in the making." By 1916, only a few years after its debut, annual sales of Crisco reached 60 million pounds. Fluffy pies, cakes, and breads stuffed with slippery, solid vegetable fats filled the stomachs of American children and adults.

Procter & Gamble gave away a free cookbook called *The Story of Crisco*, which had 615 recipes all made with Crisco. The book reflected the culinary tastes of the age and included recipes for kidney omelets, baked brains, tripe, calf's head vinaigrette, stuffed hearts, braised ox tongue, and fried pancreas and thymus sweetbreads. It stated, "America has been termed a country of dyspeptics. It is being changed to a land of healthy eaters, consequently happier individuals. Every agent responsible for this national digestive improvement must be gratefully recognized."

Other companies followed suit. Partially hydrogenated vegetable oils continued to hijack the food supply, pushing lard out of American kitchens. Many margarines and vegetable oils were made from partial hydrogenation. Butter rationing after World War II boosted the shift. Soon, these oils were recast as health foods and became some of the most important ingredients the food industry had ever encountered. Partially hydrogenated oils were much cheaper than, say, natural olive oil. They could serve as preserva-tives, yielding profitable packaged goods with long shelf lives. They could be used many times over in fast-food companies' commercial fryers for batch frying.

But during the mid-twentieth century, as research on the role of diet in chronic ailments like heart disease, cancer, obesity, and dia-betes flourished, scientists started to realize that remaking fat in a laboratory by pumping hydrogen molecules into vegetable oils pro-duces a unique substance called *trans fat*. Hydrogenation also boosts omega-6s while eliminating omega-3s.

The immune system does not tolerate trans fats, which are not simply stored as fat. Rather, they displace normal fatty acids in the membranes of every cell in our body, inextricably weaving them-selves into our physiology. When this happens, the cells do not func-tion as they should. They make an excess of volatile molecules called free radicals, which injure healthy cells. Free radicals are neces-sary to sustain life. Macrophages and other immune cells typically make free radicals as they combat germs, toxins, and more. But an overload of these molecules leads to oxidative stress, a situation in which the body cannot produce enough antioxidants to neutralize the free radicals. Oxidative stress, be it from the wrong quantity or quality of food calories—or other environmental factors—ramps up the expression of inflammatory genes. It can irreversibly dam-age proteins, lipids, genetic information, and other substances in our bodies, exciting the immune system and adversely affecting the functions of our cells. Chronic inflammation is both a cause and a consequence of oxidative stress. For example, LDL cholesterol becomes more inflammatory after it is oxidized by an excess of free radicals, activating macrophages and drawing a hostile immune response. Much of the cholesterol in an atherosclerotic plaque is made up of oxidized LDL.

Many studies have shown that trans fats are tied to chronic, low-level inflammation, with an increase in inflammatory blood markers. Trans fats turn on NF-κB, the master activator of inflammatory genes and cells, by triggering the formation of free radicals or by directly affecting innate immune receptors on cell membranes. They exacerbate the inflamed milieu of adipose tissue and atherosclerotic plaques, bullying macrophages into heightened activity. They inflame the endothelial cells lining blood vessels, causing them to make less nitric oxide, an essential gas that calms inflammation and prevents blood clots. Trans fats influence the risk of heart disease by affecting not only inflammation but also lipids. They dramatically raise LDL cholesterol levels—more than saturated fats do—as well as triglycerides, another type of fat that is linked to heart disease. And unlike saturated fat, trans fats lower HDL, a "good" cholesterol that can help prevent atherosclerosis.

By the turn of the century, studies had tied trans fats to heart disease, stroke, hypertension, obesity, diabetes, cancer, growth problems, learning disorders, and infertility. Fred Kummerow, a professor of biochemistry at the University of Illinois in Urbana-Champaign, had been ringing the alarm bells about trans fats since 1957, largely alone and ignored in his field and frequently facing the wrath of industry giants. He published hundreds of papers on the topic and continued to voice his concerns until his last days. In 2013, at the age of ninety-eight, he filed a lawsuit against the FDA and the United States Department of Health and Human Services, hoping to compel them to respond to a petition he had filed asking for a ban on partially hydrogenated oils. By June of 2015, the FDA finally ruled that trans fats were no longer "Generally Recognized as Safe (GRAS)" and called for their complete removal from the food supply. The term GRAS, which was introduced in 1958, allows companies to evaluate their own substances and deem them acceptable, after which the FDA can review the evaluation—or choose not to. GRAS allows producers to take new food additives to market without informing the FDA. By 2018, the majority of food suppliers had eliminated trans fats from their products.

The story of trans fats is a lesson in humility, demanding that we manipulate food as a surgeon wields a scalpel, with great care and an awareness of the stakes at hand. Trans fats are one of the earliest examples of how a human-made food, like a foreign organ, has the potential to be rejected by the body's immune system. But unlike the raucous inflammation that ensues when a transplanted lung, liver, or kidney is acutely rejected, an emergent, visible phenomenon captured by medical tests, the inflammation elicited by routine consumption of trans fats and other inflammatory foods can be a relative innuendo, a whisper, a subtle insinuation that accrues over time but is no less menacing.

While imbalances in the quality and quantity of fats can inflame, so, too, can an excess of sugar, refined carbohydrates, and salt. Long before the advent of systematic nutrition research like the Seven Countries Study and the Nurses' Health Studies, one nineteenth-century physician intuitively recognized this, insisting that what people put into their mouths was crucial for wellness. He soon came up with an idea that would radically disrupt the food industry—and his own life.

CHAPTER 10

Sweet, Salty, Deadly

J ohn Harvey Kellogg was the fifth son of New England pioneers who left their family home of six generations in Hadley, Massachusetts, for the wooded West. Seeking to escape a life of onerous farming on a plot of barren soil, they made their way to a homestead in the Michigan frontier, where they built a farm and labored every day from dawn until dusk to carve out a living from the wilderness. Kellogg's parents were members of the new Seventh Day Adventist religious group in the city of Battle Creek. Under the sponsorship of Seventh Day Adventist church leaders, Kellogg attended Bellevue Hospital Medical College in New York City, finishing his studies in 1875 at the age of twenty-three.

Bellevue was affiliated with the earliest and largest public hospital in America. Aspiring young physicians clamored to be let into its iron gates and sloping red brick portal and through a plethora of pavilions comprised of laboratories, wards, surgical suites, the morgue, and more. They hoped to learn from the vast array of afflictions public ward patients presented with. In Kellogg's day, suffering and destitute New Yorkers arrived at the hospital's door in swarms and filled more than a thousand beds each night. But the state of medical care was such that over 15 percent of patients perished while admitted. Effective medical treatments were scarce, and the surgeon's scalpel was more likely to kill than to cure. Health-care staff seldom washed their hands or discarded their bloody, vomit-stained clothing. Infections ran rampant.

Meanwhile, gustatory excesses defined the good life in the latter portion of the nineteenth century, particularly in the decades after the civil war. Americans indulged in food as their means allowed. For the rich, lunch would consist of meaty main courses with gravies, creamed vegetables, bread and butter, cheese, whole milk, and desserts like puddings and fruit pies. Dinner was more of the same, and all fats were thought to be tasty and healthy. In Michigan's remote forest areas, homesteaders feasted on copious amounts of cured pork as well as beef in wet brines. During festive times, veal, mutton, or beef tongue might be served, or other game animals might be shot and quickly eaten. Molasses and cane syrup stood ready to sweeten teeth. But locals, even farmers, faced a shortage of fresh produce. Finances and the season dictated availability. Fruits and vegetables were preserved by canning, pickling, and jellying. Any loss of flavor with these early attempts was atoned for by using generous quantities of salt.

Breakfast was rarely simple fare, and its preparation was time-consuming. In addition to grains and potatoes, salty cured meats like ham or bacon, fried in congealed fat, would make an appearance. "Hot beefsteak," food historian Abigail Carroll wrote, "was a dish without which a proper nineteenth-century middle-class breakfast was increasingly considered incomplete." With the torrent of salt running through the body from these foods, thirst ensued and drink was a must. Saloons serving alcohol opened early in the morning, and plenty of coffee, tea, and cocoa were also available.

The most common ailment of the day was dyspepsia, or indigestion, a term used for anything from flatulence and heartburn to diarrhea, constipation, and upset stomach. (Today, the diagnosis of dyspepsia rests on more specific criteria.) In 1858, Walt Whitman wrote that indigestion was "the great American evil." Gastrointestinal issues were endemic, endlessly discussed, and profusely covered in newspapers and magazines, akin to modern chronic diseases like obesity, heart disease, and cancer. In fact, a nineteenth-century *New England Journal of Medicine* article listed dyspepsia among the top killers of humankind, along with infectious diseases. Entrepreneurs peddled new tonics and potions meant to cure intestinal distress. But Kellogg believed that the answer could not be found in a bottle. He would tell eager audiences, "When I was a boy, we knew nothing about diet . . . I thought there was nothing more delicious than an oxtail, which had been turned to a rich brown in the oven."

When Kellogg returned to Michigan a freshly minted surgeon with a prestigious degree, his sponsors asked him to take over medical directorship of the Western Health Reform Institute. They promised him a free hand to run the institute based on the latest science, without interference from the Adventist church. Kellogg accepted, but his ideas were ambitious. Perturbed by a culinary legacy that would be described by future generations as the era of "the great American stomachache," he believed that the public needed instruction in natural and healthful ways of living. Blending his Adventist beliefs with his background in science and medicine, he renamed the institute the Battle Creek Sanitarium, later known as the "San," to reflect his vision. Under Kellogg's leadership, the San, which started out as a two-story converted home, morphed into a colossal, luxurious medical center.

Guests both rich and poor frequented the San's quarters, and it had no shortage of celebrity patronage. Kellogg was an early advocate not only of eating and exercising to promote health but of other ideas progressive for the time, including Louis Pasteur's germ theory and Joseph Lister's sterile technique in operating rooms. However, he also—particularly in his later years—supported eugenics, a popular scientific theory of his day. Although he rejected segregation of African Americans at the San, trained African American doctors and nurses on its grounds, and fostered over forty children of various racial backgrounds with his wife Ella, his vocal role in helping to spread the eugenics movement in America, which was ultimately deemed racist and unscientific, is a dark imprint on his legacy.

Kellogg was one of the first physicians to tell patients that food played an important role in health. "Eat what the monkey eats— simple food and not too much of it," he advised his patients. Menus at the San included creamed cauliflower, stewed raisins, kafir tea, pineapple sauce, and whole grain biscuits. These plain feasts were free of added sugars or animal products, a stark contrast to typical nineteenth-century meals. They also drastically reduced salt intake, quenching thirst—and inflammation.

———————

The human body needs salt, as it does water, for survival. Salt helps to regulate the body's fluid balance and allows muscles and nerves to function properly. But the dose is critical. Traditionally, high salt intake has been tied to different health problems, including heart disease, hypertension, certain cancers, swelling, and stroke.

More recently, science has shed light on salt's potential to alter immune function. Small amounts of salt accumulate in skin wounds, pushing macrophages to heal (giving new meaning to the phrase "rub salt in a wound"). But a large load of salt sends macrophages into turmoil, driving them to assemble NLRP3 and other inflammasomes, which pump out dozens of inflammatory molecules. They also produce an excess of free radicals.

A surplus of salt activates—directly and through the macrophages—inflammatory Th17 cells, which are tied to many autoimmune diseases. In petri dishes sitting on laboratory tables, the cells of the innate and adaptive immune systems spew out inflammatory markers when immersed in solutions saturated with salt. Even a few days of a high-salt diet can cause the body to become inflamed, as if it were being attacked by germs or suffering with an autoimmune disease. Inflammatory cytokines like IL-6 and IL-23 increase, while anti-inflammatory ones like IL-10 decrease. Salt disables Tregs, the regulatory white blood cells that are critical for balancing inflammation. Tregs tell the body to tolerate its own antigens, preventing autoimmunity. They make IL-10 and inhibit the inflammatory Th17 cells, helping to resolve inflammation.

Take the case of hypertension, once thought to be solely the result of blood flow gone awry. Uncontrolled hypertension leads to the thick, stiff arteries of atherosclerosis—and vice versa—as well as fatal complications like heart attacks and strokes. We now know from studies in both humans and animals that hypertension is, in part, an inflammatory disease, an idea that dates back to the 1970s. Hidden inflammation is implicated in arterial stiffening, and hypertensive patients have increased levels of inflammatory molecules in their blood. Th17 cells play an important role in the inflammation of hypertension, along with macrophages, Tregs, and other immune cells. Inflammation is one likely route through which salt contributes to hypertension. Salt may help immune cells infiltrate the blood vessels and kidneys, releasing inflammatory cytokines that make it hard for the vessels to relax or for the kidneys to remove waste—both of which lead to hypertension.

A glut of salt in the diet can promote both hidden and overt inflammation, contributing to chronic inflammatory diseases—including autoimmune diseases like rheumatoid arthritis, lupus, multiple sclerosis, and intestinal inflammation—or even organ transplant rejection. And as one researcher found, confirming a hundred-year-old observation by German pediatrician Heinrich Finkelstein in his *Textbook of Sickling Diseases*, salt can make for bad skin, prompting autoimmune conditions like childhood eczema.

While salt shaken onto home-cooked meals can be utilized in moderation with little loss of flavor, the salt hiding in processed goods or fast food is usually exorbitant. Popular restaurants can serve up, in a single meal, several *days'* worth of the recommended daily allowance of salt.

The San, with its simple meals devoid of excess salt and animal fare, was also the birthplace of the first modern breakfast cereal, one that entirely excluded added sugars or refined carbohydrates. One afternoon, when Kellogg was still a medical student, a grocer sold him a package of "steam-cooked" oatmeal. It took just as long to cook as any other hot grain cereal. Kellogg was disappointed. Why, he wondered, did preparing cereals have to be so laborious? Why was it not possible to buy them ready to eat at the store? Decades later, he would describe the experience as a turning point in his creative life, his "eureka" moment. Exhausted housewives across the nation struggled with the same problem. They woke at the crack of dawn to stand over a hot, wood-burning stove for hours, stirring gruels made of barley, cracked wheat, oats, or corn.

Years later, in the San's kitchen, where Kellogg toiled with Ella and his younger brother Will, he sought to create plant foods that were not only convenient but also healthy. A thick spread made from peanuts, initially called "nutbutter," would one day fill millions of kids' lunch boxes. Nuttose, the first commercially produced meat substitute, also relied on peanuts and tasted like "cold roast mutton." Protose was made from nuts and grains and sold as mock chicken or beef. Kellogg used a good amount of gluten to thicken and shape these fake meats. He loved gluten and saw few adverse reactions to it, even in his patients with the most fragile guts.

One day, the trio discovered that instead of pulverizing whole wheat into a flour, they could steam it and run it through heavy rolls that squeezed the grain into flakes. The flakes could be made not just from wheat but other grains as well.

Flaked cereal was an immediate hit at the San. Patients' stomach-aches improved and they had more regular bowel movements. Will suggested selling the flakes to the public, but Kellogg refused. His main goal was to improve digestion in his clinic, not to profit.

But over the years, Will secretly worked to perfect the recipe for flaked cereal. His systematic, arduous search for the right flavor and texture turned him away from wheat and toward corn, the quintessential American grain. Corn was cheap, sweet, and plentiful. Will steamed the kernels over a bubbling water boiler and cut the kernel from the oily hull and germ. He ground the leftovers into "flaking grits." The result, after toasting, was a crunchy, crispy, and golden-brown flake, the "sweetheart of corn," as admen would later proclaim. Will added precise portions of sugar and salt to give his flakes a nutty, delicious taste.

Kellogg was livid when he found out about Will's actions. In particular, he considered salt and sugar to be unhealthy in excess. Will left the San to start his own food business. The brothers stopped speaking to each other and waged a long war in court over who had the right to use the family name. Will's company turned into a stunning sensation, becoming a household name and forever changing the American breakfast landscape. In just a few years, he became a millionaire. He continued to prosper through the Great Depression. John Kellogg's Battle Creek Sanitarium, however, faced hard times. By the 1930s, the San was more than three million dollars in debt and eventually fell into receivership. Kellogg, once vibrant and lauded, tumbled into obscurity. In the ensuing years, many popular breakfast cereals filled with a surplus of sugar would populate grocery shelves, a sweet start to the day becoming an American staple.

In the 1950s, while Keys focused on saturated fat, nutrition scientist John Yudkin, his most prominent rival, spoke out against sugar. At Queen Elizabeth College in London, Yudkin worked to prove that saturated fat was not the major—or only—potential element behind the epidemic of heart disease. "There is no physiological requirement for sugar," Yudkin wrote. "If only a small fraction of what is already known about the effects of sugar were to be revealed in relation to any other material used as a food additive, that material would promptly be banned."

Yudkin was not referring to the sugar found in whole fruits but rather to the small white or brown granules created on sugar plantations in the Caribbean that were fueled by the slave trade, setting in motion the lucrative operations of the modern sugar industry. He outlined his thoughts and the scientific evidence against sugar in a 1972 book called *Pure, White and Deadly*. Yudkin explained that too much sugar in the diet probably contributed to the high rates of heart disease, diabetes, obesity, tooth decay, and some cancers— conditions that tend to show up together in individuals and populations. Sugar, he wrote, could also harm the eyes, joints, and skin. It could contribute to indigestion or even severe intestinal inflammation. The book initially did well but soon came under attack. The World Sugar Association termed it "science fiction," and The British Sugar Bureau claimed his work was full of "emotional assertions." Mild-mannered Yudkin attempted to engage in political rhetoric but ultimately failed to successfully defend his ideas.

Keys's Seven Countries Study revealed a link between sugar and heart disease, but upon more sophisticated statistical analyses, it was eclipsed by the more potent relationship between saturated fat and heart disease. And a widely accepted mechanistic explanation for how eating sugar could lead to a heart attack was absent. On the other hand, scientists knew that saturated fat clogged arteries by raising blood cholesterol. But while Keys blamed saturated fat for the outbreak of heart disease and pointed out weaknesses in Yudkin's data, he was no fan of added sugars. He wrote, "Nutritionally we would be better off if we used more natural foods to provide most of the calories we now get from sugar, a refined chemical . . . gourmets agree that a common fault of much American cooking is an excess of sugar with a resulting cloying sweetness that drowns out more delicate flavors."

Neither Keys nor Yudkin claimed that their studies on saturated fat or sugar could prove that these substances caused heart disease. Rather, they only offered clues. Both men presented valuable insights that would give rise to inflammatory debates in modern nutrition science. "Final proof," Keys wrote, "is elusive in any field. It is extraordinarily difficult with regard to prevention of a slowly developing condition such as coronary heart disease."

Despite their bitter intellectual debates, both Keys and Yudkin struggled with opposition from the food industry. But the relative quality of Keys's data as well as its place in a larger body of research led to increasing support of his ideas. Meanwhile, Yudkin was not invited to international nutrition conferences, and research journals refused his papers. But their reflections on food and health were similar in more ways than was apparent by the trajectories of their lives. The advice to avoid an overload of animal fare, salt, and sugar would have been met with approval by Kellogg.

Sugar, like salt, is not an intrinsic poison. It is the lactose in breast milk, the fructose in an apple, the glucose swimming in our blood at all hours. The body has no need for sugars isolated from whole foods, or added sugars. But it is the dose of a thing that makes the poison, as Paracelsus, the father of toxicology, pointed out in the sixteenth century. And as with salt and saturated fat, it is both the dose of sugar and the context in which it is consumed that determines how it affects the body.

The first edition of the Dietary Guidelines for Americans, released in 1980, had warned against eating too much sugar, albeit with less fervor than the case against saturated fat. Even if sugar did not contribute to heart disease, the guidelines stated, it could rot our teeth. But science is beginning to show that what happens in the mouth can affect the heart and other organs. As we know, dental diseases like gingivitis—inflammation of the gums—can lead to hidden inflammation throughout the body, a risk factor for heart disease and other chronic inflammatory diseases. Cytokines spilling from the mouth in dental diseases prompt the same in the walls of blood vessels, an effect that is exacerbated in people prone to becoming inflamed due to genes or the environment. Oral hygiene, in addition to preventing bad breath and rotting teeth, fights hidden inflammation.

In the eyes of the immune system, Kellogg's steam-pressed whole wheat flakes and whole fruit markedly contrasted with Will's corn kernels, which were stripped of their outer layers—and hence many nutrients—before being laced with added sugars. The body breaks down grains, fruits, vegetables, and other carbohydrates and absorbs their simple sugars, which provide energy for cells and organs. Glucose molecules, harnessed by the fiber in whole foods, enter the blood slowly. They prompt the pancreas to pump out insulin, which guides the glucose into fat and muscle cells. Human beings evolved to obtain limited sugar from whole foods at this leisurely pace. When they repeatedly encounter vast quantities of refined grains and table sugar, foods stripped of fiber and other nutrients, quick, steep rises in blood sugar and insulin levels—as well as dramatic falls—ensue, stressing the body and activating inflammatory gene regulators like NF-κB.

When saturated with meals filled with sugar and refined carbohydrates, endothelial cells lining the inner walls of blood vessels make less nitric oxide, and macrophages in fat tissue pump out greater numbers of inflammatory cytokines. An overload of sugar stresses the liver, which converts dietary carbohydrates to fatty acids, the simplest form of fat molecules. Excess fatty acids trigger free radical formation and inflammation throughout the body, activating NF-κB and molecules like CRP, IL-6, and TNF-α.

A habitual pastry with morning coffee, a soda at lunch, and a scoop of ice cream after dinner are enough to heighten our risk of dying from heart disease. A daily can of soda or a few slices of white bread can increase inflammatory blood markers, LDL cholesterol, and fat in the liver and around the abdominal organs—the highly inflammatory visceral fat.

In the latter half of the twentieth century, evidence implicating sugar in more than just bad teeth began to build up. Research to date shows that excess sugar and other refined carbohydrates, robbed of nutrients but replete in calories, are tied to hidden inflammation and chronic inflammatory diseases, including heart disease, obesity, hypertension, diabetes, cancer, fatty liver, neurodegenerative diseases, arthritis, and inflammatory bowel disease.

Unlike refined grains and table sugar, sweet whole fruits can be consumed with impunity in nearly any amount and tend to counter inflammation. Kellogg's pureed pineapple, for example, contains bromelain, a potent anti-inflammatory enzyme used to treat pain after physical sprains and strains. But routinely discard all but the juice of an apple, or strip the bran and germ from wheat berries, and the body begins to recoil.

For over a century, pioneering physicians and scientists have pointed to the adverse health effects of unfavorable fats as well as excess salt and sugar. But these inflammatory foods swarm the standard American diet, or the "Western" diet, which is popular in most industrialized—and, increasingly, nonindustrialized—nations. The Western diet, a powerful trigger for hidden inflammation and chronic inflammatory diseases, is filled with fatty animal fare, salt, sugar, refined carbohydrates, and processed foods. And in the twenty-first century, a close examination of this diet as it invades not only humans but also the microbes that live within us further elucidates its noxious effects.

CHAPTER 11

Feeding Germs

W hen they operated on Olivia, she missed two full weeks of work. The inflammation in her bowels had refused to respond to medications. Her body tried to contain the damage, but repairing the intestines is more complex than, say, remaking the surface of the skin. Attempts at healing brought forth fibrous scar tissue, forcing the gut to contract and distort, narrowing hollow tubes with long, grotesque strictures, obstructing the passage of food and feces. The surgeon cut out much of Olivia's intestines, halting the cycle of inflammation and scarring. He sewed the end of the bowels to the wall of her stomach, creating an ostomy—a mercy for Olivia, who had tolerated excruciating pain when defecating. She called me one day, limp with resignation, feeling she had lost jurisdiction over her body. Unopened books on food and health cluttered her apartment. Excel spreadsheets on her laptop lingered past their due dates. The surgery was still fresh, her voice low and strained. But she wanted more answers than ever before.

Macrophages are whimsical characters with wide-ranging emotions. In Olivia's intestines, they had developed unruly features. Unlike typical, tolerant intestinal macrophages that encourage Tregs and anti-inflammatory cytokines like IL-10, Olivia's macrophages triggered inflammatory Th17 cells and made reams of inflammatory cytokines. And they were not as adept at clearing bacteria or resolving inflammation. In patients with inflammatory bowel disease, an illness in which all kinds of immune cells are affected,* these wild macrophages begin to show signs of troubling behavior while they are still young and floating in the blood, before they make their way to the intestines, where they run amok and create—and respond to—intense inflammation.

Why had Olivia been burdened with these wayward creatures? The answer is only partly etched into the sinuous sugar-phosphate strands of her DNA. An integral part of the story lies in the everlasting exchanges between her immune cells and the "non-self," including food and germs. Microbes are malleable and mainly molded by lifestyle, particularly by diet. Food can affect the body directly through inflammatory or anti-inflammatory—and even pro-resolving—pathways. But it also wields its power upon inflammation through germs. How we feed the germs that live within us helps to dictate the presence or absence of inflammatory microbial species and behaviors—including the nature of microbial metabolites, the essence of their conversations with immune cells. One of the main culprits in cultivating an inflammatory microbiome is an overload of modern animal fare, which, along with other inflammatory foods in the Western diet, starves gut germs of fiber, their most critical nutrient.

———

John Kellogg had an intuitive inkling of the intricate relationship between food, gut germs, and human health. During his lectures, he would slice off a sliver of beefsteak procured from the Post Tavern, a popular Battle Creek restaurant, and take a look at the meat under a microscope. He pointed out the millions of germs on-screen. "What is the difference," he would say, "whether it

———

* Traditionally, the adaptive immune system has been considered a major player in the pathogenesis of inflammatory bowel disease. However, the innate immune system is also an important inducer of gut inflammation in these patients.

be beefsteak rotting in the butcher shop, and then you swallow it first, or whether you let it become rotten and decay after you swallow it?"

He believed that meat was difficult to digest and putrefied in the gut, leading to the production of "bad germs" that emitted all kinds of toxins. He wrote that Americans had "the feeble stomach of a primate" but tended to eat anything, including "artificial foods," concluding that "it is no wonder the human gastric machine has broken down and that dyspepsia, constipation and peristaltic woes of various description have become universal in civilized lands." Meat-heavy diets, he argued, could lead to not only the plague of gastrointestinal issues that were so common in the day but also to many other maladies. He believed that Americans ate more protein than necessary, which could harm the body. The natural diet of humans, he stressed, was a combination of whole grains, fruits, vegetables, and nuts. Kellogg often remarked that it took massive amounts of grains—which could be used to feed humans—to produce small portions of beefsteak.

Many years after John Kellogg's lectures, another scientist came to the conclusion that Americans ate too much animal protein, compelling him to speak out against foods that had been the crux of his family's livelihood for generations. Years before he became Professor Emeritus of Nutritional Biochemistry at Cornell University, T. Colin Campbell was a five-year-old boy tasked with milking cows on his family farm, which lay long miles from Washington, DC—and the best public schools. Campbell would finish his morning chores every day and sit down to a breakfast of whole milk and boiled eggs with the tops chipped off. Sometimes he had fried potatoes or even a rare, luxurious strip of bacon.

In the 1960s, Campbell's research had doggedly fixated on promoting animal protein. But as he delved into medical literature and conducted his own laboratory experiments, he began to question dogma. Early in his career, working with children in the Philippines, he noticed ties between high protein consumption and liver cancer. His subsequent studies in rats showed that animal protein, unlike plant protein, tended to foster cancer growth, especially at levels beyond about 10 percent of daily calories.

Animal protein stimulated inflammation, increasing levels of free radicals in the body, but weakened immunity, impeding natural killer cells that protect against tumors. It encouraged cells to replicate and carcinogens to enter cells and bind DNA while hindering DNA repair mechanisms. Animal protein was also more likely than dietary cholesterol to raise blood cholesterol. Revisiting Keys's Seven Countries Study, Campbell noticed—as did a few other scientists—that the link between animal protein and heart disease was stronger than that for saturated fat.

Campbell was not content with experiments in rats, nor with focusing his efforts on an isolated nutrient. In 1983, he teamed up with Chinese scientists to launch what the *New York Times* would dub the "grand prix of epidemiology," the first major research project between China and the United States, a massive undertaking that spanned decades and looked at the influence of diet on disease in six thousand rural Chinese individuals who had lived, worked, and dined in the same county for most of their lives. The team administered nutritional questionnaires, directly measured everything subjects ate over a three-day period, took blood and urine samples, and analyzed food from local marketplaces. Since most rural Chinese consumed more plant foods than the typical Westerner, Campbell was in effect comparing diets somewhat rich in plants relative to Western fare with diets *very* rich in plants.

Thousands of statistically significant associations between diet and disease emerged from the China Study, and in concert they suggested that animal foods were linked to higher rates of chronic inflammatory diseases like heart disease, diabetes, cancer, and obesity while plant foods were protective. Campbell soon gave up his childhood meals, including any dishes with meat, dairy, and eggs. He was careful to acknowledge that the China Study could not, standing alone, prove that diet caused disease. But it was an important chapter of a larger story.

In later years, all kinds of human studies found that animal protein was tied to inflammation, a higher risk of various chronic diseases, and even early death. Over a century after Kellogg's lecture at the Post Tavern, scientists understood that excessive fermentation of undigested proteins by germs in the gut, unlike that of fiber fermentation, could harm the body. This issue typically arises when we overeat animal foods, which tend to be concentrated in protein and bereft of fiber. When bacteria ferment, or "digest," protein in the gut—a process known as putrefaction—toxic substances are produced, including hydrogen sulfide, a malodorous gas that smells like rotten eggs and is tied to diseases like inflammatory bowel disease and colon cancer. Hydrogen sulfide can damage DNA and inflame the gut. To a lesser extent, fermentation of select plant foods can yield hydrogen sulfide, but its effects are attenuated by fiber and other beneficial nutrients in these foods. Microbiome scientists continue to point out that animal protein helps to create an inflammatory microbiome that fosters disease, while plant protein has the opposite effect.

Beyond certain fats and protein, the immune system responds poorly to other substances in animal foods, like N-glycolylneuraminic acid, a sugar in red meat that is tied to chronic, low-level inflammation and an increased risk of cancer. Or heme iron, as the iron in meat is labeled. Heme iron is often lauded for its facile entry into the bloodstream. But unlike iron from plants, it paradoxically bypasses the body's exquisite mechanisms that regulate iron absorption and can accumulate in excess. While adequate levels of iron are essential to maintaining body functions, too much can induce oxidative stress and inflammation. Heme iron has been linked to a variety of chronic inflammatory diseases.

The sheer amount of germs in animal foods, as Kellogg had glimpsed, may contribute to inflammation. The body encounters fresh microbes with each meal. Some of these germs are helpful to humans, while others may be indifferent or even harmful. But their number is drastically higher in meals comprised of animal foods than in meals made up of plants. Further, studies show that animal foods contain an overload of bacterial toxins called endotoxins (akin to the endotoxins of Cani's microbiome studies, except the germs do not reside within the body), which are potent triggers of the innate immune system and inflammation.

In 2011, scientists at the University of Leicester in the United Kingdom wondered why blood inflammatory markers rose significantly after meals heavy in animal foods. They suspected that foodborne germs were largely responsible for this phenomenon. They tested fruits, vegetables, meat, and dairy foods for endotoxins and found massive levels in animal products, including pork, poultry, and dairy. In petri dishes, extracts of these foods prompted innate immune activation, inducing human macrophages to secrete inflammatory signals like IL-6 and TNF-α. The endotoxins, which lingered whether bacteria were dead or alive, withstood all types of processing, even if the meat was boiled for hours, dipped in an acid bath, or exposed to digestive enzymes. Small amounts of endotoxins are tolerated in the blood of healthy humans, but studies show that some meals—like an Egg and Sausage McMuffin—carry a heavy load of dead bacteria that emit enough endotoxins to inflame the whole body—a brief flare at first, but one that turns chronic with repeated slights, sowing fertile soil for disease to take hold.

The Western diet is deadly, in large part because it inflames humans while starving them of their most essential anti-inflammatory nutrient. At the crux of the relationship between food, germs, and inflammation, the "keystone" factor that all sanguine interactions rely on is fiber.[*]

In 1980, soon after completing his studies in nutrition and gastroenterology at London University, Stephen O'Keefe embarked for Cape Town, Africa, where he began a medical practice. O'Keefe was struck by his patients' immaculate colons, which were largely free of precancerous polyps. Unlike his patients in London, the South Africans rarely developed colon cancer or other intestinal ailments, including constipation, hemorrhoids, and diverticulosis. O'Keefe remembered the Irish missionary surgeon Denis Burkitt (famous in medical circles for discovering a childhood cancer known as Burkitt's lymphoma), who had made similar observations

---

[*] The word "fiber" is used in this book to refer to both soluble and insoluble fiber as well as resistant starch.

in the 1960s while working in Uganda. Burkitt watched millions of Ugandans flourish on high-fiber diets of whole grains and tubers with little to no meat, dairy, eggs, or processed foods. They barely encountered deadly diseases common to developed nations,* like heart disease, cancer, obesity, or diabetes. In his twenty years of surgery in the country, Burkitt removed exactly one gallstone. "The health of a country's people," he wrote, "could be determined by the size of their stools and whether they floated or sank, not by their technology."

Many large-scale epidemiological studies conducted over the years confirmed and built upon Burkitt's hypothesis, tying fiber deficiency to high blood cholesterol, hypertension, heart disease, various cancers, infections, lung diseases, obesity, diabetes, and an overall higher risk of death from these and other diseases. But *how* does fiber exert its potent health effects?

Fiber represents the parts of plants that humans cannot digest and absorb, unlike protein, fats, and carbohydrates. Plant foods are rich in fiber and other nutrients, slowing the stomach down and triggering gut hormones that tell the brain to feel full, helping us to lose weight. Fiber bulks up feces and allows it to hold more water, putting pressure on colonic walls and helping to stimulate bowel movements and dispose of excess cholesterol and hormones, toxins, and other unwanted elements. It helps with both diarrhea and constipation, promoting soft stools that are easy to pass, averting issues that manifest with repeated straining during defecation, such as hemorrhoids, hernias, diverticula, varicose veins, and even intracranial bleeds. It dampens glucose and insulin spikes after meals. Fiber dilutes and binds carcinogenic compounds in the intestines, escorting them out of the body.

One of the most essential ways in which fiber can prevent or treat modern chronic diseases—inflammatory diseases—is by

---

* Burkitt's observations are only partly attributable to the fact that life expectancy is higher in Western nations and older individuals are more prone to developing chronic diseases.

manipulating the immune system. Fiber can lower inflammation by melting visceral fat, an inflammatory organ. And studies show that diets high in fiber are tied to lower levels of inflammatory markers—including CRP, IL-6 ,and TNF-α—in specific tissues and throughout the body even after controlling for weight.

By the year 2015, O'Keefe had been living in the United States for over a decade, having joined the medical faculty at the University of Pittsburgh. His patients likely devoured some of the city's most iconic dishes: oversized hotcakes drenched in butter and syrup, pizza with reams of mozzarella cheese, and sandwiches stuffed with french fries, bacon, and coleslaw that once allowed rushed 1930s steelworkers to finish their entire lunch, sides and all, in mere minutes. O'Keefe noticed that the African Americans who had grown up in Pittsburgh feasting on this Western diet, filled with animal foods, refined grains, sugar, and salt, suffered from soaring rates of colon cancer and other diseases. O'Keefe thought of his days in South Africa. What would happen to gut germs, he wondered, if the African Americans ate the food of South Africans—and vice versa—for just a few weeks? He enrolled twenty African Americans from Pittsburgh and twenty rural South Africans from KwaZulu-Natal in his study.

Before switching their diets, he took baseline measurements, collecting their blood, urine, and feces. He performed colonoscopies, sampling intestinal tissue by taking tiny biopsies. A special stain revealed that the Americans' intestinal epithelial cells were dividing more rapidly than those of the South Africans, predicting a higher risk of colon cancer. Nearly half of them—but none of the South Africans—had adenomatous polyps in their colons. When O'Keefe took a close look at participants' feces, he was not surprised to see a vibrant array of germs living in the South Africans' intestines.

Other studies, including one comparing children in Burkina Faso with those in Italy, had also found that high-fiber diets bred a robust variety of gut microbes, some of which had never been seen in Western guts. The hundreds of thousands of edible plants on the earth contain distinct types of fibers that provide a variety of health benefits. Soluble fiber,* which dissolves in water and is found in most plant foods—some more than others—is a prebiotic.† It is fermented by gut bacteria and stimulates not only a diverse microbiome but one with anti-inflammatory behaviors, a cohort that keeps inappropriate immune responses in check while adeptly crowding out or defending against pathogens. Some of the best sources of soluble fiber include legumes, oats, nuts, seeds, and certain fruits and vegetables.

Fiber and other nutrients can exert their influence on the innate and adaptive arms of the immune system directly or through gut germs. Microbial waste, like human waste, can have potent effects on immune function. When bacteria ferment soluble fiber, they make metabolites that regulate the function of nearly every type of immune cell and affect inflammation both within the gut and throughout the body. O'Keefe found high levels of a metabolite called butyrate in South Africans' feces. Butyrate, along with acetate and propionate, is a short-chain fatty acid. These substances prevent the growth of fat tissue‡ and boost gut hormones that reduce hunger and nurture the intestines. For the single layer of intestinal epithelial cells, the sprawling yet delicate barrier between human beings and a plethora of foods and germs, butyrate is salvation, their main source of nutrition and good health. It helps the gut discard carcinogenic waste and excess cholesterol. It modifies gene expression, preventing unchecked epithelial cell growth and aids in eliminating cells that may turn into cancer.

---

* Two main types of fiber are soluble and insoluble fiber (there is also a third type with properties of both, known as resistant starch). Most fiber-containing foods contain both soluble and insoluble fiber but usually have more of one type than the other. Most soluble fiber is soluble in water. It absorbs water and turns into a gel-like substance. This contributes to fecal bulk, creating stool that is soft and easy to pass. Insoluble fiber, or "roughage," does not change much when combined with water. But it, too, adds bulk to stool and absorbs water as it moves through the digestive system, which can increase pressure on colonic walls and help to stimulate bowel movements.

† It is important to eat both soluble and insoluble fiber for a variety of health reasons. Generally speaking, soluble fiber is fermented by gut bacteria (though some insoluble fiber can be fermented too) and promotes the growth and maintenance of beneficial gut bacteria.

‡ Short-chain fatty acids decrease the ability of fat cells to take up fatty acids.

Short-chain fatty acids can influence immune function in ways that hamper hidden inflammation and chronic inflammatory diseases. They force the immune system to become more tolerant of foods and germs, averting allergies and sensitivities. They induce Tregs while inhibiting NF-κB, a critical conductor of the inflammatory response that leads to the expression of hundreds of inflammatory genes. They help to breed calm macrophages, ones that produce fewer inflammatory cytokines like IL-6 and TNF-α even when faced with a potent trigger like lipopolysaccharide. They heal intestinal inflammation: butyrate enemas treat patients with inflamed guts after radiation for cancer treatment or surgery for inflammatory bowel disease and other conditions.

Short-chain fatty acids prevent—or treat—the leaky gut and endotoxemia tied to hidden inflammation. They enrich the syrupy, protective mucus lining the intestines and prompt the body to make adhesive proteins that seal gaps between intestinal epithelial cells. They keep microbial molecules like lipopolysaccharide, an endotoxin, from passing between or across these cells.

Short-chain fatty acids can travel through the intestinal wall and into the bloodstream, manipulating immune cells and subduing inflammation in distant organs. They even cross the barrier between the blood and the brain, linking gut microbes to the mind, improving learning, mood, and memory. They prevent macrophages from infiltrating fat tissue, partly impeding the inflammatory potential of fat. They weave through the placenta in pregnant women to protect the fetus against inflammatory lung and other diseases.

O'Keefe found that American excrement, in stark contrast to that of the South Africans', revealed the ravages of a Western diet: poor microbial diversity and a wealth of toxic, inflammatory metabolites rather than short-chain fatty acids. Among those inflammatory chemicals in American feces were secondary bile acids. Bile, made in the liver and stored in the gallbladder, helps break down fat. Most bile acids are reabsorbed in the small intestine, but the continuous flow of bile farther down into the bowels on a Western diet high in fat ensures that an excess reaches the colon, where it is consumed by bacteria to create secondary bile acids, potential carcinogens that can damage DNA. A surplus of secondary bile acids promotes inflammatory microbes and adversely affects immune cells and inflammation within the gut and throughout the body. Secondary bile acids have been tied to various cancers.

O'Keefe also glimpsed high levels of choline in Americans' feces and an increase in a substance called trimethylamine N-oxide in their urine. Choline and carnitine are nutrients concentrated in animal foods like meat, dairy, and eggs and found sparsely in some plant foods. When the body receives these nutrients through animal foods, gut bacteria produce metabolites that are converted by the liver to trimethylamine N-oxide, a noxious compound that activates macrophages and other immune cells, increasing inflammation throughout the body. Studies show that blood trimethylamine N-oxide levels correlate with the size of atherosclerotic plaques and predict the risk of heart attacks, strokes, and even death. High trimethylamine N-oxide levels are also tied to other chronic diseases, including cancer, dementia, diabetes, and kidney disease.

For two weeks, O'Keefe swapped his subjects' diets, feeding the Americans corn porridge, beans, and fruit while the South Africans indulged in barbecue meat, hamburgers, and french fries. He then peered into their intestines again. Remarkably, the South Africans' epithelial cells were now dividing more rapidly than the Americans', and markers of colonic inflammation, including the sheer number of intestinal macrophages, had increased. The South Africans lost gut microbial species—a decrease in diversity—as well as half of their fecal butyrate, while gaining more choline and secondary bile acids. Meanwhile, the Americans enjoyed greater microbial diversity than they previously had, as well as more than twice as much butyrate and fewer secondary bile acids.

Fiber is vital to amassing beneficial microbes. The long, winding guts of humans and other great apes processed mostly plants as they evolved over millions of years. Early humans coveted meat but subsisted largely on leaves, vegetables, nuts, flowers, and fruits from bushes and shrubs or the odd snack of a small insect. Their lengthy, saccular intestine, with its broad absorptive surface and ample room for germs, was well suited to breaking down fibrous plant matter. Our Paleolithic ancestors, who began to roam the earth around 2.5

million years ago, also feasted largely on plant foods, consuming at least 100 grams of fiber each day—with rare exceptions, as most anthropologists concede. Grains from various wild plants have been found on the surface of Stone Age tools such as mortars and pestles, and fossilized dental plaques of Paleolithic individuals reveal plant remains like barley, legumes, and tubers, suggesting that humans may have been eating these foods well before the agricultural revolution ten thousand years ago.

While the spoils of foraging made up the bulk of most Paleolithic diets, hunted flesh foods did sporadically enter the stomachs of these individuals, often in the form of insects, which are rich in nutrients and low in calories. Their larger game was wild and lean, procured from animals accustomed to roaming free and eating an assortment of green leaves. They embraced carcasses in totality, devouring not only muscle but the marrow of the bone and all types of organs. Antelope flesh, which anthropologists suggest is similar to Paleolithic meat, contains around 7 percent fat, much in the form of omega-3s and almost none saturated. In contrast, modern cuts of cattle raised on corn—and often pumped up with hormones and antibiotics—contain around 35 percent fat, most of which is saturated, and few omega-3s.

Even fish now contain fewer omega-3s and more saturated fat than their predecessors. And modern animal foods also accumulate an overload of omega-6s. Studies show that eating wild game, like kangaroo flesh, triggers less inflammation—as measured by cytokines like CRP, TNF-α, and IL-6—than beef from domesticated cattle. Modern meat, which is intrinsically distinct from its Paleolithic forebearers, is also mostly consumed in a different context: frequent, hefty portions that are not accompanied by adequate fiber in the diet or activity for the body.

Any iteration of a diet brimming with modern animal foods is linked to a higher risk of many chronic inflammatory diseases. Diets that eschew nearly all carbohydrates—including grains, legumes, many fruits, and certain vegetables—in favor of animal fare are particularly insidious. These diets may initially help us to shed some pounds and even activate a few of the same metabolic pathways triggered by healthy practices like intermittent fasting. But over a meaningful measure of time—if not months then perhaps years—the costs of flooding our body with animal foods can manifest in an increased risk of chronic illnesses, nutritional deficiencies, and even premature death. An enviable physique, as with the Finns of Ancel Keys's time, may belie the damage wrought by a poor diet upon the body and its germs as hidden inflammation seeps through tissues and organs, biding its time.

Genetics play a greater part in some chronic inflammatory illnesses—like inflammatory bowel disease—than others. But even in these diseases, environmental factors have an important role. Studies show that a Western diet and its components, particularly a high intake of animal protein, are tied to an increased risk of inflammatory bowel disease, while fiber intake—even at modest levels—is tied to a decreased risk by nearly half and lowers the rates of disease flares. Fiber feeds germs in the colon and keeps the intestinal barrier intact. In the small intestine, where Crohn's disease often starts, fiber inhibits bacteria from invading the gut wall. Olivia's story began long before the surgeons sliced through her intestines, long before that fateful day during her senior year of high school when she first stumbled into an emergency room with blood clots pouring from her rectum, her body contorted in pain. Hidden inflammation, a key trigger of autoimmune diseases, may have been drifting in her blood for months or years, stoked by environmental insults and progressing unchecked in an unfavorable genetic milieu.

Food choices, the most potent environmental influences on hidden inflammation and chronic inflammatory diseases, can help to curb the leading causes of disability and death worldwide. But despite an increasing understanding of the links between food, germs, and disease in the twentieth and early twenty-first centuries, the Western diet, fueled in part by industry interests and lenient governmental policies, managed to remain on most plates around the country and beyond. It also became unhealthier. When Kellogg had lamented the gustatory excesses of his time, some of the most inflammatory foods known to humankind had yet to enter the food supply.

CHAPTER 12

Farm Country

I n the 1960s and '70s, Americans faced a strange paradox. In 1968, South Dakotan senator George McGovern recognized a hidden scourge of hunger in America and implemented a series of landmark federal food assistance programs to protect low-income individuals against malnutrition. But he was becoming aware of a distressing new obstacle: Americans with plenty of food were getting sicker than ever.

McGovern was familiar with Ancel Keys's research. For a long year, McGovern's bipartisan Select Committee on Nutrition and Human Needs had taken a close look at all the science on food and health, reviewing the literature, conducting public hearings, and consulting with experts. The science was controversial, but McGovern's committee concluded that most deaths in the country were due to diet. In January of 1977, they released the country's first dietary guidelines, a seventy-two-page report called *Dietary Goals for the United States*. The report encouraged Americans to increase fruits, vegetables, whole grains, legumes, and nuts in their diets and decrease meat, dairy, eggs, refined sugar, and salt. McGovern hoped that the report would do for food what the surgeon general's report had done for smoking a few years before—inform people of healthier lifestyle options. "The public wants some guidance, wants to know the truth," he insisted. But despite McGovern's benevolent intentions, the country continued to embrace the Western diet rather than move away from it, consuming increasing amounts of processed foods—which are some of its most inflammatory components—rather than whole foods, ignoring the intricate language of the immune system.

Only a few weeks after the guidelines were published, McGovern and his committee faced an onslaught of criticism. The most incensed voices came from behemoth food industries. The sugar industry submitted a brief calling the report "unfortunate and ill advised," part of an "emotional anti-sucrose tidal wave which has swept industrialized nations in recent years," accusing the committee of robbing citizens of life's comforts. The National Dairy Council thought it "highly questionable" that changes in diet could affect health. The Indiana Egg Council called for the immediate withdrawal of the report. The Salt Institute stated that decreasing daily salt intake was "not an important or even significant dietary goal for the public." The meat industry demanded a separate hearing to air its grievances.

American farm country held powerful influence in Washington. By the end of the year, the McGovern committee was forced to publish a revised version of the dietary goals to placate incendiary industry protests. Recommendations to decrease eggs and whole milk were removed, and a higher allowance was provided for salt intake. Instead of telling the public to decrease meat intake, the report now advised them to "choose meats, poultry and fish which will reduce saturated fat intake." This subtle semantic shift—removing the word *decrease*—would have tectonic effects, inspiring an entirely novel way of thinking and talking about food that persists today, a language with oblique intent.

The meat industry successfully lobbied to have McGovern's committee disbanded and its functions turned over to the United States Department of Agriculture (USDA), a move that was like "sending the chickens off to live with the foxes" as the *New York Times* noted, since the USDA looked after the producers of food rather than the consumers. In the subsequent election, in 1980, the industry voted McGovern out of office, sending a warning to anyone on Capitol Hill who dared to disturb the traditional source of protein on the American plate.

A couple of years later, when the National Academy of Sciences looked at the links between diet and cancer, they spoke of nutrients rather than whole foods in order to avoid irritating influential interests. If attacking meat, dairy, and eggs was not feasible, no such industrial protections existed for the predominant nutrients in those foods, such as saturated fat, which ran through animals as distinct as salmon, chickens, and cows. Reducing foods to their invisible nutrients stripped a burgeoning language of its essential layers, as people began to speak of fats and cholesterol and carbohydrates rather than steak and eggs and apples.

Two of the thirteen panel members at the National Academy of Sciences, including Cornell nutrition biochemist T. Colin Campbell, who had embarked on the China Study, and Columbia nutritionist Joan Gussow, objected to this approach, arguing that the current science shed light on dietary patterns and whole foods, not nutrients. Campbell noted that "all of the human population studies linking dietary fat to cancer actually showed that the groups with higher cancer rates consumed not just more fats, but also more animal foods and fewer plant foods as well." He later wrote, "This meant that these cancers could just as easily be caused by animal protein, dietary cholesterol, something else exclusively found in animal-based foods, or a lack of plant-based foods." Gussow saw the same scientific reductionism with foods deemed to be healthy. "The really important message in epidemiology, which is all we had to go on, was that some vegetables and citrus fruits seemed to be protective against cancer," she said. "But those sections of the report were written as though it was the vitamin C in the citrus or the beta-carotene in the vegetables that was responsible for the effect. I kept changing the language to talk about '*foods that contain* vitamin C' and '*foods that contain* carotenes.' Because how do you know it's not one of the other things in the carrots or the broccoli? There are hundreds of carotenes."

Campbell advocated for minimally processed, whole plant foods rather than specific nutrients. He called this diet "plant-based," divorcing it from ethical, political, or other considerations that strayed from science. In his eyes, the "wholeness" of a food was paramount to health outcomes. He cautioned, for example, against an overindulgence in plant oils. Compared with whole plant food sources of fat, most oils contain few beneficial nutrients but plenty of calories rapidly absorbed by the body.

Whole foods are foods closest to their natural state, eaten raw, ground, soaked, dried, fermented, or cooked. Gentle cooking methods are best for our bodies and the germs within them: steaming, sautéing, stewing, boiling, poaching, or light baking. Whole foods provide more fibrous fuel for gut germs, which prefer plants close to their natural state: steamed or lightly sautéed vegetables, for example, instead of overcooked or fried ones; whole, intact grains like steel-cut oats over rolled, instant oats.

Foods cooked under extreme heat and with little moisture, as in grilling, deep-frying, broiling, roasting, and searing, are prone to building up a variety of toxic, inflammatory by-products. This tends to occur with many fatty or heavily processed foods, but the effect is particularly striking for animal foods like meat, cheese, butter, and eggs. In contrast, whole plant foods rich in carbohydrates—fruits, vegetables, grains, and legumes—tend to accumulate the lowest levels of these damaging compounds, an excess of which are tied to chronic inflammatory diseases, including heart disease, obesity, diabetes, cancer, and neurodegenerative diseases. Some of these substances even bypass absorption in the small intestine and feed gut germs, modifying the microbiome. Our Paleolithic ancestors often used gentle methods to cook their meat, boiling it with hot stones or in hide-lined pits. Inflammatory cooking styles can be tempered in part by using certain types of spices, acidic marinades like lemon or vinegar, and by discarding the charred parts of foods. Olive oil in the traditional Mediterranean diet was often used raw, preserving delicate flavors and nutrients, or to stew and sauté vegetables.

Whole foods can be reduced to lone nutrients or even the elemental building blocks of protein, carbohydrates, and fat: amino acids, simple sugars, and fatty acids. But this distillation neglects the hidden language that transpires at the gastrointestinal border as food is digested, dissolved, and absorbed, a translation of familiar meals into the manifold signaling pathways of the immune system, a language that is more nuanced than ever imagined, integrating signals from food and germs with a complexity we are only beginning to uncover.

That the immune system speaks a singular language is not a novel idea. Danish immunologist Neils Jerne, who first coined the term *immune system* in 1967, drew parallels between the amino acid sequence of the combining site of an antibody and the sequence of spoken words in a sentence in a 1985 article titled "The Generative Grammar of the Immune System." In 1984, three immunologists inspired by Umberto Eco's *A Theory of Semiotics* hoped that a better understanding of semiotics might complement their work in immunology. They organized a conference—attended by Eco himself—on "immunosemiotics."

> "Immunologists are forced to use unusual expressions in order to describe their observations," the semiotician Thure von Uexkull observed at the conference. "Expressions like 'memory,' 'recognition,' 'interpretation,' 'individuality,' 'reading,' 'inner picture,' 'self,' 'non self,'" he maintained, were unknown in physics or chemistry. "Atoms and molecules have no self, memory, individuality or inner pictures," he said. "They are not able to read, to recognise or to interpret anything and cannot be killed either."

The same word can refer to several different foods and yield various meanings, depending on the context in which it is interpreted. For example, "saturated fat" can represent many items, including marbled beef, lean chicken, salmon, coconuts, or even human breast milk. The context in which saturated fat is consumed further transmutes meaning: for a farmer subsisting largely on greens and legumes, it is a rare novelty the immune system tolerates. For a hungry infant, saturated fat is sustenance. For a midwestern kid lunching on refined carbohydrates and sugar, saturated fat fuels the inflammatory fire.

When I was a child, my plate was often filled with Western staples, including many foods far removed from their whole forms. I grew up in a small Indiana town called Valparaiso, known to locals as Valpo. The town is nested in the American heartland about a half hour from Gary, the infamous city of steel mills and the Jackson 5. Valpo stretches across level terrain in the midst of endless cornfields. One movie theater, a public library, and a quaint downtown area with a handful of shops and restaurants provided most of our entertainment. Three railroads and four highways run through the town. It stands on the Native American Sauk Trail, once a horse path frequented by hunters and traders.

Like many midwestern towns originally defined by industries that no longer exist, Valpo was largely unknown to the nation—but most of its food was not. As a kid, I attended local schools where lunch was served daily in the cafeteria: rectangular slices of pizza with red sauce and mozzarella cheese on pale crusts, french fries and tater tots, fried chicken fillets between soft buns that would crumble to the touch, and chocolate or strawberry milk. On school holidays I would celebrate at a fast-food joint, downing shakes so thick and tall I would not be hungry again for ages. I was a picky eater and small for my age, so my parents gave up trying to limit what I ate. Back then, cell phones and the Internet had not yet made their ubiquitous appearance. I spent hours roaming the sylvan underworld of childhood, the quiet spaces outside and within my home, comfortably trapped in the tangible but insulated from the world at large in our sleepy town. I knew little about where the food on my plate had come from or what it meant for my health. And I did not have easy access to the roads that could solve the mystery, illuminating the past to explain the present.

Over the years, our food had undergone a transformation. During the 1980s, whole foods began to cower from grocery shelves. Supermarket perimeters shrank while their central aisles ballooned. Industries had capitalized on the post-McGovern era of nutritionism, creating thousands of processed foods that contained few "bad" nutrients—like saturated fat—and many "good" ones—like vitamins. Low-fat products multiplied: SnackWell's cookies, Entenmann's cakes, potato chips filled with olestra, Slim-Fast shakes, and many more. Although people shied away from saturated fat, they weren't really eating *less* overall fat, only a smaller percentage of fat relative to refined carbohydrates. Industries with brilliant marketing campaigns replaced saturated fats with other types of fats, like partially hydrogenated vegetable oils teeming with trans fats, and plenty of sugar and salt. Simple foods traditionally made with just a few ingredients—bread or yogurt, for example—now had long lists of additives. We did not really think these foods were better for us than fresh food made from whole ingredients. But they were tasty, convenient, and seemed to be mostly in line with the government's dietary guidelines—so how bad could they be? Meanwhile, an epidemic of obesity was underway in Valpo and much of the country, and other chronic inflammatory diseases continued to balloon.

Food processing occurs in degrees. Milling and polishing brown rice yields white rice, which loses some fiber and many vitamins and minerals but retains a semblance of its original form. Stripping the bran and germ from wheat berries before pulverizing them into a pale pastry flour, on the other hand, or extracting golden raw crystals from the sugarcane plant, entirely transforms these foods. The most profound metamorphosis involves heavily processed industrial foods, which are made from whole food derivatives and extracts—including loads of sugar, refined carbohydrates, salt, and fat—as well as additives, all of which make food look, feel, and taste better. These foods are painstakingly designed to be addictive and inflammatory, quickly spiking blood sugar and insulin levels before prompting a deep crash that leads to keen hunger, making processed food companies some of the most profitable in the world. Processed foods are harmful for human health. Studies show that they increase our risk of getting sick or dying from any cause.

A sweetener known as high-fructose corn syrup, a common ingredient in all kinds of processed foods, often made its way into my childhood meals. High-fructose corn syrup is created by crushing whole corn in mills to yield cornstarch and transforming it with laboratory enzymes into a viscous liquid. Table sugar, or sucrose (which is made up of glucose and fructose molecules), and high-fructose corn syrup are the two main types of added sugars in the Western diet. Both are inflammatory, but high-fructose corn syrup is sweeter and cheaper. And unlike glucose, fructose does not suppress hunger hormones or stimulate insulin. We could drink cans and cans of pop (for most midwesterners, all soda is "pop") or eat whole bags of pretzels without our brains ever realizing it, keeping us from feeling full.

High-fructose corn syrup and other added sugars filled our sports drinks and fruit juices. On fairgrounds, they suffused funnel cakes, fried dough, and reams of cotton candy, hiding in batters or gripping sticky oil, saturating our hands and mouths. They turned breakfast cereals into candy. They lurked in many savory foods, too, like ketchup, pasta sauce, and salad dressings. Nothing in our evolutionary history had prepared us for this onslaught of sweetness in modern life.

Unlike high-fructose corn syrup, whole corn on the cob, which my little brother used to pluck straight off a farm near our house on occasion for an after-school snack, can be a healthy treat. So can plain, lightly salted popcorn—an important food in Valpo's history. Orville Redenbacher, America's "popcorn king," lived in Valpo for decades, building his life and dreams there, experimenting with corn hybrids in search of the perfect popping kernel. In his iconic red suspenders and an oversized bow tie, he became the face of microwavable popcorn, lending a homespun, wholesome quality to his brand. At Valpo's annual, spirited Popcorn Festival, which began in homage to Redenbacher (he attended the festival every year until his death in 1995), prom queens ride elaborate floats and people enjoy popcorn, live entertainment, craft vendors, and more.

Not only are processed foods full of empty, unhealthy calories and lacking in fiber and other important nutrients, they also contain additives that are uniquely inflammatory. Artificial sweeteners, for example, are meant to sweeten without adding calories. Since they are barely absorbed into the bloodstream, they should not raise blood sugar. But although artificial sweeteners bypass the blood, they inevitably encounter gut germs. Studies show that artificial sweeteners induce changes in the gut microbiome that promote inflammation and insulin resistance. They also nourish a taste for intensely sweet foods.

Emulsifiers, or detergent-like molecules that are added to most processed foods, extend the shelf life and improve the texture of these foods. Common emulsifiers like carboxymethylcellulose, polysorbate-80, maltodextrin, and carrageenan keep ice cream and peanut butter thick and creamy, improve the mouthfeel of cookies and cakes, and help oil and vinegar in a salad dressing stay blended. But emulsifiers may interact poorly with germs, adversely affecting the immune system.

In a 2015 study published in *Nature*, microbiome scientists found that consumption of carboxymethylcellulose and polysorbate-80 for several weeks—at the doses one would expect to eat—inflamed healthy mice, altering not only the composition of gut microbes—*who* they were—but where they lived and how they behaved. Mice fed emulsifiers had greater numbers of inflammatory microbial species and behaviors and fewer anti-inflammatory ones, with an overall reduction in microbial diversity. More microbes dared to encroach upon the intestinal wall, ironically thinning out its mucus while their host enjoyed thickly textured food. Hidden inflammation weaved through the intestines of these mice. They became ravenous, gaining weight and excess blood sugar, features of the metabolic syndrome that presages chronic inflammatory diseases like diabetes and heart disease. A fecal transplant from these mice into germ-free ones transferred an inflammatory microbiome, low-level inflammation, and metabolic syndrome. In mice genetically susceptible to developing inflamed intestines, carboxymethylcellulose and polysorbate-80 induced dramatic, obvious intestinal inflammation—typical of inflammatory bowel disease patients like Olivia. Laboratory studies on human intestinal tissue have found that emulsifiers like polysorbate-80 and maltodextrin help invasive microbes thought to play a part in inflammatory bowel disease tunnel through the gut wall, while some soluble plant fibers—like those found in plantains and broccoli—prevent this effect.

Food industrialization was born in America, where major historical challenges included feeding a growing population while moving food across an expansive country. Market forces, lobbying, and oversights in government regulation gave rise to cheap, convenient, heavily processed foods. Today, as in the 1990s, these inflammatory foods make up a substantial portion of the Western diet, which continues to spread around the world via globalization, displacing traditional foods in burgeoning economies.

In the last decades, the number of additives in the American food supply shot up to over ten thousand, with most breezing through the GRAS loophole. More than 99 percent of these substances have not been studied, and robust human studies are often lacking. When testing is performed, it typically occurs in animal models designed to detect acute toxicity or a potential for promoting cancer. But this type of research does not always capture subtle yet costly effects on gut germs and hidden inflammation.

Foods far removed from anything the human body has ever encountered in its biological history are often considered innocent until proven guilty. This practice is antithetical to the immune system's wary handling of the environment as it expertly wavers between tolerance and fury, emboldened by a wisdom born of countless conversations with food and germs through millennia, a reflection of a greater understanding: that language is essential for the translation of food into fuel, of science into policy, of public rhetoric into private truths. A slight shift in diction, a faint rewording, a poorly translated phrase—just like the addition of a few hydrogen atoms into a vegetable oil—can have catastrophic, enduring effects.

The immune system responds to the Western diet, loaded with animal fare, sugar, salt, refined carbohydrates, and processed foods, as it would a noxious germ. Studies in both humans and animals show that this diet can directly activate the immune system, stressing cells in our body and inciting an angry response from macrophages and other immune cells, which produce an overload of inflammatory cytokines and fewer anti-inflammatory ones. The Western diet can irrevocably alter normal cells, generating misfolded and misplaced molecules, malfunctioning mitochondria, senescent cells, and other unwanted clutter that can inflame. It may hasten the degradation of telomeres, the structures that keep chromosomes from fraying.

Telomeres are essential to a long and healthy life. Many studies have connected diets replete in fiber with longer telomeres and higher amounts of telomerase, the enzyme that rebuilds telomeres.

The Western diet is tied to altered expression of inflammatory genes and high blood levels of inflammatory markers like CRP and IL-6. With recurrent exposures, the immune system may *remember* a noxious diet and react more aggressively to it during future encounters. Innate immune memory, which evolved to fine-tune responses against germs, is elicited by other triggers as well, like food and microbial metabolites. In mice, a Western diet rewires gene expression in macrophages so that they produce more intense inflammatory responses to all kinds of stimuli, a tendency that persists even weeks after the mice resume a normal diet.

A Western diet also wields its influence through microbes. It starves gut germs, transforming their behavior and shrinking some species into oblivion. These extinctions, as animal studies suggest, can be passed to host progeny and persist even when younger generations improve their diets. Bereft of fiber for a little or long while, microbes inch closer to the mucus lining the intestinal barrier, dining on its sugars and thinning it out, forcing the immune system to react. They express inflammatory molecules like lipopolysaccharide, triggering an insidious cycle of hidden inflammation, leaky gut, and a warped, inflammatory microbiome. Obese people tend to have relatively high levels of both lipopolysaccharide and inflammatory cytokines—like TNF-α and IL-6—in their blood.

When my former colleague Carrie had come to me seeking advice on how to lose weight, her dilemma consisted not simply of physical flesh and mental will but something more explicit: the chronic, hidden inflammation that seeped from her fat and the germs in her gut. It altered hormone signaling and skewed the liaisons between her gut germs and brain, keeping her insatiable and fostering cravings for the sweet, salty, and fatty foods—high in calories and low in nutrients—of a Western diet, causing her to accrue even more inflammatory fat. Dysbiotic microbes tinkered with her genes, increasing her tendency to gain weight, and extracted extra calories from her food. Encouraged by a lack of microbial diversity in the gut, they devoted more of their energy toward manipulating Carrie's eating behaviors rather than competing with other germs. Carrie was caged in a circular trap much like Cani's mice, with food, germs, and inflammation colluding to keep her unwell.

Far from the Western world, while McGovern reflected on how to reform policy to better feed America, earlier generations of my family enjoyed a much healthier diet than my own childhood diet. In the summer of 1971, soon after India's first female prime minister had been reelected, my mother, then fifteen, imagined herself pedaling through the hard dirt roads of her town, the wind fanning her face and countering the stultifying heat as she soared past street vendors, stonecutters, and her favorite temple. But in her rural farming village in Andhra Pradesh, a coastal state in southern India, the wish was impossible. Bicycles were rare commodities only meant for boys or men. In a land where farmers suffered from crop failures, large debts, floods, water shortages, and unmarried daughters, luck and drudgery kept my grandfather afloat. He toiled on a few acres of rich black soil nourished by canal water from a nearby river, plowing the land with oxen and growing legumes, rice, millet, sorghum, sweet potatoes, peppers, and an array of other vegetables. Against difficult odds, he managed to feed his family of nine. As a child, my mother learned to speak *Telugu*, the official language of the state. Its lilting, melodious tones, with nearly every word ending in a vowel, inspired fifteenth-century explorers to declare it the "Italian of the east." This native tongue would forever translate the full range and nuances of her thoughts and emotions as no other.

My mother spent her days in a house built from clay and bamboo. She poured water from nearby ponds and wells into massive copper pots, which served as a natural purification system. She lit the flat wicks of kerosene lamps housed in slender glass chimneys. She brought sticks of wood and crop waste to fuel a small, traditional mud stove coated with clay and cow dung. Sitting next to this stove with her mother and sisters, she learned how to expertly transform the foods from her father's farm. The inflammatory

staples of a Western diet were unknown to her, including processed foods.

She stewed curries of eggplant, okra, pumpkin, and the oblong
bitter gourd with its jagged ridges. She cooked rice largely intact,
after discarding only the outer hull, avoiding the expense of a pol-
ish. She hand-ground pearl and finger millets, sorghum, and lentils, fermenting the dough for days before creating crepe-like *dosas*
stuffed with spicy onions or steamed round *idlis* dipped in coconut
chutney. She made a multitude of legume soups, or *daals*, doused
with garlic, ginger, cumin, and fragrant curry leaves and peppers
cut fresh in the field. My mother worked with classic and popular
leafy greens of the day, fashioning chutneys from the tart *gongura*
plant or lightly spicing and smashing green amaranth. *This* language of hers—expressed through food—was shared and understood in many other parts of the world. In fact, it was yet another
rendition of Keys's observations in rural Naples, a way of eating that
would prove to align with the preferences of the immune system.

CHAPTER 13

Mangiafoglia

I n the 1950s, a steady pulse had emerged during Keys's months in the Mediterranean. The food captured his imagination and altered the course of his personal and scientific life, spurring him to action upon his return to America while drawing him back to Naples in later years. But the traditional Mediterranean diet, filled with beneficial nutrients and whole foods, would be difficult to duplicate in America. Nutrition science research, historically challenging to conduct and interpret, would prove to be a controversial topic, its overarching messages on how to eat for health often muddled.

In Naples, frequent flat tires and other car troubles on the difficult roads of the day had not dampened Keys's enthusiasm. He found much to marvel at beyond favorable fats, including ample legumes, loads of fresh produce, and a variety of whole grains, including "a much darker and more solid bread" than American versions that was particularly common in rural districts and came in "big round loaves weighing nearly five pounds." The loaves were baked at night and arrived in shops "still warm and gloriously fragrant, by eight o' clock in the morning." His coffee was always strong and fresh, made from "darkly roasted, newly ground beans." Espresso machines flooded bars and restaurants, and cream was largely absent from coffee and tea. Locals united not only in their love for coffee but also in their affinity for mineral water, the only source of hydration through most of human history, an essential beverage that helps to fight inflammation by keeping cells and organs fit and by flushing toxins out of the body.

As he wandered the Neapolitan countryside sampling fare in local kitchens, Keys watched cooks grate their own nutmeg and insist on cinnamon bark and whole cloves rather than the dried powders. In the proper kitchen, peppercorns, too, were freshly ground. Cooks gathered aromatic wild herbs by the side of the road or visited grocers who provided fresh sweet basil, mint, chervil, tarragon, sprigs of rosemary, small bunches of parsley, and celery—bred for flavor rather than appearance or texture—and many wild herbs with names that did not appear in any dictionary. These herbs, collectively called *odori* in Italian, were responsible for much of the characteristic flavors of Mediterranean dishes.

Neapolitans peeled and bottled tomatoes into sauces or sliced them raw with a splash of vinegar or lemon juice. Cooking without tomatoes in any part of the Mediterranean, Keys noted, was unimaginable. Locals were also generous with the use of garlic and onions and brought wild mushrooms to life in the spring and fall—including the flavorful, sponge-bottomed *boletus*—through a good soak. Mushrooms, garlic, and onions are prebiotic foods that benefit gut germs. They not only stimulate immunity and prevent infections but also discourage chronic inflammation. Human clinical trials have shown that cheap, unassuming white button mushrooms boost salivary levels of IgA, the antibody that bars toxins and infectious germs in the gut, lungs, and other mucous barriers throughout the body and floats in spit, sweat, and tears. Even single-celled mushrooms—like a daily sprinkle of the cheese-flavored nutritional yeast—may lower the rates of upper respiratory tract infections.

Mushrooms are a natural source of vitamin D, which is traditionally known for its role in helping the body to absorb calcium and build strong bones. But vitamin D is also integral to the health of the immune system. It plays an important role in immunity, enhancing our body's ability to defend against infections. Vitamin D deficiency is tied to many chronic inflammatory diseases, including obesity, heart disease, diabetes, inflammatory bowel disease, multiple sclerosis, rheumatoid arthritis, lupus, and cancers of the colon, breast, and prostate. Receptors for vitamin D, which express many genes relevant to immunity and inflammation, are found in tissues throughout the body—including the brain, heart, and muscles—and in immune cells. Vitamin D modulates macrophages, preventing them from producing too many inflammatory cytokines, and induces Tregs.

Keys was struck by the plentiful citrus trees of lemons, oranges, and tangerines and the heavy loads of cauliflower, lettuce, and artichokes headed to markets on farmers' carts. Through summer and fall, heaps of fresh tomatoes, red peppers, and eggplant hung in balconies or appeared on tables. Keys made note of all the new, seasonal vegetables almost unknown in the United States, like rape florets (*broccoli di rape*), finocchi, fava beans, and dozens of wild greens typically gathered by the roadside. "No main meal in the Mediterranean countries is replete without lots of *dure* (greens)," he wrote. "*Mangiafoglia* is the Italian word for "to eat leaves" and that is a key part of the good Mediterranean diet." Dark leafy greens have some of the highest measured antioxidant potential of any food. They are an especially rich source of many vitamins and minerals that regulate the immune system, including vitamins A, E, K, and C and folate, magnesium, and iron.

Keys had immersed himself in a pattern of foods that was notable not only for its abundance of fiber, vitamins, and minerals from whole plants but also for an array of substances intrinsic to a plant's survival. Phytochemicals are chemicals produced by plants that help them to overcome predators, pathogens, and even the harsh glare of the sun. They are a plant's primary defense, laboriously crafted with a wealth of precious energy in an organism without a sophisticated immune system. Many phytochemicals benefit human health, uniquely affecting germs, inflammation, and disease.

Cruciferous vegetables—which include broccoli, kale, cauliflower, cabbage, brussels sprouts, collards, arugula, and bok choy—are the main dietary source of a unique class of phytochemicals called isothiocyanates. Isothiocyanates, which evolved to defend plants, can also protect the humans who eat them. These molecules remove toxins, prevent DNA damage, and kill cancer cells and infectious germs.

Research shows that cruciferous vegetables also offer powerful protection against chronic inflammatory diseases. Germs play an important role here. In the gut, microbes feast on cruciferous vegetables and secrete enzymes that activate isothiocyanates. They ferment fiber, phytochemicals, and other nutrients, creating short-chain fatty acids and many unique metabolites that help to establish tolerant macrophages. Some of these metabolites interact with a special molecule called the aryl hydrocarbon receptor, which is found on immune cells and recognizes patterns on food, germs, and other environmental stimuli. The aryl hydrocarbon receptor is a major anti-inflammatory hub that integrates signals from food and germs, setting off chain reactions that influence genes controlling inflammation and immunity. It helps to shape an anti-inflammatory microbiome, boosts immunity, and prevents the intestines from becoming inflamed and leaky (aryl hydrocarbon receptor expression is decreased in inflammatory bowel disease patients). It is activated both indirectly—by microbes and their metabolites—or directly, by specific components of cruciferous vegetables (and, to a lesser extent, plant polyphenols and other phytochemicals).

Polyphenols are perhaps the best studied phytochemicals to date. They influence inflammation through multiple pathways. Polyphenols, ubiquitous in plants and plant foods, are responsible for the varied hues of fresh produce, the bitter taste of raw cacao, coffee, and red wine, and the arresting appearance of autumn foliage or flowers that captivate honeybees. The quercetin in apples and onions, hesperidin in citrus fruits, resveratrol in grapes (where it is found in the skins at higher concentrations than in red wine), catechins in green tea, and anthocyanins in red, black, blue, and purple foods are a few of many polyphenol aliases. They are found in whole grains, legumes, nuts, seeds, and soy foods. Unlike vitamin and mineral deficiencies, a polyphenol deficiency has not yet been identified. But they are biologically active molecules with potent effects on the immune system. Studies show that plant foods high in polyphenols tend to lower markers of inflammation in the blood—including CRP—and help to prevent various chronic inflammatory diseases. While polyphenols abound in dark leafy greens, colorful berries, and teas, spices and herbs contain some of the highest concentrations.

Polyphenols, which adeptly subdue free radicals formed in plants during grueling sun exposure, function similarly in the human body, where they act as powerful antioxidants and even enhance the body's intrinsic ability to produce antioxidants. Beyond their roles in combating free radicals, polyphenols dampen an overreactive immune response by regulating proteins like NF-κB, affecting inflammatory genes, enzymes, and cytokines, including TNF-α, IL-1β, and IL-6. They also activate anti-inflammatory gene transcription factors.

Over 90 percent of polyphenols make their way down to the colon, where—like fiber—they act as prebiotics. They are fermented by microbes to yield beneficial metabolites that alter gene expression and modulate the immune system. They stimulate the growth of anti-inflammatory bacteria, like certain *Clostridia* species, *Lactobacilli*, and *Bifidobacteria*, and have been shown to improve endotoxemia after meals. Polyphenols are potent antimicrobial agents, keeping infectious germs at bay in humans as they do in plants.

Keys popularized the "Mediterranean diet" in a 1959 cookbook he wrote with his wife, *Eat Well and Stay Well*. They introduced middle America to gnocchi, gazpacho, and paella, largely basing their recipes and lifestyle tips on travels to Greece, Italy, and the Mediterranean portions of France and Spain. The Seven Countries Study provided the earliest scientific support for the Mediterranean diet, the nation's first anti-inflammatory diet. In Karelia, Finland, public policy shifted to reflect the advice in Keys's book. Over time, the incidence of heart disease dropped by 80 percent and life expectancy increased by more than ten years.

But in America, even toward the end of Keys's life, the traditional Mediterranean diet failed to truly take hold. As the years wore on, Keys and his wife, Margaret, made frequent trips to Pioppi, a small fishing village south of Naples. Eventually, they settled there for most seasons, escaping frigid Minnesotan winters. They basked in a villa with a sun-drenched terrace on the sea and a citrus garden, growing tangerines, apricots, pears, plums, kumquats, and chinotto—a citrus fruit that produced delectable marmalade. In his old age, as travel became increasingly difficult, Keys relinquished the villa and moved into an assisted living facility in Minneapolis. "My room is filled with books I cannot read," he wrote in his final months as his vision faded. "The words are only spots that spoil the page." In 2004, just shy of his 101st birthday, Keys passed away.

In the years following his death, revisionist histories undermining his work cropped up in online blogs and in popular media. The idea of a "Mediterranean" diet, which had never been properly implemented in America, largely veered from tradition—even in the Mediterranean itself—as whole plant foods, once meal centerpieces, shrunk to small side dishes. Meanwhile, the ostensibly simple question that persisted for the general public was an old one: could nutrition science reliably guide people on what to eat for disease prevention or treatment?

In another era, during and after World War II, Keys and Yudkin had scoured international mortality data. Keys noticed that deaths from heart disease dropped dramatically across Europe during wartime food shortages of meat, dairy, and eggs. But Yudkin stressed that sugar was also scant during the war. Other researchers pointed to gasoline shortages, which led people to inhale fewer exhaust fumes and exercise more by walking or cycling. Did any of these factors, which seemed to be linked to heart attack deaths, actually *cause* the disease? How could hazy population-based science translate into public health guidelines for individuals? Data also tied ownership of television sets to a higher risk of heart disease, but it was nonsensical to think that those objects could clog arteries. Rich people with television sets, however, had more access to fatty animal foods, sugar, cigarettes, and vehicles.

Epidemiology, a branch of medical science that studies the distribution of diseases in humans, attempts to understand the exposures related to illness and to answer questions that can improve public health. In the first half of the twentieth century, nutrition science had largely encompassed malnutrition and the discovery of various vitamins and minerals. However, it expanded along with economic development and increased low-cost food production in high-income countries to address the rising burden of diet-related chronic diseases. The Seven Countries Study, which helped to standardize nutrition research, also heralded the end of intellectual innocence in the field as scientists engaged in vitriolic disputes on the verity of relationships between diet and disease brought to light by complex mathematics.

People pointed out that observational studies like the Seven Countries Study could not *prove* that diet caused disease. A research method known as the randomized controlled trial, on the other hand, could do so. In the 1940s, when new antibiotics made their debut and physicians wanted to use them to treat patients, English statistician Bradford Hill devised the randomized controlled trial to test their efficacy. Hill proposed that patients *randomly* be assigned to a treatment or a control group, thus obviating any of the physicians' inherent biases that could skew a study's results. Subjects would then get the antibiotic or an inactive placebo pill, not knowing which they had gotten.

Randomized controlled trials became the gold standard of epidemiological study design, the means by which scientists could evaluate new therapies free of any prejudice. They became enshrined in medical education. Randomized controlled trials were especially suited to studying drug therapies, and interest in them grew as the pharmaceutical industry allied with medical establishments. Unlike observational studies, which could only provide correlations, randomized controlled trials were experimental studies that could prove causation.

But observational studies, when evaluated alongside the totality of evidence on a topic, can provide indispensable information, as seen in the case of smoking and lung cancer. By the 1950s, the average American adult smoked half a pack of cigarettes a day. Tobacco companies claimed that smoking could help us stay slim, aid in digestion, and prevent disease. They toyed with the idea of producing fruit-flavored cigarettes for young people. Many doctors smoked, and free cigarettes could be found at medical conferences.

As scientists debated on the potential of food to cause disease, similar arguments raged on lung cancer. Young medical trainees were intuitively making connections between smoking and soot-stained airways, but the notion that smoking could cause lung cancer seemed absurd. Such cause-and-effect relationships, epide-miologists argued, only made sense for infectious diseases, in which a known germ invaded the body, overcoming immune defenses and causing illness. Chronic, noninfectious conditions like heart disease, diabetes, and cancer were too complex for a single cause or even multiple causes to be isolated.

Hill was one of the researchers who began to systematically study the risk factors for lung cancer. Along with his physician colleague Richard Doll, he surveyed over forty thousand doctors on their smoking habits and tracked their health. The results that poured in over the years, published in 1956, painted a strong statistical relationship between cigarette smoking and lung cancer. Frightened by the findings, Doll threw out his cigarettes. As additional observational studies on the dangers of smoking were published, tobacco companies repeatedly emphasized that correlation did not equal causation. Public health organizations suggested moderation rather than cessation of tobacco intake.

Hill knew that randomized controlled trials on smoking and lung cancer would be hard to conduct—and ethically irresponsible. He was disturbed by the semantics surrounding the word *cause,* which made it easy to dismiss observational data. The idea of causality in disease had originated with nineteenth-century microbiologist Robert Koch, who described four criteria an infectious germ must fulfill to be construed as the cause of a disease.* Hill thought it archaic to repurpose such rules for noninfectious exposures and diseases, like smoking and lung cancer. Epidemiology, he decided, needed a revamping of the word *cause.* He came up with a set of principles that scientists could use to infer a causal relationship between an exposure and a disease.

Studies had to show that a correlation was strong and consistent. With smoking and lung cancer, various studies in distinct

---

* Developed in the nineteenth century, Koch's four criteria were designed to assess whether a germ caused disease: 1. The germ had to be found in diseased rather than in healthy animals. 2. The germ had to be isolated from the diseased animal. 3. The germ had to be capable of transmitting disease when introduced into a healthy host. 4. The germ had to be re-isolated from the diseased individual and matched to the original germ.

populations had come up with the same robust association. The exposure had to come before the effect, with greater exposures yielding worse outcomes. Smoking preceded lung cancer, and the more one smoked, the higher the risk of lung cancer. The correlation should be specific but act similarly in comparable situations. Tobacco smoke led to a cancer of the lungs, the site where smoke entered the body, but it had also been linked to cancers of the mouth and esophagus. Hill decided that a potential mechanistic link—*how* smoking led to lung cancer—would be helpful, even if it was limited by current knowledge.

The physical experiments transpiring in basic science laboratories, including studies performed in animals—like painting mice with cigarette tar—lived in realms far removed from the abstract number crunching of epidemiologists and statisticians. But Hill thought they could shed light on the reliability of epidemiological evidence, showing how the cause-and-effect relationship played out on a biological level. He established the idea of *biological plausibility*, explaining that coherence between laboratory and epidemiological findings increased the chance of causality. The more of Hill's criteria a hypothesis fulfilled, the more that elusive notion of *cause* arising from epidemiology could be made concrete, a cohesive truth resting not on the strength and mechanics of one study—or studies—but on a collection of intuitions that construed a flexible framework, one that could push for action on public health measures. Still, as noted in the 1964 report of the surgeon general, it took thousands of studies and countless deaths before a consensus was reached in the medical community on smoking and lung cancer.

Randomized controlled trials are not ideal for many of the questions that emerge in nutrition science. Imagine randomly assigning an individual to a pattern of foods as opposed to a pill in a drug trial. It is impossible to hide the nature of the treatment with a "placebo" comparison, and generic labels—like vegan, gluten-free, low-fat or low-carb—can represent heterogeneous ways of eating that ignore the proportions of whole foods in the diet and how they were prepared, important considerations for inflammation and disease, including effects on gut germs. While most randomized controlled trials span a few months to years, dietary influences operate across a range of time. A teenager's food habits, for example, may affect cancer risk in middle age. A randomized controlled trial lasting decades would drain finances or soon be ethically impractical.

Unlike many drugs, which are designed to target a single system or pathway in disease and yield dramatic outcomes that play out in a relatively short period of time, nutrients behave more discreetly but engage in countless interactions with each other and in a variety of mechanisms throughout the body for long durations. The benefit of a single food is related not only to its intrinsic qualities but also to what the food displaces in the diet. Thus, an egg may appear healthful when compared with a processed doughnut, but not so when set alongside a bowl of whole grain oats. Saturated fats do better than trans fats, but not as well as unsaturated plant fats.

As nutrition science struggles against imperfect methodology, it demands that researchers ask honest and useful questions. The primary value of Keys's Seven Countries Study may lie not in the links between specific nutrients and heart disease but in something far more profound: his contention that diet does indeed influence the risk of chronic diseases, posited in an era when most were thought to be an inevitable consequence of aging.

The wealth of nutrition studies conducted to date encompass various methods. Experiments in test tubes and animals explore mechanisms of action and support biological plausibility, even if they cannot dictate what will happen in people over time. Randomized controlled trials exist alongside observational studies—like the China Study and the Nurses' Health Studies—which are prone to design limitations but can look at entire populations over decades, life spans, or even generations. Human intervention studies without a control group may occasionally offer stunning real-world outcomes that matter not only because of the merits of the individual studies but also because of their place in a larger body of knowledge. Anecdotal data from clinical experience and an eye to history and human evolution also add invaluable information.

Nutritional epidemiology, rich with subtext, guides us away from deliberating on foods that may be harmful to harmless at best and toward those that can actively *heal*, balancing inflammation. Therapy, an essential branch of medicine, initially bloomed on empirical grounds: accidental observations led to useful cures, many of which derived from plants. Botanical medicine is older than humanity itself, relying on extraordinarily complex molecules designed through eons of evolution. But it is difficult to imagine using the Peruvian cinchona tree to treat malarial fevers rather than synthetic quinine, or the opium poppy plant in place of modern-day morphine. Our historical tendency to distill and the use of synthetic chemistry has served us well, leading to the development of potent lifesaving treatments, including vaccines, antibiotics, anesthetics, chemotherapy, and anti-inflammatory and immune-modulating drugs. To do the same with food by isolating beneficial nutrients from their whole foods, turning them into supplements, is a deep-seated instinct—one that often fails. Some isolated nutrients do show promise as therapeutics, including roles as anti-inflammatory agents, but this tends to be the exception rather than the rule. The seemingly pedestrian practice of embracing a pattern of whole foods, on the other hand, routinely influences inflammation in ways that drugs cannot. Food, readily available and free of unwanted hazards, can be a powerful antidote to the protean nature of hidden inflammation, with nutrients working synergistically along numerous pathways to moderate the course of inflammation over time. Certain traditional ways of eating around the world, including—but not limited to—the Mediterranean diet, help to illustrate the vigorous therapeutic potential of food.

———————

The traditional Okinawan diet once inspired a bright young surgeon to forgo his scalpel for a fork. In 1968, shortly after spending a year in Vietnam as an army surgeon, Caldwell Esselstyn joined the general surgery team at the Cleveland Clinic. By 1985, after performing numerous surgeries, he had become increasingly disheartened by America's unrelenting chronic killers, including heart disease and cancer. As he excised tumors from breast cancer victims, he wondered why places like rural Japan boasted such low rates of these illnesses. But when Japanese women migrated to the United States, their daughters and granddaughters developed the same risk of breast cancer as Caucasian Americans—similar to Keys's observations on heart disease in Japanese migrants.

Over the years, as Esselstyn scoured scientific databases for information on diet and health, he began to treat a cadre of heart disease patients with food, an unexpected trajectory for someone accustomed to healing with a scalpel. In the late twentieth century, while the embryonic idea of hidden inflammation in heart disease began to take shape, the practical lifestyle changes prescribed by Esselstyn and other pioneers helped to counter this root cause.

Esselstyn became obsessed with protecting and preserving the delicate, single-layered endothelial cells lining the inner walls of arteries that supplied blood to the heart and brain. Endothelial cells injured by LDL cholesterol or other risk factors for heart disease morph into inflammatory powerhouses, recruiting immune cells and churning out inflammatory mediators. They make substances that constrict blood vessels and fail to secrete enough nitric oxide.

Esselstyn asked patients to avoid processed foods, meat, dairy, eggs, refined carbohydrates, oils, and excess sugar and salt. He pointed to human experiments showing that these foods— particularly those of animal origin—injured and inflamed endothelial cells, an effect that could last for hours, right up until the next meal. He advised patients to focus on foods that would best fight the "cauldron of inflammation" in their arteries and restore the capacity of endothelial cells to make nitric oxide, filling their plates with a medley of whole plant foods, including leafy greens, vegetables, fruits, legumes, whole grains, and a couple of daily tablespoons of flaxseeds. What emerged was an iteration of the Okinawan diet of rural Japan, which had sustained some of the healthiest and longest-lived people in the world for generations. Okinawans subsisted largely on leafy greens, root vegetables, legumes, whole soy foods, and whole grains. Over 96 percent of their calories came from plants, with few animal foods, added sugars, or oils.

Purple and orange sweet potatoes are staples of the traditional Okinawan diet. The typhoons that troubled the islands would destroy many crops but leave these underground tubers unharmed. Sweet potatoes, originally seen as a poor person's food that sustained farmers, fishermen, and the like, are rich in nutrients that manage inflammation and enhance immunity, including fiber, polyphenols, and vitamins C, E, B$_6$, and folate. They are one of the best sources of beta-carotene, a precursor to vitamin A, which, like vitamin D, is especially important for the health of the immune system. Potato peels may be even more essential than their inner flesh, and the intensity of a potato's color predicts superior nutrition. In humans, eating a purple potato each day has been shown to decrease inflammatory markers like CRP and IL-6. In modern Okinawa, however, where the availability of animal foods has grown and the inhabitants have begun to adopt Western habits, the traditional Okinawan diet is disappearing, along with its health benefits.

In 2014, Esselstyn published dietary intervention data on nearly two hundred patients. At baseline, all suffered with coronary artery disease, except for three patients with blockages in the arteries leading to the brain or legs. Around half had already succumbed to a heart attack or stroke. A few had even failed their first or second coronary artery bypass graft or balloon angioplasty. During a coronary artery bypass procedure, surgeons rewire arteries or harvest veins from patients' legs in order to bypass blockages in vessels serving the heart. Minimally invasive techniques used to open up blood vessels include balloon angioplasty and, more commonly today, stents. But treating one blockage at a time in piecemeal fashion to combat a widespread disease may fail to provide a durable solution. Addressing the root causes of heart disease, including elevated levels of blood cholesterol and inflammation, is essential.

After implementing a diet of whole plant foods, Esselstyn tracked these patients for an average of almost four years. An encouraging 89 percent stuck with the new diet. Their blood cholesterol levels plummeted and their chest pains disappeared. Only one patient in this group experienced a cardiac event related to disease progression—a stroke—yielding a remarkably low recurrent event rate of 0.6 percent. Meanwhile, among the patients who did not comply with the diet, 62 percent experienced at least one cardiac event, including heart attack, stroke, and sudden cardiac death. Esselstyn found that the diet had not only arrested inflammatory disease in most but also *reversed* it in some: imaging tests revealed once constricted coronary arteries harboring a lighter plaque burden, regaining in part the configurations of their youth.

––––––––––

Okinawan, Mediterranean, and other traditional diets are uniquely adept at not only dampening but also resolving inflammation. Nutrients work synergistically to resolve inflammation, engaging immune cells and germs and participating in a variety of biological pathways throughout the body. The resolution of inflammation involves not only omega-3s but an array of substances found in plant foods. Polyphenols, for example, trigger genes that promote the healing of damaged tissues. The bitter yellow salicylic acid crystals that chemists once extracted from willow bark to create aspirin—one of the few modern anti-inflammatory drugs that not only reduces but also helps to resolve inflammation—lurk in many other plants, defending them against germs and stressors. Salicylic acid is ubiquitous in fruits and vegetables, but the highest concentrations are found in spices and herbs, including chili powder, turmeric, paprika, and—one of the richest sources—cumin. Studies show that people with diets centered around whole plant foods tend to have low levels of salicylic acid coursing through their blood, comparable to those individuals taking a small daily dose of aspirin—but, unlike aspirin, with no known adverse side effects.

In costal South India, many miles from Mediterranean and Okinawan cultures, my mother's culinary concoctions also boasted a wealth of whole plant foods, including a medley of spices and herbs filled with polyphenols and other nutrients. Amla fruit, the Indian gooseberry, was one of her favorite snacks. She ate it raw or cooked, dried or pickled. Human clinical trials show that amla, which is packed with polyphenols and has two hundred times the antioxidant capacity of blueberries, lowers blood sugars and LDL cholesterol, improves the function of blood vessels. and dramatically reduces inflammation, cutting CRP levels in half. It also promotes immunity.

Spices, in addition to furnishing the iconic flavors of a culture, historically served as medicines and food preservatives. Drinking chai, the most popular beverage in India, is a ritual that eclipses boundaries, a daily practice in rural villages, city slums, and lavish restaurants. Chai is suffused with a multitude of spices—like cloves, cinnamon, cardamom, and ginger—that favorably modulate the immune system. The sweet, pungent ginger root has been used to treat digestive symptoms since ancient times. It soothes the intestines, allaying its spasms. Ginger bolsters immunity and combats inflammation. Studies in humans point to its utility for muscle pains, arthritis, diabetes, obesity, fatty liver, nausea, and even painful menses and migraine headaches.

One of the best-studied spices is turmeric, which my mother cultivated. She would dig up tubers as the plant's foliage began to discolor, drying them in the sun and crushing them into a fine, golden powder with a bitter, earthy taste. The powder's hue comes from curcumin, a polyphenol found in turmeric. Although turmeric has a four-thousand-year history of medicinal use, thousands of articles on curcumin made their way into medical literature only in the last couple of decades, including dozens of human clinical trials since the turn of the century.

Curcumin inhibits many inflammatory pathways, including some targeted by major anti-inflammatory drugs—but without any reports of serious side effects or deaths. It prevents NF-κB and cyclooxygenase activation and suppresses inflammatory cytokines like TNF-α, IL-1, IL-6, and IL-1β. Curcumin converses with a wide range of cells, from macrophages and other immune cells to those in the liver, pancreas, heart, and fat tissue. It calms inflammation in a variety of autoimmune diseases, like rheumatoid arthritis, inflammatory bowel disease, psoriasis, and inflammation of the eyes. It holds promise for treating postsurgical pain and osteoarthritis and for helping to prevent or treat memory problems, Alzheimer's disease, and cancer. In inflammatory bowel disease patients, curcumin improves both subjective gastrointestinal symptoms and obvious intestinal inflammation, helping to heal flares and lower the rate of recurrent disease. It strengthens the intestinal barrier, preventing the seepage of harmful bacterial metabolites into the blood. It discourages infectious germs and tumors from gaining a foothold in the body.

While most clinical trials have studied curcumin supplements, recent research suggests that whole turmeric stripped of its curcumin continues to exhibit potent anti-inflammatory activity. A curry with typical levels of turmeric used in cooking sends only a fraction of curcumin into the bloodstream, but adding a little black pepper (an anti-inflammatory spice in its own right) to the curry boosts the bioavailability of curcumin by 2,000 percent, a testament to the importance of diversity in the diet. Spices promise gains out of proportion to effort. Throwing a pinch of cloves and cinnamon—two of the most antioxidant natural substances on earth—into mashed sweet potatoes or a hot tea or tossing some rosemary, oregano, and cayenne into a soup can enhance the anti-inflammatory flavor of these foods.

An important therapeutic effect of Okinawan, Mediterranean, and other traditional diets derives not only from the presence of fiber but also from its sheer quantity and diversity. A few weeks of giving up a Western diet and living on mostly plants, as studies reveal, is but a bare initiation, the dawn of a deep and enduring transformation of gut microbes with ongoing food choices. A diet filled with enough fiber from whole foods progressively molds the microbiome and selects for bacteria that specialize in fermenting fiber, training the body to tolerate large amounts of fiber. It also nudges existing gut germs into changing their behavior, decreasing inflammation. Indeed, while many bacteria are capable of fermenting fiber to produce short-chain fatty acids and other useful metabolites, they often lack the chance to do so.

Both ancient hunter-gatherers and modern farming populations with low rates of chronic disease—as Burkitt and O'Keefe had observed—consume vast quantities of fiber, around 100 grams a day or more. The dose of fiber can determine its utility. Burkitt and O'Keefe had recommended, based on human studies, at least 50 grams of fiber per day to prevent colon cancer. But the more short-chain fatty acids present, the more potent their effects on immune function and inflammation. And a gut flooded with short-chain fatty acids develops a lower pH, which inhibits the growth of inflammatory and potentially infectious bacteria like *Salmonella* and *Escherichia coli*. Short-chain fatty acids can also directly suppress these strains.

Studies show that fiber supplements fail to provide the same health benefits in healthy people as fiber from whole foods. Each plant contains unique fibers, and each fiber may feed one or more kinds of germs, creating beneficial metabolites yet uncharted. The Hadza of Tanzania, one of the last remaining hunter-gatherer tribes in the world, incorporate hundreds of varieties of plants into their diet during any given year, including the dry and chalky baobab fruit and wild tubers so fibrous that they are chewed up and spit out rather than swallowed. Their microbiomes are incredibly diverse and stable ecosystems that can withstand the onslaught of parasites or a seasonally dependent food supply. The most critical factors in designing an anti-inflammatory microbiome—one linked to health rather than disease, including robust immunity—are the quantity and diversity of plants in the diet.

A few nutrition studies taken out of context can lend support to almost any dietary choice. But the aggregate data, including growing evidence on food, germs, and inflammation, provide overwhelming evidence for the following hypothesis: a diverse diet made up largely or entirely of whole plant foods is the best way of eating to prevent— or, in many cases, to treat—most chronic inflammatory diseases that plague modern humankind. This pattern of foods can avert, suppress, and even resolve inflammation while boosting immunity.

Cell, animal, and human studies of all types point to the inflammatory or anti-inflammatory effects of a variety of foods. Research groups have drawn on this large body of literature, taking into account the number and quality of studies, to "score" nutrients, foods, and dietary patterns on how likely they are to impede or promote inflammation based on common blood markers like CRP, IL-6, IL-1β, TNF-α, and IL-10. Saturated fats, red and processed meats, and refined carbohydrates—all well represented in Western diets—tend to receive inflammatory scores, which have been tied to metabolic syndrome, heart disease, stroke, diabetes, dementia, inflammatory bowel disease, cancer, and other chronic inflammatory diseases. On the other hand, foods with notable anti-inflammatory scores include leafy greens, fruits, vegetables, spices, herbs, tea, soy, whole grains, legumes, nuts, and seeds. Polyphenols, carotenoids, and other phytochemicals found in these foods also receive anti-inflammatory scores, as do unsaturated plant fats—particularly omega-3s. Most or all of these foods and nutrients are plentiful in Mediterranean, Okinawan, and other traditional diets around the world.

Randomized controlled trials point out that a diet of mostly plants helps to ward off or heal a variety of ailments, including metabolic syndrome, obesity, diabetes, and heart disease. It improves the makeup of gut microbes and lowers blood markers of inflammation as well as cholesterol levels. Observational studies that have enrolled over a million people all over the world in total and span several decades link this type of diet to lower rates of many chronic inflammatory diseases—including obesity, diabetes, heart disease, hypertension, some cancers, autoimmunity, and neurodegenerative diseases—as well as a lower risk of death from any cause. The science will continue to be shaped over time as new evidence surfaces, a slow molding that will rely on incremental gains, innovations rather than sensationalist inventions attempting to abolish history and current context.

CHAPTER 14

Shaping Sustenance

B y the time she came to see me, Emily had given up her favorite bagels—multigrain, with sesame seeds, garlic, and onions—from the deli across from her apartment and the easy sandwiches she had lunched on for the better part of a year while working on her dissertation in libraries and coffeeshops. She threw out most of her beers, crackers, and cereals and started to cook pastas made of legumes or rice. Gluten, she claimed, made her mind murky and her body weary. She would get headaches and vague pains in her belly and bones and often be drawn to the toilet with bouts of diarrhea.

Celiac disease is an autoimmune condition in which gluten, a protein found in grains like wheat, barley, and rye, can cause intestinal inflammation leading to various issues both within and outside of the intestines, including stomach pains, altered bowel habits, weight loss, vitamin and mineral deficiencies, fatigue, skin rashes, osteoporosis, an increased risk of some cancers, and even neurological or psychiatric symptoms. Emily did not have celiac disease. Her blood was free of telling antibodies and her intestines gleamed pink and unscathed. Nor, as studies revealed, did she have a classic allergy to wheat or to any other grains. What, then, could account for her suffering—or help to alleviate it?

The answer to Emily's questions begins with a close inspection of whole plant foods and their relative inflammatory potential when compared with other foods. Various iterations of anti-inflammatory diets may exclude certain types of plant foods, including grains, legumes, and "nightshade" vegetables like tomatoes, eggplants, peppers, and potatoes. For most people, however, these foods, replete in Mediterranean and other traditional diets, are not truly inflammatory. Further, how we prepare and choose plants— including the use of potent ancient techniques like fermentation— can affect our tolerance of them and influence the intertwined relationship between food, germs, and inflammation.

Eating, as Elie Metchnikoff once observed, is an inflammatory act. He wrote that "in man, as well as in several other mammals, the number of white blood cells increases some time after meals," and he wondered if the digestion of food might also be a type of infection. A meal, any meal, prompts an immune response and inflammation, however modest. This humble, transient phenomenon, an adaptive reaction to one of humankind's most primal activities, differs from the excessive, persistent inflammation—triggered by particular dietary patterns and other environmental factors—that leads to chronic inflammatory diseases.

Plants, like all organisms, retain the imprints of their evolutionary struggle to survive and reproduce, a slow, laborious ascent toward imperfect perfection. Lectins, for example, are proteins found in small amounts in most foods, particularly grains, legumes, and nightshade vegetables. While some lectins evolved to protect plants from predators and can be poisonous in isolated, excessive doses, eliciting an immune response in rodents, others are actually anti-inflammatory and may hold promise for cancer therapy and gastrointestinal health.

If all meals incite at least a flash of inflammation, and if no food is entirely bereft of elements that may inflame, then anti-inflammatory diets and inflammatory diets are comparative labels: How *much* inflammation does a particular food trigger, and in what context? Will it help to fuel or hinder chronic, insidious inflammation? Answering these questions means considering not only specific nutrients but also whole foods, dietary patterns, and food selection and preparation methods. We must also consider other lifestyle factors that can modulate inflammation. Severe stress, for example, may muffle the beneficial effects of an anti-inflammatory diet.

The relatively minute potential of whole plant foods to inflame a healthy individual is eclipsed by their robust ability to reduce and resolve inflammation. They saturate the body with fiber, favorable fats, vitamins, minerals, and phytochemicals like polyphenols, all of which engage in countless interactions with the body and its germs to influence inflammation.

All kinds of studies in both humans and animals decisively show that whole grains—including those containing gluten—and legumes dramatically lower the risk of hidden inflammation and chronic inflammatory diseases. Randomized controlled trials reveal that whole grains and legumes cause CRP and other markers of inflammation to plummet. They contain more fiber than many fruits and vegetables and craft a diverse gut microbiome, promoting the growth of beneficial bacteria that produce short-chain fatty acids while inhibiting inflammatory bacterial species and behaviors.

Legumes, brimming with even more fiber than whole grains, slow the entry of glucose molecules into the blood accordingly, barely spiking blood sugar. Lentils and chickpeas, two of the most anti-inflammatory legumes, are also some of the best tolerated. Mung beans, a staple in traditional Asian cuisines, show promise in early studies for calming inflammation and boosting immunity.

Whole grains and legumes are packed with phytochemicals, some of which do not exist anywhere else in our food supply, as well as a plethora of vitamins and minerals that help to support immunity and lower inflammation, like magnesium, zinc, selenium, and an array of B vitamins, including B_6. Deficiencies in these micronutrients can contribute to hidden inflammation and chronic inflammatory diseases. Nightshade vegetables, found alongside whole grains and legumes in the traditional Mediterranean diet, are similarly favorable for immune health. Human clinical trials indicate that tomatoes, for example, increase immunity and reduce low-level inflammation in the body.

The variable magnitude and duration of inflammation after meals can be apparent during medical testing. Studies reveal that within hours of eating a typical Western breakfast—a small sausage, eggs, and a slice of cheese atop an English muffin, all coated with salted butter—low-level inflammation sears through our body, measurably stiffening and crippling the blood vessels and inflaming the airways. Many components of this meal prompt a raging response from macrophages and other immune cells; they include an excess of unfavorable fats, animal proteins, endotoxins, refined carbohydrates, salt, and a bevy of environmental toxins that tend to concentrate in organisms that escape the bottom rungs of a food chain and are cooked under high heat. And eventually, that iconic breakfast of sausage and eggs will make its way down to the colon, jarring gut germs. Recurring injuries will foster a dysbiotic, inflammatory microbiome and chronic disease.

The inflammation generated by meals rich in whole plant foods, as opposed to those filled with animal foods and refined carbohydrates, is typically meager and swiftly resolved, failing to result in bodily harm. In fact, plants can help to temper the inflammation created by other foods. A load of vegetables—peppers, tomatoes, and carrots in one experiment—added to sausages and eggs stirred in butter slightly blunts the ensuing increase in inflammatory cytokines and impairment of endothelial cells lining blood vessels. A half an avocado atop a burger does the same. A blend of spices—including turmeric, ginger, black pepper, cumin, cinnamon, coriander and oregano, rosemary, and thyme—can dampen the inflammation from a dinner of chicken and biscuits. And a handful of berries or nuts will diminish the heat from white bread and processed breakfast cereals. This protective effect can carry over to the next meal: lentils eaten for breakfast can reduce the blood sugar rise from refined carbohydrates at lunch.

An immune snapshot of one meal, however, cannot portend health outcomes. It is the overall design of the diet that plays a crucial role in the body's intrinsic ability to adeptly manage inflammation, deploying or dousing it as needed, warding off hidden inflammation and chronic diseases. After an identical meal, levels of inflammation can vary significantly between healthy adults. For example, when people with gut germs bred on all or mostly plants eat a rare slab of steak, they make little to no trimethylamine-N-oxide—the noxious substance that people typically make after eating a steak—that activates macrophages and other immune cells and induces inflammation throughout the body.

The manner in which a food is prepared can also influence inflammation and plays an important part in nourishing microbes. The wheat in Emily's bagels and sandwiches is far removed from its ancestral iterations. Einkorn and emmer wheats, first cultivated around ten thousand years ago, contain more fiber, carotenoids, and polyphenols and nearly three times as much protein as modern wheat, but much less gluten. Selective breeding, genetic modifications, and modern agricultural practices have rendered wheat and other grains abundant, resistant to pests, and malleable in baking. But a rise in gluten, amylase-trypsin inhibitors—which protect plants from pests and parasites but can activate the innate immune system and inflame the intestines—and other proteins as well as novel chemicals ensued. Modern grains are speedily transformed, largely bypassing an important process before they are cooked: fermentation.

Humans have been preparing and eating fermented foods since the birth of civilization. Before refrigerators and iceboxes or even the underground snow-filled pits of the ancient Greeks and Romans, our ancestors—notably those living in tropical climates—fermented food to keep it edible, safe, and healthy. In nearly all ancient cultures, these mythical metamorphoses kept food from spoiling, be it Egyptian beer, Japanese miso and natto, or the leavened bread of the Gauls.

In 1856, Louis Pasteur showed that living yeast germs in grape juice transformed sugars into alcohol in the absence of oxygen, a process he called fermentation, or "respiration without air."* He found that some bacteria could also ferment foods, turning sugars into lactic acid, souring milk into buttermilk or yogurt. Microbes tend to degrade food, causing it to rot, but tweaking the bacteria that live naturally on foods can preserve a dish while radically transforming its taste, look, and intrinsic elements. Fermented foods can be probiotics, introducing live, helpful bacteria—and more nutritious food—into the intestines. Microbes consume sugars and produce acid, alcohol, and gases, initiating the process of digesting food. Little effort is needed to ignite their laborious feats: some chopping and immersion, watching and waiting; a few simple, elegant maneuvers, often amenable to slapdash, imprecise recipes that yield unique outcomes with every attempt.

In the early twentieth century, fermented foods became Metchnikoff's popular microbial legacy. His work on microbes coincided with the height of the "autointoxication" movement, when several European physicians had begun to talk, in vague terms, of poor digestion leading to toxic intestinal waste that could poison the body. Surgeons, including the famous Englishman William Lane, regularly removed patients' colons in order to treat this malady. But Metchnikoff considered this cure to be worse than the disease. If it was possible to fall ill after ingesting "bad" bacteria, he reasoned, could the opposite hold true as well? Maybe the key to improving digestion and balancing gut germs lay not in the surgeon's scalpel but in the human diet, in ingesting "good" germs.

Germs that ferment inhibit the growth of those that cause food to rot. Metchnikoff noted that lactic acid killed putrefying microbes in a laboratory dish and wondered if "good" germs could displace "bad" ones in the gut. The spry Bulgarian peasants, who boasted long lives and regularly ate yogurt—a largely

---

* Two of the most common types of fermentation are lactic acid and alcoholic fermentation. In lactic acid fermentation, bacteria convert sugars into lactic acid, yielding foods like sauerkraut, kimchi, and yogurt. In alcoholic fermentation, yeast transform sugars to produce the ethanol found in beer, wine, and other foods and beverages.

unfamiliar food at the time—caught his eye. He isolated pure cultures from a sample of Bulgarian fermented milk, including *Lactobacillus bulgaricus*, a species still used to make yogurt today, and reported that the bacteria had a favorable effect on the gut flora of mice.

Metchnikoff's findings, which he initially only presented as a hypothesis, led to a frenzy of hyperbolic media reports and launched an international demand for yogurt and Bulgarian sour milk cultures, which were sold as tablets, powders, and bouillons—the precursors of modern probiotics. Although Metchnikoff often countered exaggerated claims stemming from the misuse of his research, he held fast to core notions on fermented foods and beneficial bacteria: "A reader may be surprised by my recommendation to absorb large quantities of microbes, as the general belief is that microbes are all harmful. This belief, however, is erroneous."

Doctors from around the world, including John Kellogg, traveled to Paris to hear Metchnikoff speak on sour milk. But Kellogg refined Metchnikoff's advice. Yogurt, he thought, was not the only or even the main method by which to introduce helpful bacteria into the gut. The key lay in a vegetarian diet favoring high-fiber foods. He also believed that changes in intestinal flora could be sped up by enemas seeded with helpful bacteria. When Metchnikoff's theories on gut flora fell out of favor due to a lack of hard scientific evidence, Kellogg continued to voice his support, writing that Metchnikoff had "placed the whole world under obligation to him" for discovering that intestinal germs needed cultivation.

People around the world continue to ferment nearly any food, including grains, fruits, vegetables, legumes, dairy, and meat. In Iceland, shark flesh is stored for months in gravel-filled hillside holes to produce a traditional dish called *hakarl*. South Indian cuisine is replete in fermented plant foods, many born of economic necessity and coaxed by unrelenting heat. In the nineteenth and twentieth centuries, new ways to keep foods from spoiling by managing microbes included canning, cooling, pasteurization, and the addition of preservatives. But unlike fermentation, these methods did not always make food healthier—and sometimes did the opposite.

———————

Scientist Justin Sonnenburg regularly ferments some of his food. His bread sits in warm air for hours to days, the batter bubbling and turning tart before it enters the oven, where fresh sourdough loaves finally rise. Old tea fizzles like soda, morphing into kombucha. Chopped cabbage and green tomatoes sluiced in salt water and germs fill glass jars that line his kitchen counter. Sonnenburg, who directs the Center for Human Microbiome Studies at Stanford University, researches how dietary interventions can alter gut germs and the immune system. He had become captivated by gut microbes around the turn of the century, when the field was still "a quirky, beautiful thing in biology, not yet functionally crucial or at the center of the biomedical revolution."

In a 2021 study, Sonnenburg and his team randomly assigned a group of people to a diet high in fermented foods such as kimchi, kombucha, and vegetable brine drinks for ten weeks. They collected samples of subjects' blood and feces before, during, and after the dietary intervention. Back in the laboratory, they took a close look at gut germs. Those eating fermented foods had acquired amicable anti-inflammatory bacteria, like *Lactobacilli*, typically found in such foods. New species, including many unrelated to the fermented foods, bloomed in their microbiomes, increasing diversity. Next, Sonnenburg and his team executed an ambitious analysis of the subjects' immune systems, evaluating around 350 parameters with each blood draw, providing a multidimensional portrait of their inflammatory status as they reshaped their diet and gut microbes. Sonnenburg's team measured the levels of eighty cytokines. They collected blood immune cells, largely innate immune cells like macrophages but also B and T lymphocytes, and irked them in petri dishes to gauge their inflammatory potential: were they patient or prone to anger? By the end of the experiment, the people who had been eating fermented foods fared well on these tests, experiencing an enormous decline in markers of inflammation.[*]

How does turning cabbage into sauerkraut calm inflammation? When the leaves are deprived of oxygen and soaked in brine, bacteria consume sugars and make healthy acids, like lactic acid, that sours the solution and lowers its pH—similar to how short-chain fatty acids lower the pH in the gut. This changes the balance of germs, selecting for the acid producers while inhibiting inflammatory, infectious species or those that degrade food. Acid-loving bacteria like *Lactobacillus* and *Bifidobacteria* thrive. Beyond its ability to curate an anti-inflammatory cohort of bacteria, lactic acid directly decreases inflammation in the gut and prevents macrophages from secreting inflammatory cytokines, with more potent effects as the dose of acid rises. Even small amounts of lactic or other acids in fermented foods that make their way into the intestines may benefit the body, as Metchnikoff once proposed.

Microbes can enhance the inherent anti-inflammatory properties of a food, laboriously unpacking these qualities like a sculptor with rasps and a chisel. As they devour simple sugars in fermenting foods, they curtail the ability of a food to spike blood sugar—and thus inflammation. Sourdough bread, for example, raises blood sugar less than regular whole wheat bread. Fermentation also reduces gluten, amylase-trypsin inhibitors, and poorly absorbed carbohydrates[†] and removes most potentially harmful lectins. A

---

[*] In this study, those randomized to a diet high in fermented foods saw a decline in nineteen different inflammatory proteins and thirteen different immune-cell-signaling parameters. Meanwhile, those randomized to a diet of 45 grams of fiber experienced a decrease in inflammation only if their microbiome was diverse at baseline.

[†] The acronym FODMAPs describes a group of previously unrelated short-chain carbohydrates and sugar alcohols. These include fructose, lactose, fructo- and galacto-oligosaccharides (fructans and galactans), and polyols (such as sorbitol, mannitol, xylitol, and maltitol). In those with gastrointestinal issues, FODMAPs may be poorly absorbed in the small intestine for a variety of reasons. For example, fructose does not move efficiently across the brush border due to poor transport mechanisms across the epithelium. The activity of lactase, the enzyme responsible for metabolizing lactose, decreases after weaning. Due to their poor absorption as well as their small

traditional sourdough bread ferments for hours to days before baking as microbes dismantle problematic substances. Many patients with irritable bowels or a sensitivity to wheat are better able to tolerate fermented grains, some of which have less gluten than products labeled "gluten-free" in grocery stores. These packaged goods often contain inflammatory substances, including refined carbohydrates, excess sugar, and additives.

During fermentation, microbes can also increase the amounts of certain nutrients in foods that directly affect inflammation, including polyphenols, many types of vitamins and minerals, and other biologically active metabolites that favorably modulate the immune system. Kombucha, which is best consumed in moderation*—a few ounces a day—teems with healthy acids, polyphenols, vitamins (including vitamins C and B_{12}, thiamine, and pyridoxine), and the anti-inflammatory amino sugar glucosamine.

Fermented plant foods can be uniquely anti-inflammatory, both prebiotic—full of the fiber beloved by gut germs—and probiotic, bringing live, valuable bacteria into the intestines. Many of these bacteria survive the journey through the digestive tract, although they might choose to visit rather than remain, especially if they are unable to find the prebiotics that sustain them. But their transient tenure in the gut may affect the immune system in ways beyond their sheer presence, as they converse with indigenous microbes and make anti-inflammatory metabolites. Even when fermented foods are cooked with heat and no longer contain many viable bacteria, the foods have been inherently transformed by the process of fermentation. These foods may also retain microbial metabolites and dead microbes. All of these factors can favorably influence the immune system.

---

size, FODMAPs are osmotically active and increase intestinal water content. They are rapidly fermented by gut bacteria, leading to increased gas production. Greater intestinal water content and gas production leads to bowel distension, which translates into adverse gastrointestinal symptoms such as abdominal pain, bloating, and motility changes.

* While kombucha tea is generally considered a healthy beverage, some evidence links it to rare yet serious health risks, including life-threatening lactic acidosis and liver toxicity, perhaps a result of contaminated batches or overconsumption.

Probiotic supplements are distinct from fermented foods. Both may contain live bacteria, but a probiotic supplement holds a few highly concentrated species—the bespoke organisms of synthetic biology—while fermented foods contain lower bacterial counts but a greater variety of species, allowing us to ingest cooperative families of naturally bred microbes, taking advantage of bacterial synergies and increasing the chance of encountering a helpful microbe. Probiotic supplements can have anti-inflammatory effects, but they may be best used in disease rather than in health at present, aiding patients with a variety of specific conditions, including inflammatory bowel disease, diarrheal illnesses, and irritable bowel syndrome.

Every fermented food is singular, the product of a particular time and place in which certain foods and germs intermingle. Fermenting in small batches yields safe, high-quality products that are usually made in home kitchens or found in farmers' markets and other specialty shops. Large-scale commercial enterprises, on the other hand, may sell barely fermented foods with few viable bacteria and an excess of added salt or sugar. Most pickles, once naturally fermented, are now simply preserved in vinegar and suffocated with heat to create sterile, immaculate condiments.

A return to traditional grains and preparation methods can help to favorably modulate inflammation. Incorporating a diversity of whole grains into the diet, as with any plant food, is ideal. Ancient grains beyond einkorn and emmer wheat abound, including spelt, kamut, barley, bulgur, rye, and farro. Some are entirely free of gluten, like millet, sorghum, quinoa, amaranth, buckwheat, and teff. Research in petri dishes and in people suggests that ancient grains may suppress inflammation more easily than their modern counterparts. Beyond fermentation, traditional food preparation methods like drying, soaking, and sprouting can also improve the digestion and absorption of nutrients. Soaking some nuts, seeds, legumes, or grains in water for a few hours lessens any bloating and indigestion from these foods. And soaking, like cooking, inactivates most lectins. Sprouting can significantly inflate some of the nutrients in foods. Broccoli sprouts, for example, contain ten times as much of the powerful anti-inflammatory phytochemical sulforaphane—which is an isothiocyanate—as the florets and have been found to counter the inflammatory effects of air pollutants like diesel exhaust. They also increase immunity and dampen inflammation induced by viruses.

Simply eating a few foods in their raw, unadulterated state can be profitable for humans and the germs living within them. Raw foods, including fruits, vegetables, nuts, and seeds, are often higher in fiber, phytochemicals, and other nutrients than their cooked analogues. Enzymes in raw broccoli—which are inactivated by cooking—drastically heighten sulforaphane levels even before the vegetable makes it way down to gut microbes that do the same. Raw fruits and vegetables may be tied to a decreased risk of chronic inflammatory diseases, including heart disease and cancer, as well as symptomatic improvement in rheumatoid arthritis and other autoimmune diseases. Cooking destroys some nutrients but also natural plant toxins. And it can make certain nutrients even more available. Simmering tomato sauce for a long while in the traditional Italian method triples its load of lycopene (a carotenoid phytochemical). For most people, a judicious inclusion of both raw and cooked foods in the diet is optimal.

Emily's symptoms improved—but did not entirely resolve—as she embraced new ways of preparing her food. She was diagnosed with a condition called *wheat sensitivity*. In patients with celiac disease, inflammation can be obvious during an endoscopy or it can be microscopic. Pathologists comment on classic findings, like the infiltration of certain immune cells into the intestinal wall or changes in the structure of the small intestine. In patients with wheat sensitivity, however, none of these abnormalities are found.

Chronic, uncontrolled inflammation in the gastrointestinal tract can be seen with the eye during endoscopies, or it may be picked up only when small pieces of tissue are magnified and examined by a pathologist who looks for inflammatory cells. But sometimes it is so unobtrusive that it may entirely elude the tests we use to catch it. Emily was perhaps silently inflamed.

While gluten largely elicits an adaptive immune response in patients with celiac disease, data suggest that the innate immune response drives symptoms in patients with non-celiac wheat sensitivity. Studies show that people with non-celiac wheat sensitivity tend to have low-level intestinal inflammation and a robust rise in markers of innate immune inflammation and intestinal damage floating in the blood.*

Hidden inflammation can play a part in other gastrointestinal diseases for which no apparent cause has been pinpointed, like irritable bowel syndrome, a condition in which the bowels become easily angered and give rise to troublesome digestive symptoms, or dyspepsia, the boundless nineteenth-century malady that now refers to certain cases of an upset stomach. In these diseases, inflammation may affect the nerves and muscles of the stomach and intestines, helping to alter pain thresholds and movements in the gut, contributing to issues like bloating, stomach pain, diarrhea, or constipation. Even a transient spell of inflammation in the gut, as with an infection like traveler's diarrhea, can lead to changes in the way the gut functions that persist for months or even years in some people, prompting ongoing gastrointestinal distress and even sensitivities or intolerances to specific foods. A healthy gut, however, inevitably harbors a low hum of incessant inflammation as it deals with the flood of food, germs, and other matter from the outside world.

Emily's burdensome symptoms disappeared entirely after she discarded wheat and other gluten-containing grains. But while she

---

* While celiac disease largely involves a highly specific adaptive immune response against gluten, innate-type components of the immune system contribute as well. Similarly, while research to date mostly implicates innate immune responses in non-celiac wheat sensitivity, an adaptive immune response has been found as well. Non-celiac wheat sensitivity patients have high levels of IgG (similar to celiac disease) and, to a lesser extent, IgA (though less so than in celiac disease) and IgM antibodies to gluten. The IgG response in non-celiac wheat sensitivity is different from that found in celiac disease, and while there is indeed an adaptive immune response in non-celiac wheat sensitivity, it remains to be seen whether it has a pathogenic role. There is just as much of an increase in circulating markers of epithelial cell injury in non-celiac wheat sensitivity as in celiac disease, which suggests that there exist inflammatory responses in non-celiac wheat sensitivity that have not yet been identified.

had blamed gluten for her woes, the true story may be more nuanced than expected. Gluten is the most plentiful—but not the only—protein found in wheat and related grains like barley and rye. These grains contain other proteins that may induce symptoms in patients with celiac disease, wheat allergy, or wheat sensitivity, including amylase-trypsin inhibitors. Small amounts of certain carbohydrates in these grains may also be poorly absorbed by people with wheat sensitivity, contributing to gastrointestinal problems.

To bar any of the whole plant foods from the diet, given their vigorous, synergistic net effects on inflammation, the burden of proof for harm must be high. Avoiding all gluten is appropriate for patients with celiac disease, as is renouncing wheat for those suffering from an allergy or, like Emily, sensitivity to the grain. Gluten avoidance may rarely be useful for other conditions, though evidence supporting the practice in these cases is not established. For the vast majority of the population, gluten-containing whole grains help to prevent or treat chronic inflammatory diseases.

Many patients who claim they cannot tolerate certain foods are not likely to be inflamed, silently or otherwise, from these foods. While food allergies and sensitivities involve immunological reactions against food antigens, food intolerances generally relate to digestive and other woes from a lack of adequate enzymes needed to break food down (as in lactose and fructose intolerances or pancreatic disease), problems with the structure and function of the gastrointestinal tract, or other issues unrelated to the immune system. Fermentation and other food preparation techniques can help to improve both food sensitivities and intolerances.

Often, intolerances crop up simply because the body needs time to adapt to a certain quality or quantity of plants. Those unaccustomed to an abundance of diverse plants may initially struggle with exaggerated gastrointestinal issues as they embark on this dietary pattern, particularly with foods highest in fiber like legumes and whole grains. Bloating, diarrhea, or discomfort in the belly do not necessarily suggest inflammation but rather a burgeoning discourse between food and germs as gut microbes ferment fiber and other nutrients. Over time, the gut alters its secretions and contractions to better process plants and becomes enriched in microbes that help to digest them. Humans produce fewer than twenty enzymes that target carbohydrates, but microbes wield tens of thousands. Crafting a healthy gut microbiome allows us to tolerate all kinds of plants, and any transient displeasure from this process soon yields to optimal fitness throughout the gut and beyond. Banning one or more plants from the diet without due cause will only heighten intolerances.

Prescribing a diet that excludes certain foods is a science that considers the stage and severity of a patient's illness and any coexisting conditions, among other factors. A sensitivity to wheat, for example, may disappear when underlying issues like an overgrowth of bacteria in the intestines—which can silently inflame the intestines and the body at large—or microscopic inflammation of the colon are addressed. Stress, an unhealthy diet, or other lifestyle factors that inflame can also affect how our bodies handle specific foods.

Many exclusion diets are meant to be ephemeral. The elemental diet, which consists of the building blocks of protein, carbohydrates, and fat distilled into their most basic elements and thus lacks the plethora of antigens from whole foods, treats raw, inflamed intestines during a flare of inflammatory bowel disease. But after the gut heals, the elemental diet will outgrow its use, failing to favorably mold germs and inflammation to ward off additional flares or to prevent chronic diseases. Some patients with an irritable bowel or other gastrointestinal worries find it difficult to adequately absorb certain carbohydrates and may need to avoid them for a short period. Many of these foods are indispensable to microbiome and immune health* and should be gradually reinstated in a few weeks as tolerated.

To manage inflammation in health or illness, we must be mindful not only of how we prepare plants but also of how we choose them. Modern plants have been molded by humans. Paleolithic fruits and vegetables were small, bitter, and fibrous, packed with nutrients

---

* Healthy FODMAP foods include cruciferous vegetables, garlic, onions, mushrooms, blackberries, apples, beans, peas, whole wheat, barley, rye, nuts, avocados, and teas.

and an occasional excess of toxins. A Paleolithic forager might find tomatoes the size of berries, apples the flavor of tart potatoes, bananas crammed with seeds, and olives clinging to thin skin. Over generations—and, beginning with twentieth-century science, over years or even hours—humans bred plants to be more palatable and digestible, selecting for sweet, ample flesh and sparse natural toxins. Wild, straggly teosinte grass gave rise to corn. A single plant species painstakingly sculpted through time, *Brassica oleracea*, yielded cabbage, broccoli, cauliflower, brussels sprouts, and kale.

Taming wild plants comes with a cost for immune health. The ancient leafy green purslane contains fourteen times more omega-3 fatty acids and six times more vitamin E than spinach. Wild blueberries are packed with twice as much fiber as plump modern versions. One native tomato has fifteen times more lycopene than those found in the supermarket. And a few ounces of a wild apple growing in Nepal provides the same amount of phytochemicals as six large Gala apples. Cultivating the sweetest, mildest, most attractive plants means a dramatic loss of fiber, favorable fats, vitamins, minerals, and polyphenols and other phytochemicals.

To recoup this loss, at least in part, we must select the most nutritious plants, like those that are fresh, ripe, and in season. Simply increasing the diversity of plants in the diet predicts lower body inflammation, even without changes in quantity. And while nearly all plants are salutary, some have more pronounced effects on the immune system than others. In general, red, purple, red-brown, or green plants are stuffed with the most phytochemicals. Studies show that berries and greens smother systemic inflammation better than equal servings of bananas and lettuce. Blue corn has thirty times as many phytochemicals as white corn, and the dark and diminutive grape or currant tomatoes are richest in both flavor and lycopene. Some plants approach the immune-modulating power of their ancestors, like the leafy green kale, or dark-skinned potatoes with deeply colored flesh, or scallions, which have 140 times more phytochemicals than common white onions and even resemble their wild counterparts. Some of the healthiest plants are not sweet and mild but rather sour and bitter.

The appearance of a food may belie its nutritional value. Overly handsome produce may have been touched up with inflammatory chemical additives, the culinary equivalents of lipstick and mascara that are often made from the same pigments. On the other hand, around a quarter of fruits and vegetables in the United States are thrown out because they are considered too unattractive to be consumed, contributing to food waste. A crooked carrot, a pockmarked apple, an oddly curved cucumber, and many other plants with unexpected blemishes and dimples are cast to the wayside. But these foods might be even more adept at fighting inflammation than their attractive counterparts, their misshapen bodies a testament to their resilience. As plants battle heat and pests, they divert their energies to making polyphenols and other phytochemicals rather than sugars. Produce grown with fewer pesticides build up more polyphenols as well as salicylic acid. A plant visibly marred by the scars of its struggle to survive is perhaps uniquely precious to human health.

Foods choices are habits that move beyond health considerations. We demand much more from meals than we do from medicines. If the growth of a civilization, as British philosopher A. C. Grayling writes, is measured by an increasing distance from the immediacies of survival, giving rise to a cultivation of pleasure for its own sake, then eating is an essential player in this modern order. The way a food looks, smells, and feels, and even how and with whom we consume it, can affect its taste and potential to trigger our brain's pleasure centers. Some of the same receptors on our tongue that recognize bitter, sweet, savory, and umami are also present in the gut, where they sense food and germs and can prompt immune responses. While we are biologically hardwired to crave sweet, salt, and fat, an excess of these elements, especially in artificial industrial foods, exploits our natural inclinations and desensitizes our taste receptors. But we can redress this damage in mere weeks, shifting our notions of pleasure. Humans are designed to enjoy whole foods close to their natural form. Summer peaches plucked in the field. A crisp herb salad drizzled with lemon and vinegar. Warm brown bread torn by hand. A subtle simplicity with potent effects upon germs, inflammation, and disease.

Beyond the ways in which we eat, including how we prepare and select food, the ways in which we live can also help to prevent or treat inflammation and illness. Of these seemingly mundane yet pivotal decisions of daily life, some of the most critical ones relate to our relationships with germs.

CHAPTER 15

Dirty Cures

I have no memory of this, but before my first birthday, my mother pierced my ears, shaved my head, and lined my lower eyelids with dark kohl. She carried me to the Ganges, a river in northern India near the hospital where I had been born, and dipped me in its waters from head to toe, an ablution she would repeat many times over the next few months—until she and my father packed their bags to make a new life in America.

The city of Varanasi sits on the banks of the Ganges and is considered the holiest city in the land, the spiritual capital of the country. Mark Twain wrote in 1897 that Varanasi "is older than history, older than tradition, older even than legend, and looks twice as old as all of them put together." Every year, millions of religious pilgrims flock to Varanasi, descending its ghats to bathe in the Ganges. As legend has it, the Ganges is not only a body of water but also the goddess Ganga, worshipped by Hindus as the deity of purification and forgiveness, who was brought down to earth from her home in the stars by Lord Shiva. Her mystical water could wash away all sins and prevent disease. The sixteenth-century Mogul emperor Akbar deemed it the "water of immortality" and served it in his court.

High in the Himalayas, the Ganges drains out of Gangotri Glacier, its ice-cold stream colored a murky gray by rock silt. It begins its journey biologically immaculate, free of most germs. But for much of its course, as it runs through Varanasi and over 1,500 miles to the Bay of Bengal, serving nearly a third of India's population with drinking water and irrigation, it becomes one of the most polluted rivers on the planet. I grew up in an American suburb and swam in Indiana Dunes, which clasps the southern shore of Lake Michigan, or in chlorinated local pools. Meanwhile, people continued to dump all kinds of things into the Ganges, including human and animal waste, industrial metals and pesticides, raw sewage, and cremated corpses. Fecal coliform counts in the river reached astronomical levels. That a holy river with legendary healing properties might breed disease weighed heavily on the conscience of a nation, both a literal and a metaphysical defilement that drew loud calls from activist groups to clean up the waters. The Ganges remains benevolent at its core, attempting to regain its balance in an ecosystem disrupted by humans. Modern research is beginning to reveal part of the science behind her healing potential: the intimate relationship between germs, inflammation, and disease. The microbial friends I left behind in her waters are characters in this story—I never found replacements for them in America.

Louis Pasteur's 1850s germ theory, which German physician Robert Koch later broadened, proved that organisms too small to be seen with the naked eye could infect the body and cause disease. Germ theory, considered the cornerstone of modern medicine, provoked a radical shift in humanity's relationship with germs, drastically reducing deaths and increasing life expectancy.

Pasteur patented a process to kill bacteria in milk and other liquids by using high-temperature heating, which he called *pasteurization*, preventing routine infant fatalities from impure milk. In operating rooms, British surgeon Joseph Lister envisioned Pasteur's germs running through infectious, gangrenous limbs. At the time, the same probes were used for all patient wounds to look for pockets of pus, and surgical instruments were cleaned only before storage. The death rates from infections propelled movements to ban surgery in all hospitals. Fighting legendary skepticism from his colleagues, Lister pioneered antiseptic surgery, the creation of a sterile surgical environment, leading to a dramatic decrease in postoperative infections. His work laid the foundation for modern surgery.

Germ theory altered how people viewed their bodies and surroundings. Cleaner food and water as well as the introduction of toilets, sewer systems, and garbage collection saved millions of lives. Americans learned that "the higher life is everywhere interpenetrated by the lower life," in the words of one scientist, and that their damp cellars and dank water closets could breed disease. People washed their hands and used disinfectants to help get rid of germs. They started to take daily rather than occasional baths.

In Scotland, on a fated fall day in September of 1928, biologist Alexander Fleming returned from a holiday with his family to find a petri dish of *Staphylococci* bacteria he had left on a laboratory bench killed by a blue-green fungus. Fleming found that the fungus could destroy many different types of bacteria. He named it *penicillin*. His discovery would lead to the world's first antibiotic and hail a new era in the treatment of infectious diseases, a "war on germs" that would push Metchnikoff's probiotics to the scientific periphery. But today, scientists understand that certain germs are indispensable to health. Adequately incorporating these germs into our lives means redefining the notion of hygiene—and our individual identities.

—————

Compare Varanasi in the late twentieth century with a city in the United States, where the muddy countryside gives way to concrete and glass; where filth is managed by toilets and sewer systems, water is chlorinated, and food is sanitized; where humans live ensconced by brick and plaster rather than dirt and thatch, shielded from natural flora and fauna. In modern Western cities, fatal infections are overshadowed by allergies, autoimmune diseases, and other inflammatory conditions. In these locales, immune systems are more likely to rage when faced with common foods—a peanut or a shellfish—or harmless germs, and to churn out hidden inflammation that weaves through the body at all hours of the day.

The diversity of species in the human microbiome, a marker of its health, has been steadily shrinking over time, particularly since the Industrial Revolution. Fossilized feces reveal the rich, varied microbiomes of ancient populations. Even modern descendants of aboriginal hunter-gatherer tribes, like the Hadza of Tanzania, embrace microbial species absent from city dwellers. The loss of primal microbes that evolved alongside humans is not merely a nominal casualty but the loss of old contextual milieus, of patterns and functions that are essential to shaping the immune system. Helminths, parasites once universally present in human intestines, are now rarely encountered in industrialized nations. But these worms honed their skills through thousands of years of intimate contact with humans and can modulate multiple immune pathways. They are powerful inducers of tolerant macrophages, Tregs, and anti-inflammatory cytokines like IL-10. Helminths have been used to treat inflammatory bowel disease and other autoimmune conditions, with patients swallowing microscopic live eggs that hatch in the intestines.

Urban city life is witnessing an explosion of chronic inflammatory diseases. Meanwhile, locations beset by poor sanitary conditions struggle with infectious illnesses. For example, helminths may benefit immune function at large, but they can also cause acute diarrheal infections in some people that lead to malnutrition and anemia. But we do not need to choose between these opposing misfortunes. Most meetings between microbes and immune cells are benign yet illuminating. In order to encourage encounters with the right quality and quantity of microbes—helping to avoid both infections and chronic inflammatory diseases—we must reform our relationship with the germs living on, inside, and around us, cultivating habits that begin at birth and endure through life.

A pregnant woman exposes her fetus to microbial metabolites in utero. They cross the placenta and make their way into fetal tissues, driving immune development and giving the baby a taste of the mother's microbes before it faces the massive onslaught of germs at delivery. The mode of birth matters: babies born via cesarean section pick up a motley collection of microbes from the mother's skin and the hospital room rather than from her vagina, which contains a group of germs optimized by evolution to train the nascent

---

* While some authors suggest that helminth-mediated immune regulation might have evolved to become a physiological necessity, others disagree.

neonatal immune system. Breast milk provides sugars that nourish the baby's gut microbes, helping to raise an anti-inflammatory class, and delivers antibodies that protect against infections.

A vaginal birth followed by breastfeeding sets a baby's immune system on an advantageous trajectory. But medical issues, the demands of modern life, and the fact that the number and gender of partners caring for children may vary often do not allow for this combination of events. Fortunately, the mode of birth and early feedings are not the only opportunities for favorably shaping the microbiome, nor are they the main causes of its demise. A wealth of choices we make beyond infancy and throughout adulthood allow us to continually influence our microbiomes.

In 1987, British epidemiologist David Strachan noticed that children with older siblings had less hay fever and other allergies. He came up with his famous "hygiene hypothesis," which contends that children in industrialized countries lack infectious exposures due to small family size or other factors and grow up with jumpy, maladjusted immune systems. The hygiene hypothesis outlined a trade-off between childhood infections and ensuing chronic inflammatory disorders, leading to the notion that overzealous personal and household cleanliness was largely responsible for inadequate microbial influences. But this picture is incomplete, and our understanding of hygiene today warrants revision.

Hygiene is officially defined as "conditions and practices conducive to health." Its roots lie in the nineteenth-century combat against contagion by Pasteur, Lister, and other scientists. Personal and societal hygienic practices like hand washing, careful food storage and preparation, citywide garbage collection, and the introduction of sewage systems have saved countless lives. Hygiene remains essential in both industrialized and nonindustrialized nations for preventing epidemics and pandemics, antibiotic resistance, food- and waterborne poisonings, and hospital-acquired infections. It protects vulnerable populations, including infants, the elderly, and the immunocompromised. For much of the world's population, particularly in poorer nations, access to clean water for drinking and washing is a major public health challenge that costs lives.

The words "cleanliness" and "hygiene" are often used interchangeably, but they are not identical concepts. As early as 1939, German sociologist Norbert Elias argued in his book *The Civilizing Process* that cleanliness must be distinguished from hygiene. In many instances, cleanliness does not contribute to health and thus fails to align with science-backed hygiene. How clean we keep ourselves—or our homes—often relates to aesthetic and social considerations rather than to preventing disease. The rising standards of cleanliness typical of civilized society, philosopher Olli Lagerspetz writes, are "of no obvious utility for either biological survival or for the general quality of life." The distinction between "clean" and "dirty," he continues, is a guiding principle in human societies, like right and wrong.

Dirt is both a tangible, worldly element and a symbol of a cultural order. In America, a country without an aristocracy, cleanliness was a mark of status in the years after the Civil War. It became a personal and social responsibility—and eventually an obsession that transcended preventive health. Symbolic associations, rather than scientific hygiene, dictate our tastes. Scatological topics in particular elicit profound disgust, with excrement seen as the paragon of filth and kept out of sight during day-to-day life. And yet the world is doused in a patina of feces, invisible intestinal microbes with remarkable potential to both heal and harm that swarm over most things we touch and taste, making the idea of absolute cleanliness— and the echelons it engenders—a delusion.

After birth, a baby continues to acquire microbes from the environment. By age two, trillions of germs populate the toddler's gut. Exposure to a wide variety of germs in the first few days and years of life is essential to cultivating a diverse, anti-inflammatory microbiome. Studies show that when children are exposed to greater numbers of microbes during infancy, their risk of hidden, chronic inflammation in adulthood may decrease. In the critical window of early life, microbes have more power to influence the immune system than at any other junction, teaching it to react only when faced with authentic threats, to resolve inflammation in a timely manner, and to tolerate an eclectic mix of germs. Without these timely conversations, children may go on to develop immune systems that overreact to benign germs, foods, and other matter, such as pollen or household dust. They have a higher risk of becoming silently inflamed and succumbing to chronic inflammatory illnesses, including overt autoimmune diseases.

A child needs to interact not only with the right *quantity* of microbes but also with the right *quality*. Studies show that childhood infections do not protect against allergies and other chronic inflammatory diseases. In 2003, at the University College of London, microbiology professor Graham Rook considered a new hypothesis. Rook knew that common childhood infections—colds, flus, measles, and others—manifested relatively recently in human evolution, after the Neolithic agricultural revolution in 10,000 BCE, when human populations grew large and close. These "crowd infections" had not taken hold in hunter-gatherer communities, where they either killed an individual or induced quick immunity. In contrast, ancient microbes that had formed mutually beneficial partnerships with humans, evolving alongside them in mud and water and rotting vegetation, are crucial for molding the immune system. These "old friends"—as Rook proposed—are indispensable to optimal immune function. They activate a variety of immune pathways, including Tregs, which dampen inflammatory responses. They prevent the body from attacking its own tissues or harmless airborne particles like dust, dander, and pollen. The most vital exposures for children are not infections—exceptions rather than the rule in microbial interactions with the immune system—but rather the missing conversations with old friends.

How can humans of all ages find old friends in modern environments? One of the keys is early and regular contact with the natural world, including the dizzying array of plants and microbes rooted in the soil and water of the earth. Poets and artists have intuited throughout history that engaging with nature has immeasurable benefits. In 1986, when twenty-year-old Christopher Knight drove into the woods of Maine's Belgrade Lakes—not to "suck the marrow out of life" as Thoreau, whom he considered a dilettante, had once mused, but to fulfill a deep, gnawing hunger for solitude—he remained the "North Pond hermit" for nearly three decades. He thieved in order to stockpile supplies and survive brutal winters until the police caught him red-handed and took him into custody.

Gone were the predawn hikes with misty sunrises or the musky, earthy smell of petrichor in the air after a rain. But nature had altered his brain, gifting him with a photographic memory as well as a penchant for deep reflection and focus.

The idea of "forest bathing," or *shinrin-yoku*, emerged in Japan in the 1980s. It refers to connecting with the natural world through sight, hearing, taste, smell, and touch. Forest bathing increases creativity, eases stress and aggression, and improves mood. It helps create a healthy microbiome and a reflective immune system that responds appropriately to triggers. It affects not only microbes in the gut but those on the skin and in the airways, which also play a role in immune health. Fresh forest air teems with bacteria, viruses, fungi, pollen, and plant biomass. Inhaled phytoncides, essential oils emitted by trees and other plants (including fruits and vegetables) to protect themselves from insects, boost the immune system's ability to fight infections. Even a few hours—or a few minutes— spent in nature is restorative. As children immerse themselves in nature, they gain not only old friends but, as Italian physician Maria Montessori noted, endless opportunities for freedom and experiential learning that promote all aspects of human development.

We can also gain old friends from other humans and animals. Children amass microbes through every interaction with family members and friends and through schools, day care, and sports groups. A single dog—or cat—can have an enormous effect on their microbiomes, acting as a conduit between outdoor and indoor environments and bringing unfamiliar germs into their worlds. Many studies show that keeping a pet or farm animal early in life can lower the risk of allergies and asthma.

The science to date on germs, inflammation, and disease demands that we reconsider what it means to be "hygienic" in our daily lives. What is hygienic is not always clean, and what is clean is not inevitably hygienic. Cleanliness veers from hygiene when it fails to offer health benefits, while its obverse—playing in dirt, for example—can be supremely hygienic. Protective biological instincts keep us away from excessive dirt and feces—behaviors that allowed for early human survival—but a degree of exposure to both the natural world and to the microbes living on and inside of other people and animals is critical to hygiene, helping to prevent hidden inflammation and chronic inflammatory diseases. A dip in the Ganges will never quite become "clean," though it might become more hygienic as purification efforts progress.

Rethinking personal and household cleanliness can play a part in meeting old friends, although to a lesser extent than other measures. Choosing to keep the home and body immaculate is a labor-intensive aesthetic or cultural preference rather than scientific hygiene, which permits us to discard harsh detergents—for the home or body—in favor of natural or homemade ones, with the goal of diluting the concentration of germs rather than eradicating them. Or we might invest in a simple, portable bidet and take less frequent showers, preserving the oils and bacteria that keep our skin and hair healthy. Lowering the standards for personal and household cleanliness can increase leisure and hence the quality of life. But a balance must be struck: if we neglect to clean entirely, our homes become damp and moldy, invaded by microbes we have not encountered in our evolutionary past, since modern homes are built with modern materials. And our bodies will swim in filth that provides fertile ground for infections to arise.

Perhaps the best way to make the home more hygienic is to increase its contact with the natural world, aspiring in small ways to mimic traditional farm environments, which are tied to greater indoor and outdoor microbial diversity and help to protect against chronic inflammatory diseases, including hay fever, asthma, allergies, and autoimmune diseases. Growing a garden in a yard—or even in an apartment, with potted plants and airy windows—is one way to do this, as long as the soil is free of human-made toxins. Regenerative organic agriculture, which is rooted in ancient traditions, prioritizes soil health as well as that of the planet and its inhabitants, helping to fight climate change. It avoids the use of antibiotics and synthetic inputs, focusing on traditional techniques like composting and crop rotation. It fuels symbiotic relationships between plants and soil microbes. The soil of these sustainable organic farms, fertilized with manure from farm animals, contains a diverse wealth of microbes that draw nutrients from the earth. Plants grown in this type of soil are likely to be more nutritious than their conventionally grown counterparts, with higher levels of antioxidants and other micronutrients. On the other hand, soil containing antibiotic residues and certain types of pesticides and herbicides loses biodiversity, helps to breed antibiotic-resistant germs, and can harm plants, humans, and their helpful microbes.

The practical tenets of hygiene are not immutable. They are fluid, shape-shifting across time and space, but clinging consistently to science. Deadly infections, including antibiotic-resistant germs, remain a worldwide threat, especially for vulnerable populations. What is hygienic for one person at a certain time and place may not be so for another. An old man living through a pandemic and at high risk for contracting a serious infection warrants heightened personal and household cleanliness when he encounters potential sources of the pathogen. But he must also maintain exposures to microbial friends in nature, which will enhance immunity. A young child hiking in the woods, playing with mud in the yard, or walking the family dog may not need to wash his hands before dinner, provided the soil is free of pesticides and other chemicals (akin to the dirt our ancestors traversed barefoot most hours of the day, digging for edible tubers). But a visit to a hospital, petting zoo, or picture-perfect, weed-free public lawn demands hand washing—the key to preventing many infectious diseases passed between people.

Hygiene, which encompasses the broad range of essential behaviors that foster healthy relationships with the germs in our world, includes food and drug choices. Medications—including antibiotics, ibuprofen, and antacids—can adversely affect gut germs. Even a single course of antibiotics can profoundly alter the microbiome, decreasing its diversity and killing off old friends. Once the antibiotics are stopped, the microbiome returns to largely—but not exactly—its old self with proper care, but each additional course of antibiotics brings fresh insults. Antibiotic use in the first year of life, which can wipe out critical bacteria that help to calibrate the immune system, is tied to a higher risk of chronic inflammatory diseases, including asthma, eczema, inflammatory bowel disease, and obesity. Of course, antibiotics have saved many lives since their introduction in the mid-twentieth century, preventing routine surgeries or common infections like pneumonia from turning fatal. Organ transplant patients, who take immune-modulating medications, are often given long courses of antibiotics to thwart infections. But antibiotics are often used on a whim, without good reason.

In the early 1950s, scientists started to pay attention to the negative effects of antibiotics, including the advent of superinfections like *Clostridium difficile* and rising antibiotic resistance, calling to mind the prospect of an era in which antibiotics would become futile relics of the past. A judicious use of antibiotics is hygienic, taking into consideration the natural course of an illness and the risks and benefits of treatment. And thoughtful management of gut germs with dietary and other lifestyle choices before, during, and after a course of antibiotics can minimize the collateral damage.

From birth to death, the quantity and quality of our conversations with microbes are influenced by lifestyle factors such as food, drugs, and our tactile relationships with other humans, animals, and the germs we encounter as we mingle with the natural world, including its soil, air, and water. These conversations relate not simply to the science of immunology: they help to shape the very notion of individual identity.

The microbes that live within us and on us blur the boundaries between the causes and consequences of disease and also between the individual self and non-self. Metchnikoff's youthful discovery of macrophages had launched a language of warfare into immunology. The Darwinian struggle for life that manifested under his microscope depicted conflict, invasion, and bloodshed. But macrophages and other immune cells, in defending an organism, also *defined* it. Implicit in their actions stood the idea of self, that the rigid yet ever-evolving boundaries of a human being uniquely separated the individual from the external environment, or non-self. Immunity, a response to the violation of these borders, not only created identity but also preserved its integrity.

In his 1941 book *The Production of Antibodies*, MacFarlane Burnet explicitly introduced the words "self" and "non-self" into immunology literature. He used this newfound language to describe autoimmune diseases, immune tolerance in organ transplantation, and other conditions. By the 1970s, the self versus non-self dichotomy became deeply embedded in immunology, the defining dogma of the discipline.

Burnet expanded the common view of infection and immunity as a life-and-death struggle. Infectious germs, he argued, needed a healthy host for food, shelter, and reproduction (with the exception of germs that spread from dead bodies or those that survive indefinitely outside of the body, like anthrax spores). Because a germ that killed its host also engineered its own annihilation, death from infectious disease was an anomaly, due to perhaps the extraordinary virulence of the germ or exceptional weakness of the host (or both) and in the best interest of neither species. A low-level infection with minimal or absent symptoms, on the other hand, would allow the germ to prosper indefinitely, find a new home, or even aid the host in some manner.

While the immune system has chiefly been defined by an ability to distinguish self from non-self, with insinuations of individuality and insularity, this metaphor—as philosopher Alfred Tauber argues—does not suffice. Immunity evokes images of bodily defense against catastrophic infections, but most meetings between immune cells and germs are harmless, a truce rather than a war. The state of being infected is not absolute: like health and disease, it rests on a continuum. Living in the expansive gray area is a quiet tolerance of food, germs, and other matter. And many mundane yet essential immune functions maintain the body's own cells. For example, autoimmunity is a harmful reaction toward healthy "self" tissues. But it exists in benign form, with natural antibodies marking normal body parts, outlining an "immune homunculus." The immune system also scavenges the body for damaged or dead cells, as Metchnikoff once pointed out, or precancerous cells, getting rid of these elements as it would poisonous germs.

Microbes help steer immune identity away from its origins and into its broadest biological context. This identity rests not only on self versus non-self discrimination but on *dialogue*, the conversations immune cells are having with each other and with cells inside and outside of the body, at all hours, weakening the notion of an isolated, circumscribed self. The identity that emerges is sophisticated, entwined with the environment, with dialogue that partakes in a larger goal: to process information, to recognize, remember, and learn. In labeling a food or germ a friend or foe—or somewhere in between—context matters, since the type and volume of an immune response is dictated by multiple influences, challenging the old view that a mechanical lock-and-key match between an antibody and a noxious antigen, a mere recognition, is in itself enough to ignite a specific immune response. The nature of immunity is contextual and rests on a continuum.

Metchnikoff did not live to see the research of his youth meld with the musings of his old age to yield essential findings that redefined immunology. He did not imagine that germs, which succumbed to macrophages, would also reinvent their identity, that the health of the immune system would become a proxy for the fitness of an ecosystem. But his curiosity persisted into his final moments on earth. Minutes before he took his last breath, he reminded a colleague to look "carefully" into his intestines after his death.

Present-day technological prowess might have allowed the colleague to understand that the microbiome is highly mutable and largely shaped by lifestyle factors. These forces spark and sustain conversations between microbes and immune cells, sketching ever-evolving organismal boundaries. The array of hygienic practices that are healthful for our bodies and the germs within them is broad. And it is perhaps best exemplified by some of the longest-lived individuals in the world.

CHAPTER 16

Easter Island

N
o one can pinpoint exactly where my grandfather went on his last walk, at the age of ninety. He surely walked for miles, as he had each day of his life since his youth, in a body nimble and unobtrusive despite the whimsical freckles and furrows that marked the swift passage of time. He came home before sundown to a modest meal with many of his children and their children. Afterward, he retired to a corner of the house where, sitting cross-legged on the floor, he pursued an education that had been interrupted at age five. In his old age, having learned to read and write, he embarked on a quest to better understand all the religions of the world. He pored over translated texts while taking notes in a native script. So when, soon after that final walk, blood suddenly stopped flowing to parts of his brain, his work was yet unfinished, with death manifesting as an abrupt—but ultimate—interruption in his thoughts.

The global population of individuals over the age of sixty-five is growing at an unprecedented rate. As we age, biological order relents to disarray: the heart and lungs begin to fail, kidneys falter, bones and muscles weaken, and the mind slowly descends into darkness. But evolutionary biologists point out that the elderly are vaults of cultural wisdom, yielding societal benefits despite their dwindling health and inability to reproduce. Not every old person suffers an identical fate. Some retain robust physical and mental fitness, allow-ing for economic productivity, creative pursuits, and other plea-sures of life through their last days. But beyond genetic material and sheer luck, is there something more that determines how long and—more importantly—how *well* we will live in our later years, allowing for acuity of the mind and an insouciant bodily ignorance that are youth's greatest gifts?

———————

In 2004, demographers Michel Poulain and Giovanni Mario Pes, along with physicians Luigi Ferrucci and Claudio Franceschi and other colleagues, published a paper on a cluster of villages high in the mountains of Sardinia, Italy, with an incredible concentra-tion of centenarians, more so than in any other part of the island. In early speculations, the researchers had casually used blue ink to mark the area on a map and called it a "blue zone." Around the same time, journalist Dan Buettner, under the aegis of the *National Geographic*, set out to identify areas around the world with high life expectancies. He wanted to know *why* these individuals lived not only long but well, largely free of the physical and mental plagues of old age. Genes contributed, of course, but studies had shown that most differences in the length and quality of human life owed to lifestyle.

Buettner enlisted a team of medical researchers, including Poulain and Pes, as well as anthropologists and epidemiologists, to search for evidence-based environmental similarities between the regions. The team identified additional blue zones around the world, places with the highest average life expectancies, with many inhabitants living for nearly a century or more. Aside from Sardinia, they featured the Greek island of Ikaria in the Aegean Sea, as well as Okinawa, Japan; Nicoya, Costa Rica; and Loma Linda, California—home to one of the largest concentrations of Seventh Day Adventists in the world.

Blue zone dwellers consume a diet of around 95 to 100 percent whole plant foods eaten both raw and cooked. Their food is gently transformed, often with just a handful of ingredients, and at times fermented. People in blue zones enjoy a variety of seasonal vegetables and fruits, including those straight from the garden. Legumes loom large: cheap, versatile, and amenable to endless creations, from hearty Mediterranean minestrone or Nicoyan black beans and rice—*gallo pinto*, the national dish—to Okinawan soybeans squeezed or fermented into extra-firm tofu, tempeh, natto, and other products, all of which contain potent anti-inflammatory soy phytochemicals like isoflavones. Soy often arrives with seaweed in Okinawa, as in a hearty miso soup, the *wakame* and *kombu* suffusing the brew with not only iodine and B$_{12}$ but prebiotic fibers like agar and ulvans, unknown in terrestrial plants. Blue zone bread is typically whole wheat, rye, barley, or sourdough, and each day a small handful or two of nuts make their way into the meals. Drink is mostly water and perhaps a morning coffee or teas of various shades—black, white, or green—or simply wild herbs steeped in scalding water, all flooded with catechins, or tea polyphenols. Catechins are most bountiful in green teas, which Okinawans tend to nurse throughout the day.

If animal foods are included in the diet, they mostly function as flavorings or side dishes. Minimal dairy, a couple of eggs, and a meager serving of lean chicken, lamb, or pork—less than the size of a deck of cards—each week is typical. Fish, too, is eaten somewhat sparingly, around two modest weekly portions, with a focus in most areas on small fish like sardines and anchovies, which are less prone to building up heavy metals like mercury or toxic industrial chemicals like polychlorinated biphenyls, all of which are known to adversely affect human health.

The blue zone diet, filled with fiber and other essential nutrients from whole plants, helps to cultivate salutary gut germs, tempering the bruises evoked by time in an aging microbiome, which is prone to becoming dysbiotic and inflammatory over the years and can compound inflammaging.

Okinawan elders follow the rule of *hara hachi bu*, an old Confucian adage that reminds them to stop eating when their stomach is 80 percent full. Eating this way means being mindful of how food affects the senses and other internal cues rather than external ones—like the empty dinner plate or the end of a television show—to signal the end of a meal. In Okinawa and in other blue zones, people savor their last meal of the day late in the afternoon or early in the evening, allowing their intestines an extended rest before breakfast the next morning, a facile form of fasting seamlessly interwoven into daily routines.

The sporadic access to food for much of human history—particularly for hunter-gatherers—contrasts with modern habits of feeding and snacking throughout the day and well into the night. Various methods of fasting have been shown to affect inflammation and chronic inflammatory diseases, in both animal research and a growing number of human studies. Fasting may slow aging and help to prevent or even treat conditions like obesity, hypertension, diabetes, heart disease, cancer, memory issues, bone loss, and autoimmune conditions.

During a fast, the body depletes its supply of glucose and starts to burn fat. Throughout organs and tissues, inflammation wanes—with a downregulation of NF-κB and a decrease in the expression of inflammatory genes and cytokines, including CRP, TNF-α, IL-6, and IL-1β—but immunity remains intact. Hormones and enzymes that push cells to grow and divide, notably insulin-like growth factor and the mammalian target of rapamycin (TOR), begin to back down. Insulin-like growth factor and TOR, helpful for growth and development in childhood, are harmful in excess during adulthood, fostering aging and age-related diseases. They thrive on inflammatory diets high in animal protein but are reined in by plant foods, particularly phytochemicals like polyphenols, which can inhibit TOR.

TOR-driven aging is akin to an engine on a race car speeding without brakes. We evolved mechanisms to ensure that our bodies run at full speed during youth, so that reproduction occurs before death. In the wild, many organisms do not live long enough to age and thus have no use for breaks. In seventeenth-century London, most people did not reach their thirtieth birthday. But problems arise in modern societies when the car continues to speed during adulthood. The drug rapamycin, discovered in the 1970s in Easter Island in the southeast Pacific Ocean, suppresses the immune system and is given to transplant patients to prevent organ rejection. It also inhibits TOR. But slowing TOR down with rapamycin, unlike dietary changes, risks unwanted medical side effects.

The mild stress of a fast shifts the body's attention away from growth and toward repair and reform as it clears or recycles molecular garbage, mends DNA, and renews cells, further dampening inflammation and fortifying the body against a broad range of potential future irritants. Eating all the day's calories in a window of around eight to ten hours is one of the simplest ways to fast, reaping some of the same benefits as lengthier fasts and allaying the inflammatory actions of macrophages.

People in blue zones stress their bodies not only with a bit of hunger each day but also with physical labor, frequent natural movements born of a reluctance to use modern conveniences as they garden, work around the home, and take long walks. Research shows that regular, moderate exercise—like brisk walking, bicycling, jogging, and strength training—helps to prevent all kinds of chronic inflammatory diseases, while inactive lifestyles feed them.

During a solitary bout of exercise, inflammatory cytokines rise and peak briefly before returning to baseline levels. In exercise, the body needs inflammation just as it does when faced with an injury or a germ. In fact, it is how we build muscle. Strength training damages muscle tissue, eliciting an acute inflammatory response. Afterward, the inflammation resolves and yields to muscle repair and growth—processes that are stimulated by resolvins and other pro-resolving mediators. Taking NSAIDs to suppress inflammation disturbs this natural chain of events, as they also dampen the healing process. In exercising muscle, as in other tissues, inflammation aims to arrive quickly and flee after the task has been completed, rather than overstay its welcome.

For a body unaccustomed to exercise, this initial surge of inflammation is quite high and muscles may be sore for days. But soon, the acute inflammation incited by each session declines. So does ongoing inflammation throughout the body. Dozens of human clinical trials across age-groups show that regular exercise tones down chronic, low-level inflammation, reducing markers like CRP, IL-6, and TNF-α while increasing Tregs and cytokines that counter inflammation, like IL-10.

Exercise calms inflammation through many routes. It influences most of the hallmarks of aging, including genomic instability, telomere attrition, and the buildup of senescent cells. It melts inflammatory abdominal fat and, even in the absence of weight loss, lowers the numbers of macrophages that infiltrate fat tissue and churn out inflammatory cytokines. It manipulates microglial behavior, preserving brain function. It shrinks inflammatory fat around blood vessels and alters macrophages in atherosclerotic plaques, helping to ward off heart disease and strokes. It improves gut microbial diversity. But exercising too much, or in the wrong way, injures and inflames. Long, intense periods of exercise, particularly for those unaccustomed to such rigor, can increase the risk of chronic, hidden inflammation.

Habitual movement of most kinds benefits the body. Even simple stretching may mitigate inflammation in laboratory animals, with human trials underway. Scientists hoping to find a better way to prevent or treat human back pain, especially in light of the growing issues with anti-inflammatory drugs, have experimented with "rat yoga." During rat yoga, a laboratory rat is set down on a platform. It is gently lifted by its tail, bringing it off its hind legs, and cajoled into grasping the edge of the platform. The tail is pulled back while the rat, back arched, wordlessly holds its pose.

In one study of rat yoga, a chemical called carrageenan (which is found in many processed foods) was injected into rats' backs, creating local inflammation. Typically, when inflammation arises in the muscles of the back, pain ensues and mobility is limited. But when some of the injected rats were subjected to yoga twice a day for two weeks, they behaved as though they had taken an anti-inflammatory drug. These rats were able to walk better, with less back pain and lower numbers of macrophages in their back tissue. Another study found that yoga helped rats to make more resolvins. Further, an injection of resolvins under their skin mimicked the effects of yoga poses. Incredibly, stretching triggered nature's healing response to inflammation.

Animal studies suggest that stimulation of the vagus nerve—which is known to boost anti-inflammatory effects in the body—induces the release of resolvins. Lifestyle factors with emerging links to vagal stimulation include yoga, tai chi, meditation, deep and slow breathing, laughter, massage, fasting, social connections, singing, chanting, and even listening to certain types of music.

While periods of hunger and movement stress the body in ways that, over time, blunt hidden inflammation, some habitual stressors are incendiary. Our bodies evolved to fight or flee when faced with frightening threats. If we happen upon a tiger hunting for a meal, cortisol, a major stress hormone, floods our blood and readies us to run or resist. The heart quickens its beat and breathing becomes brisk and shallow. The stress soon passes—the tiger devours or retreats—and we either perish or reclaim equanimity. In the modern world, it is not the rare deadly predator but frequent spells of familiar stressors, real or imagined, that collude to damage health. The loss or illness of a loved one. A harrowing divorce. An irate, bullying boss or burnout on the job. Excessive, exhausting friction with family members or friends. Poverty and its multitude of handicaps. Or loneliness, an intense stressor for a species adapted to tribal tendencies for survival. These and other misfortunes, which may not mark the body with immediate, glaring wounds, tend to spur hidden inflammation, a central embodiment of chronic stress that is likely one mechanistic link between stress and an increased risk of many illnesses, including heart disease, obesity, diabetes, cancer, autoimmune conditions, neurodegenerative diseases, depression, and anxiety.

Stress alters the actions of immune cells. Macrophages become angry and maladaptive, pumping out greater numbers of inflammatory cytokines. Stressful situations—even ostensibly moderate ones, like public speaking—are tied to a rise in blood markers of inflammation like CRP, IL-1β, TNF-α, and IL-6. Often, multiple stressors operate in concert to compound inflammatory effects. A lonely individual, for example, is more likely to become inflamed when beset by challenges at work than someone rich in social support. And stress directly affects immunity, weakening the ability of immune cells to effectively engulf or kill germs.

Blue zone dwellers find ways to counter stress each day. They begin each morning after ample rest. Sleep loss, a type of chronic stress, can feed hidden inflammation and disease. So can disruption of circadian rhythms, which happens with certain changes in sleep patterns or when facing artificial light—particularly blue light—long after sundown. They also engage in rituals like meditation or various forms of yoga, which have been shown to lower inflammation, and remain close to the natural world, fostering a healthy relationship with germs. Many centenarians often share meals and conversations with friends or family. In Okinawa, for example, childhood companions create traditional *moais*, small groups of people who offer each other support that endures a lifetime.

These individuals think and move through their last days, forgoing the idea of a formal retirement. Rather, a deep sense of purpose—the Okinawan *ikigai*—careens them through the minutes and hours, through the inevitable sufferings and sorrows of life, evolving with time as well as with their physical and mental abilities, imbuing each day with hope for the next.

Broad environmental factors beyond lifestyle can also affect inflammation. In 1677, long before the blue zones were officially identified, Ikarian bishop Joseph Georgirenes attempted to capture their essence in his book, writing, "The most commendable thing of this island is the air and water, both so healthful that the people are very long lived . . . a small island, the poorest and yet, the happiest of the whole Aegean Sea."

Air quality, from pollutants and smoking to the onslaught of chemicals that suffuse modern goods, can inflame both old and young. All parts of polluted air are harmful to humans, but fine particles from sources like vehicle emissions, industrial processes, or wildfires—which can fling these particles far from the site of the flame—are particularly noxious. In the home, high-heat cooking fumes can pollute our bodies. The fumes from biomass fuels, a byproduct in some rural households where wood charcoal, dung, or crop residues are used in open stoves, are especially hazardous. Fine pollutants can penetrate our lungs, enter our blood, and land in organs all around the body. They can alter the germs within us and even our genes.

Hidden inflammation is a central mechanism by which pollution leads to poor health. The immune system recognizes and responds to pollutants as it does to germs. Macrophages, our initial defense against the air we breathe, rest in the tiny air sacs of the lungs and become incensed upon encountering pollutants. Particles from cigarette smoke can still be found inside these macrophages even years after a person quits the habit. Chronic exposure to air pollution is linked to elevated levels of inflammatory markers like CRP, IL-6, IL-1β, and TNF-α as well as a higher risk of chronic inflammatory diseases—not only lung diseases like chronic obstructive pulmonary disease and asthma, in which inflammation is an important factor, but others as well, including heart disease, hypertension, diabetes, obesity, cancer, allergies, and autoimmune and neurodegenerative diseases. Pollution ages our skin and bones and portends premature death. It can affect nearly every organ system in the body.

Each year, beyond pollutants from the air, thousands of new chemicals make their way into the items we ingest or use, infiltrating food, clothing, medications, personal care products, and cleaning supplies. An overload of pesticides, phthalates, flame retardants, polycyclic aromatic hydrocarbons, bisphenols, and more nourishes hidden inflammation though multiple routes and increases the risk of chronic inflammatory diseases. A few simple lifestyle changes can yield great gains in minimizing these exposures.

To be old is to be, one day, inflamed. Many robust centenarians fail to escape hidden inflammation in their final years. But they also manage to retain an abundance of factors that counter inflammation—including resolvins, which typically decline as the years mount.* To be inflamed in youth or middle age, however, is to quicken the passage of time, to harbor old age and diseases that lurk unseen within the confines of the body but stand ready to manifest at any hour. Children with inflammatory illnesses like obesity are, unbeknownst to them, aging briskly. Senescent cells pile up in their visceral fat and biological debris begins to clutter their bodies. Children faced with recurrent violence, bullying, neglect, or other forms of physical or psychological stress develop hidden inflammation that may persist into adulthood, even after shedding the stressors. The immune system, which memorizes infections

---

* Experiments in mice have shown that resolvins naturally decrease with age.

and vaccinations, also recalls severe threats to the survival of a nascent self. But the story starts even earlier. Environmental exposures that inflame a pregnant woman—including foods, toxins, pollution, infections, inactivity and stress—can modify genes and enable her to bequeath an inflammatory "code" as she would curly hair or a round face, giving the baby a higher risk of hidden inflammation and chronic inflammatory diseases in childhood and beyond.

Aging, then, begins before birth. The arc of hidden inflammation is unique to each individual, from its heterogeneous triggers and the precise hour of its emergence to the intensity of its peaks and valleys across a lifetime. This ongoing molding of inflammation, an *immunobiography*, is a narrative with plots and subplots, foils and mirrors, as the immune system maps and marks chronic disease, disability, and death. Genes are not blameless in this story, but their role is relatively minor, as evinced by the drastic global rise of chronic inflammatory diseases in the past decades. At the heart of hidden inflammation lies an incongruity between the ecological niche we evolved to inhabit and the one most of us currently occupy. Industrialization enhanced life and longevity in many ways, including through modern medical technology, sanitation, vaccination, and other public health measures. But it also fostered radical shifts in our food, air, movements, sleep, stressors, relationships, and more, factors that affect our bodies and the germs it hosts.

The burden of hidden inflammation and chronic inflammatory diseases forces us to reckon with all elements of our modern environment, to view health not simply as the presence or absence of one or multiple distinct diseases but as a state of being that allows for maximal fulfillment of human potential. As our environment continues to transform, the interdependence at the core of this equation is increasingly manifest: that the health of humans, microbes, and the planet are inextricable.

CHAPTER 17

Human Chimeras

After her surgery for Crohn's disease, Olivia was left with a mere 80 inches of her intestines rather than the typical 25 feet. Any food she chewed and swallowed would quickly morph into pale brown sludge that poured out of her ostomy, a surgical opening through the wall of her stomach that allowed the intestines to dump its contents into a plastic bag. She doubled her daily intake of calories but her body remained thinner than ever. I drew for Olivia a diagram of her new intestines. Her gut length was now closer to that of a tiger, I explained, rather than that of a human being. The shrunken intestines struggled to absorb enough calories and nutrients to keep her healthy. If the pattern persisted, she would need liquid nutrition pumped through her veins, putting her at risk for life-threatening infections, liver disease, and other problems. And perhaps she would eventually require a radical surgery: transplantation of the intestines and other organs into her body.

Along with the anatomy of her gut, the greatest threats to Olivia's life had transformed. Chronic inflammatory diseases like cancer, obesity, and heart disease lurked in the shadows of more primal, immediate killers like malnutrition and fatal infections. For a small handful of patients with rare medical issues—which are usually, but not always, related to digestion—an exclusive diet of whole plant foods may not be optimal. While some patients who are missing much of their intestines can tolerate loads of fiber and subsist entirely on plants, others, like Olivia, require at least some concentrated sources of animal protein and fat in the diet. I advised her toward eggs, seafood, lean poultry, and plain fermented dairy and away from red and processed meats. Still, the aim was to saturate her gut with as many plant foods as it could bear, focusing on food preparation techniques that could maximize absorption.

Plants powerfully affect inflammation, helping to ward off chronic diseases, including those that patients with short guts are uniquely predisposed to, like liver disease. They are also especially adept at aiding an unusual process unfolding in Olivia's gut called *adaptation*, whereby her slashed intestines would slowly increase their absorptive capacity over the next couple of years in an attempt to recoup some lost bowel function. Nutrients like omega-3s and prebiotic fibers—which feed gut germs—stimulate special hormones that facilitate adaptation. The "tiger gut" would become a bit bigger and longer and its muscles thicker. Microscopic, finger-like protrusions on the intestinal cells of the inner wall, called villi, would grow and multiply, easing the passage of fluid and nutrients into the bloodstream. This remarkable process reflects the relationship between eating and the world at large: food choices could help Olivia manage her inflammation and adapt to the features of a radically altered environment that existed not only within her body but also outside of it, from the rise of stunning medical advances to a changing climate and novel, deadly infections.

———

To be human is to adapt to an evolving environment. Life was brief and brutish for our Paleolithic ancestors, who frequently succumbed to infectious diseases, food shortages, predators, wars, and accidents. Babies often did not grow to become adults, and few people made it into their forties. Survival meant molding oneself to the whims of the surroundings.

To survive, humans evolved a hyperactive immune system, insulin-resistant bodies adept at storing fat, and sticky blood prone to clotting after trauma. Most Paleolithic diets emphasized plant foods, but exceptions existed. In the Arctic, for example, where frosty tundra discouraged the growth of plants, the forebearers of the Indigenous Inuit subsisted largely on the flesh of fish, seals, and whales. They evolved to tolerate this way of eating, with a genetic mutation (still present in more than 80 percent of Greenland and Canadian Inuit) that could curb ketosis, a state that emerges when the body is deprived of glucose and forced to burn fat. The Inuit are not known for long lives or exemplary health—as studies reveal—but for their ability to endure in a harsh natural climate.

Humans will attempt to find a meal in any habitat, from deserts and grasslands to forests and frozen wastelands. The foremost goal in the natural world is to sustain life until reproduction, the focal point by which evolution acts, the point after which a hungry drive for preservation deteriorates. With the advent of agriculture, some populations evolved the capacity to better digest starchy foods or even dairy products. The microbes in our guts adapted as well. In the Japanese, for example, they borrowed genes from other organisms that could break down seaweed. By the time antibiotics became prevalent, old threats had largely faded.

Yet the legacy of our past remains: bodies prone to inflammation, insulin resistance, and blood clots. The epidemic of chronic inflammatory diseases that tend to show up together in individuals—heart disease, stroke, cancer, diabetes, obesity, and neurodegenerative conditions—is also more likely to emerge during the aging process. These ailments, which are a part of our biological heritage, stem from deeply embedded evolutionary vulnerabilities. We were designed to survive historic killers and produce babies, not to live on for decades. Beyond this, we have transformed our environment, including the food we eat, the air we breathe, our relationships with germs and other people, and how we move or rest. Our immune system, handed down from ancestral times, is exquisitely sensitive to the new triggers in this world.

Historically, the cost of provoking an inflammatory response was usually a reasonable price to pay for the gains. A dire infection elicits a fury of acute inflammation that harms healthy tissue for the benefit of destroying a deadly germ. The cost of hidden inflammation triggered by environmental elements, however, initially seems slim. We cannot see or feel the inflammation, and it appears to cause no overt damage to our bodies. But the true price of this constant, low-level stimulation of the immune system—one that keeps our bodies in a perpetual state of readiness for a threat that never arrives—manifests years or even decades later, with maladies ranging from deadly heart attacks and cancers to the debilitating chronic diseases of old age.

Our bodies retain the marks of not one but many evolutionary histories. The most essential advice this history offers us on how to eat and live today is the notion that we must continue to adapt to our environment. Much as 80 inches of intestines will attempt the seemingly impossible task of compensating for 25 feet, so, too, will humans struggle to live in harmony with the ecosystems within and around them—or perish in the process.

A major feature of our modern environment is the surge in extraordinary medical and surgical treatments. The emerging links between food, germs, and inflammation, which influence the prevention and treatment of common chronic diseases, may also affect therapeutic medicine at large. One intriguing example of this phenomenon lies within the field of organ transplantation, a discipline in which manipulating the immune system is essential to successful patient outcomes.

In 1597, when the acclaimed surgeon Gaspare Tagliacozzi of Bologna, Italy, wrote of organ transplantation that "the singular character of the individual entirely dissuades us from attempting this work," he might have marveled at the feats of twenty-first-century medicine. Today, few surgical interventions are as elaborate or as ethically significant as removing an organ from one body and transplanting it into another. Transplantation is often the only treatment for end-stage organ failure. One donated body can save or improve several lives, allowing tragedy to incite unimaginable hope and healing. Today, surgeons are able to transplant not one but several abdominal organs in concert. These multiple-organ transplants, which include the intestines, uniquely illustrate the interaction between food, germs, and the immune system—as well as our emerging understanding of the intricate language that passes between them.

Tomoaki Kato, or "Tom" as he is known to his colleagues at Columbia Medical Center, is a surgeon who performs the types of complex transplants that patients like Olivia might one day require. Kato, a legendary pioneer in the field of multiple-organ transplantation as well as ex vivo surgery—a special method used to resect tumors—is slim and soft-spoken, with keen eyes and an unfailing work ethic. He has a reputation for successfully taking on cases other surgeons deem inoperable, spending long hours painstakingly peeling away cancerous tumors clinging to gossamer tissues. As a college student in Japan, he had aspired to become a molecular biologist. But one day, as he rode a bullet train from Tokyo to Kyoto, the conductor called for a doctor on board to assist with an ill passenger. Kato felt a visceral urge to help the man. In a flash, he decided to change his major.

One of Kato's toughest surgeries came early in his career. The patient, Jamie, was only twenty years old, but she was one of the sickest patients he had ever encountered. She had been a healthy child who loved to get lost in books and ballet lessons. But during her teenage years, an autoimmune liver disease prompted a liver transplant, and soon after a clot manifested in a major artery and destroyed all of her intestines. For months, she languished in an intensive care unit, tubes in every orifice and fed through veins while her organs declined and she battled frequent infections. Doctors told her parents the case was futile. Jamie had mere days, perhaps hours, left to live.

Over twenty-two grueling hours, Kato transplanted five organs into Jamie's body: a liver, stomach, pancreas, kidney, and intestines. The organs had been procured from an eighteen-year-old brain-dead donor. During the operation, 100 pints of fresh blood flowed through a catheter and into Jamie's veins. Small, meticulous sutures embedded the tissues, some dotted with cream-colored fat, into her hollowed frame. When he finally left the room and discarded his scrubs, Kato knew the battle was far from over.

For a patient as sick as Jamie, the chance of surviving such an operation was akin to a coin flip. The question of how she would fare suffused the air.

If Jamie survived the surgery, her fate would chiefly rest on her immune system's handling of the newly implanted organs. How badly it reacted against them as foreign bodies would depend in part on how genetically distinct the organs were from her own. An organ from a cat or cow sewn into a human being will elicit much more fury than an organ transplanted between members of the same species. Even among humans, organ donors and recipients are assessed for favorable genetic compatibility prior to the transplant. After the operation, potent drugs are deployed—usually indefinitely—to suppress the immune system's instinctive urge to expel unfamiliar tissues and organs. A fine balance must be struck in dosing these medications: too little and the immune system destroys the new organ; too much and the body succumbs to infections and cancers or other side effects of the drugs themselves.

Intestinal transplants can evoke a particularly robust immune response. Unlike organs considered to be mostly sterile, like the liver, kidney, or heart, a transplanted gut brings along a bevy of native germs. One receiving the organ also inherits a sizable portion of the donor's immune system residing in the intestines. Recipient immune cells eventually swarm the area, but some immune cells from donor intestines have been found to survive for nearly a decade in their new home.

The body's rejection of a foreign organ largely involves the adaptive branch of the immune system, as killer T lymphocytes cause cells in donor tissues to commit suicide and B lymphocytes begin to make antibodies against them. But the innate immune system, Elie Metchnikoff's legacy, also plays a role in organ rejection. After a transplant, low-level inflammation from an irked innate immune system, even if it cannot cast the organ out by its own efforts, may nudge adaptive immune cells to react badly to the graft.

Some of this inflammation is, of course, inevitable. The very act of wrenching an organ from a brain-dead donor, cutting its blood supply and hence oxygen, transporting it in cold ice, and finally nourishing it in a new body triggers inflammation. Bruises and nicks to tissues during the surgery itself can also inflame, as can infections or even bacteria burrowing into the intestinal wall, which tends to occur more often in sick patients as they fight to recover from their operations in an intensive care unit.

Donor grafts that embrace the external environment, like the lungs and intestines, are directly exposed to food, germs, air, and more. Antigens that are inhaled or aspirated may irritate the macrophages dwelling in the tiny air sacs of the lung, which can orchestrate an innate immune response that portends poor outcomes after a lung transplant. For a patient who undergoes an intestinal transplant, the first oral feeding generates a sudden onslaught of fresh antigens for the innate immune system, heightening the risk of rejection.

The special immune cells called Tregs play a crucial part in one's ability to tolerate a transplanted organ. Tregs suppress excessive immune reactions throughout the body, helping to prevent unwanted, lingering inflammation, lethal autoimmunity, or organ rejection. Increased Tregs in the blood or in the donor graft portend better tolerance of the transplant.

Tregs may help to enhance a unique state in transplant recipients called chimerism. A chimera, according to ancient Greek mythology, is a monstrous, fire-breathing, hybrid creature, "a thing of immortal make," as Homer says in the *Iliad*, "not human, lion-fronted and snake behind, a goat in the middle." In modern medicine, chimeras are humans or animals that contain the cells of two or more genetically distinct individuals, or two sets of DNA. This can happen naturally, as when a few cells from a fetus pass through the placenta and spread through the mother's body or when a fraternal twin absorbs his dead sibling in utero.

Human-made chimeras include patients with bone marrow and solid organ transplants. These patients retain their own genetic material along with—to varying degrees—that of the donor. At times, this may create problems. Donor immune cells may react against healthy host tissues in what is known as graft-versus-host disease. But at yet undefined tipping points, chimerism may make patients more tolerant of foreign organs as a medley of donor and recipient immune cells peacefully mingle in a single body. Studies in multivisceral transplant patients have shown that robust blood chimerism of certain immune cells is linked to lower rates of organ rejection. And, as case reports reveal, it can even help an organ transplant recipient to entirely wean off medications that suppress the immune system.

Typical transplant drugs used to subdue the immune system do not address Treg levels, but burgeoning science points to a promising role for Treg therapy—a treatment that may be less toxic than traditional drugs—in organ transplantation. In addition, Treg levels may naturally shift with lifestyle changes. Both food antigens and microbial metabolites can influence Tregs. The short-chain fatty acids yielded when gut germs ferment fiber from whole plant foods increase Tregs both in the intestines and throughout the body. In humans, diets high in fiber have been tied to a measurable rise in Tregs. Some microbes, like the anti-inflammatory *Clostridial* clusters, induce young T cells to morph into Tregs. Microbes can also indirectly spur immune cells, like macrophages, to expand Treg numbers. Beyond fiber, food components that dampen or resolve inflammation—including polyphenols, omega-3s, and immune-modulating vitamins (particularly vitamins D and A)—hold promise for boosting Tregs, as do adequate amounts of exercise, stress relief, social connections, and sleep. Meanwhile, inflammatory ways of eating and living drive a dysbiotic microbiome that can help disable Tregs and promote inflammatory T cells. Early studies in organ transplant recipients show that diets fostering an optimal relationship between food, germs, and inflammation—like the traditional Mediterranean diet—are tied to lower rates of graft failure and loss.

Keeping silent inflammation at bay may encourage the body to better accept the donated organ. In patients with kidney transplants, high blood levels of inflammatory markers are tied to worse graft outcomes. It may also help to forestall chronic inflammatory diseases, which organ transplant patients are more likely to develop compared with healthy individuals. In a dangerous cycle, chronic inflammatory diseases like obesity may increase the risk of organ rejection. The cytokines spewed out by fat tissue can heighten T cell reactions against the organ or even subvert Tregs. The dysbiotic microbiome of an obese individual, which alters both innate and adaptive immune cells and churns out constant, hidden inflammation, may sway the immune system against the organ. In organ transplant patients, a loss of gut microbial diversity is linked to worse outcomes, including graft-versus-host disease, rejection, and lower survival rates.

Jamie survived both her surgery and an arduous recovery. She went on to complete her bachelor's degree in psychology, graduating with honors, and then attended medical school, fulfilling a life-long aspiration to become a physician. She eventually specialized in pediatric transplant hepatology, spending her days caring for children suffering with liver diseases.

A unique, innovative treatment like Jamie's holds the potential to reach far beyond salvaging a single life. By contesting boundaries in uncharted territory, it may redefine the field in which it lies and perhaps even medicine at large, reverberating across disciplines. And it is yet another element of modern life that evinces the dynamic nature of our environment, compelling us to rethink the very notion of what it means to be human.

In contrast to life-altering medical advances offering us hope and improved health, we continue to face a collusion of harmful threats in our current environment. Climate change, if left unchecked, is poised to forever transform our lives in ways we cannot fully anticipate. Infectious pandemics may prove to be equally devastating. Both are heavily fueled by our food practices, particularly by the unsavory and unsustainable methods used to meet a voracious taste for animal foods. Here, too, an understanding of the relationship between food, germs, and inflammation may aid us in adapting to these realities.

Epidemics and pandemics did not exist for most of human evolution. But around ten thousand years ago, with the advent of agrarian lifestyles, humans began to domesticate animals and catch their germs, like measles from cattle, pertussis from pigs, typhoid from chickens, and leprosy from water buffalo. To date, germs have killed more people than have natural disasters and wars. With the arrival of better sanitation, vaccination, and finally antibiotics, the death toll from infectious diseases markedly declined by the mid-twentieth century. "To write about infectious disease," MacFarlane Burnet noted in 1962, "is almost to write of something that has passed into history." But by the late twentieth century, the trend had begun to reverse itself. Death from infectious diseases grew once again, with new germs emerging at alarming rates. Scientists warn that infectious pandemics may pose as much of a threat to humanity as climate change. Infections, like chronic diseases, have declared themselves a modern problem—one exacerbated by hidden inflammation, which can impair immunity. Hidden inflammation increases the risk of infections and weakens the response to vaccines, undermining inflammation's primordial functions. It can also prompt our immune systems to overreact to infections, leading to bleak outcomes.

In 2019, when the coronavirus pandemic took hold as the most destructive global pandemic since the 1918 Spanish flu, an urgent question soon emerged: why did some individuals experience worse outcomes than others? A few patterns made intuitive sense. The elderly tended to fare worse. So did those with certain health conditions, including obesity, heart disease, hypertension, diabetes, and lung or kidney disease. In some cases, men appeared more likely to fall ill and die compared with women of the same age. Women, inured to fighting pathogens that threaten their unborn children, may mount stronger and faster initial immune responses against germs (conversely, they tend to have a higher risk of auto-immunity). How much of the germ a person encountered also seemed to matter. Health-care workers, who were exposed to high loads while caring for patients, were more susceptible than the general public. The germ was more likely to replicate and rapidly spread in their bodies before the immune system could wrestle it under control.

But severe infections were not exclusive to the elderly, those with preexisting health conditions, or health-care and other essential workers. Frequent outliers loomed large, like the woman in her twenties with no medical history who ended up in an intensive care unit on heart and lung support, or the marathon runner in his forties who passed away. The motley group of patients who succumbed to severe infections was made up of people both old and young, sick and healthy. As scientists scrambled to better understand why this was so—long after Virchow had struggled to understand the forces behind an 1848 typhus epidemic in Prussia, subsequently initiating his intricate study of inflammation, or Metchnikoff, charged by his spirited discovery of macrophages in Messina, had mused on their ties to microbes during an 1890 cholera pandemic—inflammation and the immune system appeared to play a fundamental role.

When a germ first makes its way into the human body, the innate immune system reacts within minutes. This primeval response, triggered by protein structures common to many pathogens, is broad and rash, indiscriminately lashing out at anything that appears alarming, atoning for a lack of precision with speed. The innate immune system aims to temper the infection, preventing the brawl from getting unruly before the adaptive arm is able to join the fray.

Most germs attempt to foil the immune system in order to foster their own survival. A virus may muffle interferon, an early cytokine that interferes with viral replication, buying itself more time to spread stealthily throughout the body. A virus can invade cells and organs, causing harm. But unless the infection yields immediate failure in vital organs or blood vessels—as with the Ebola virus*—the risk of death typically rests on how the immune system responds to the virus. We must fear, even more so than a new germ, the caprice of our familiar bodies.

Initially, the germ itself or a measured immune reaction to it triggers most symptoms—like a fever, which alerts the body to the attack, or the coughs and loose stools that expel microscopic infectious particles. But if the immune system does not succeed in managing the germ with contained violence, it can later resort to a haphazard, unfettered flood of inflammation, where immune cells churn out reams of cytokines in a frantic attempt to fight the germ, catching an excess of healthy tissues in the cross fire. If the immune

---

* Even with the Ebola virus, the way in which the immune system responds to the virus (including its ability to generate excessive inflammation) plays a role in outcomes.

system is anxious and excitable at baseline, it may react in this way even if it easily clears the germ. Macrophages are some of the main culprits in this pathological, hyperactive immune response, which begins with the innate arm of the immune system and eventually encompasses the adaptive one as well. The condition has been labeled "cytokine storm" or "macrophage activation syndrome," but no single consensus definition exists at present. And it can be seen, in various iterations, even as a complication of noninfectious illnesses, including select autoimmune disorders.

Many people in intensive care units succumb to this burst of irrational inflammation rather than to the germ itself. The lungs and airways become inflamed, peppered with "ground-glass opacities," hazy gray areas on radiographic imaging that resemble translucent glass on a shower door. The lungs fill up with fluid and fail to oxygenate, warranting the use of a ventilator. Blood clots tend to form more easily, depriving essential organs of oxygen. Heart attacks, fueled in part by inflammation, may occur even in the absence of major plaque buildup in arteries. Inflamed heart muscle struggles to pump blood around the body and is prone to developing erratic rhythms. The intestines become porous, allowing bacteria to seep into the bloodstream, and the brain may surrender to fatal inflammation. The storm can contribute to the demise of nearly every organ. As the patient's condition worsens, blood pressure drops while the breathing rate and beating of the heart quicken. The patient becomes feverish or—paradoxically—frigid. The mind floats, unmoored from its ordinary functions. The body is prone to both clots and catastrophic hemorrhage. Multiple organs fail, and death eventually ensues. If a germ can maim, the immune system, despite benevolent intent, can kill.

In an infectious viral illness, the line between helpful and harmful inflammation can be blurry and hard to define. After all, a certain amount of inflammation is indispensable to combating germs. And the deluge of cytokines—numbering in the hundreds, including both inflammatory and anti-inflammatory types—are interdependent, driving and suppressing each other in complex feedback loops. Although universal diagnostic criteria for this disastrous inflammation are lacking, delineating the threshold between productive and pathological inflammation can have important implications. It may allow doctors to intervene early, before the storm gets out of hand, and to deploy drugs that target inflammation in a timely manner.

An overzealous immune reaction might help to explain why some young and ostensibly healthy individuals suffer with severe illness and death during epidemics and pandemics, such as the 2019 coronavirus pandemic, a respiratory illness at its core with ravages that reached far beyond the lungs; the SARS, MERS, and H1N1 epidemics; or the 1918 Spanish flu pandemic, when an influenza virus sprang from wild birds* and spread to humans, killing an estimated fifty million people around the world. In addition, many who survive a severe infectious disease complain of lasting issues related to the infection. Spending more than a week in the intensive care unit can be comparable to experiencing a major head injury, and damaged organs may not entirely recoup their lost function. Long after recovery, lingering inflammation can contribute to ongoing ailments like fatigue, headaches, insomnia, loss of taste or smell, brain fog, body aches, and heart or lung problems. For some people, the damage is irreversible.

Inflammatory storms may occur because the immune system is impetuous, keeping a searing flame alight even after an infection has died down, or because the germ is replicating too quickly, causing immune defenses to spiral out of control. Perhaps the initial infectious load was large or the patient's immunity weak, allowing the germ to disseminate with ease. Genes can also help to explain why people might react differently to the same germ.†

---

* The 1918 influenza virus was thought to be of avian origin, but whether there were intermediate carriers (like pigs) is unclear. The avian viral ancestor has not been found. Many scientists now believe that it was a direct avian to human introduction, an idea that is widely but not universally accepted. Similarly, the question of how SARS-CoV-2 and other coronaviruses came to humans is still unresolved.

† For example, a handful of genetic variants in severe cases of certain viral illnesses are linked to weak innate immunity or hyperactive inflammation during infections. *IFITM3* is a gene stimulated by interferon, and the protein it codes for interferes with influenza virus entry into cells. Nonfunctional gene variants of *IFITM3* are present in around one in four hundred Europeans and are also especially common in Japanese and Chinese people. These individuals may be at higher risk for severe illness from influenza. In addition, people hospitalized for an influenza infection are more likely to have a

But hidden inflammation may also play an important role in heterogeneous immune responses. The silently inflamed tend to have
both sluggish immunity and impulsive immune systems and are
more likely to generate overwhelming, inappropriate inflammation
during infections. Many of those with an elevated risk of poor outcomes from infections suffer with hidden inflammation and chronic
inflammatory diseases.* Obesity, for example, is generally a major
risk factor for worse outcomes during infectious diseases. Some
of this risk relates to mechanics. For instance, large amounts of fat
can compress the lungs, increase airway resistance, and impair gas
exchange. But obese people are also significantly inflamed, both
within the lungs and throughout the body. Their fat is an immune
organ that churns out inflammatory cytokines and hormones at all
hours. Their immune systems may overreact to germs, and signals
that limit or resolve inflammation are impaired.

In the elderly, another vulnerable group, senescent cells embedded in the lungs and elsewhere and bodywide inflammaging
heighten the likelihood of an inflammatory storm. Age affects all
types of immune cells. Macrophages, much like their creator, suffer the marks of time. Those in aging bodies, tasked with fighting
threats and disposing of litter over a lifetime, become less adept at
swallowing up and defending against germs, garbage, or other matter, which then remain free to continually irk the immune system
and inflame. Macrophages also lose artful skills in repairing tissues
and resolving inflammation. And older macrophages stumble and
stammer when introducing foreign substances to lymphocytes,
making for sloppy discourse. In the elderly, fresh T cells that hunt
down new pathogens, or B cells that produce specific antibodies,
begin to dwindle. Hence, the aging immune system may remember
how to combat germs from prior encounters, but new germs become
much more of a challenge, creating fertile ground for chronic infections to flourish and elicit hidden inflammation.

---

nonfunctional *IFITM3* gene. Regardless, most people with an aberrant *IFITM3* gene do
not have issues with combating influenza.

* A chronic disease, even independently of inflammation, may lower the threshold for
organ dysfunction and reduce immunity.

Immune defenses naturally decline with age, but hidden inflammation compounds this. Be it burning in the young or old, it hinders immunity. It weakens both innate and adaptive immune responses to an infection. Hidden inflammation is also a decoy that crowds the scene as a germ enters the body, making it more difficult for immune cells to mount a timely, effective attack against the intruder. This delay—a key window of time that affects disease outcomes—means that the germ gains a foothold in the fight, madly replicating before drawing too much attention to itself, destroying tissues and eliciting further inflammation. Immune cells then go berserk in trying to make up for lost time, driving uncontrolled bouts of inflammation throughout the body. When immunity suffers, both the infection and the immune system may run amok, yielding profuse collateral damage.

In addition, silently inflamed individuals, such as the obese or elderly, are less likely to mount robust immune responses to vaccines. A vaccine mimics a natural infection, generating memory T and B cells that will recognize and destroy the germ if it dares revisit the body. Immunologists are learning that some vaccines may train not only adaptive immunity but also innate immunity. Chronic, low-level inflammation can impair the immune system's ability to store intelligence about past encounters with pathogens, dampening the efficacy of vaccines.

Hidden inflammation may be implicated in unfavorable responses to infections. But food choices, which help to regulate hidden inflammation, affect not only our ability to combat germs responsible for epidemics and pandemics but also our propensity to breed these germs in the first place. Evolutionary biologists blame, in large part, intensive animal husbandry—driven by a steep global demand for animal foods—for the rise of particularly vicious germs that have led to novel, deadly infections in humans. Making enough meat for the masses inevitably drives production methods that create inflamed, diseased animals. Animals slaughter and devour one another in the natural world, but in the face of a constant order: prey nearly always vastly outnumber their predators. Tigers, for instance, are solitary creatures.

The rising consumption of animal foods around the world—which doctors and scientists have long decried as being deleterious to human health—also alters a natural order between humans and microbes, transforming harmless germs into harmful ones. A bird flu virus, for example, begins with innocuous intent, making its home in the bowels of aquatic birds like swans. It replicates and lives in harmony with its host until it is regurgitated into the waters and ingested by another swan, a cycle that has persisted for millions of years. But then the swan is dragged out of the lake and stuffed into dark, dank sheds along with hundreds of thousands of other animals. Squatting in its filth, it becomes bruised and immobile, burned by its own secretions. It becomes chronically inflamed and diseased, with meek and broken immunity. The swan's feces, rife with virus, drop onto the face of, say, a chicken, and make their way into its intestines. The virus must then adapt to its new host, a land bird, by mutating. It masters new modes of migration and spreads to various organs. It may reach the lungs and become airborne, learning to survive outside of its host. The bird flu virus, once largely harmless to both birds and people, mutates into a strain that kills over half the chickens it infects and patiently awaits its chance to trigger a human pandemic.

Most interactions between immune cells and germs are not meant to ignite a war, and the state of being infected rests on a continuum. But factory farms, live animal markets, and other places in which similar conditions are found warp this evolutionary inclination toward cooperation. They breed easy prey: crowded, sick, immobile animals lacking air or sun. The germs, emboldened, may morph into organisms capable of unnatural violence. Their victims are so close and plentiful that the death of a host is no longer the deterrent it once was. Agricultural animals also use up most of the antibiotics sold in the United States—which are often deployed merely to fatten them up—and breed formidable, multidrug-resistant germs.

Factory farms, where most of the animals eaten in the United States are raised, not only nurture but also disseminate disease. They scatter pollutants and waste into local communities, contaminating the air, soil, and water and directly affecting flora and fauna. Germs that escape their walls, born of incessant suffering, are unleashed upon the world to feed a cycle of anguish.

Our relationship with germs is often depicted as a relentless struggle, a battle of wits between disparate species trying to outsmart each other. Evolutionary biologist Leigh Van Valen, in his Red Queen hypothesis, captured this idea of an "arms race" between coevolving species, referencing Alice's predicament in Lewis Carroll's *Through the Looking-Glass* when the Red Queen tells her she must run merely to remain in place, as the world is rapidly changing. But we need to do even more than outrun germs: we must also learn to live alongside them, cultivating tolerant, perhaps even symbiotic, relationships with the microbes inside, atop, and around us, behaving in ways that enhance rather than destroy natural truces between humans and microbes. The immune response, molded over millennia by evolutionary forces to combat germs, finds not only new foes in food but also new friends in germs. Germs help to shape our defenses against them, and their behaviors—both within and outside our bodies—are influenced by our own.

Human and planetary health are also intimately bound by food and inflammation. As a college student in the 1960s, the scientist Walter Willett—whose nutrition studies would support Ancel Keys's historic observations in the Mediterranean—had been struck by a class he took with professor Georg Borgstrom, author of The *Hungry Planet*, which hinted at the then embryonic idea of climate change. Over fifty years later, Willett became cochair of a commission on food, health, and the environment called EAT-Lancet, which brought together thirty-seven leading scientists from sixteen countries with expertise ranging from human health and agriculture to political science and environmental sustainability. Climate change had ballooned into the most urgent dilemma of the century.

As the earth continues its insidious ascent toward scorching temperatures, it will be increasingly burdened by burning forests and flooded cities, heat waves and droughts. Many more plant and animal species will tumble into oblivion, further reducing the diversity—and hence the health—of ecosystems. Climate change is not only destroying our capacity to produce enough food for nearly ten billion people by 2050: it is also irrevocably altering the food itself. "Every leaf and every grass blade on earth makes more and more sugars as carbon dioxide levels keep rising," mathematician Irakli Loladze notes, "diluting other nutrients in our food supply."

The 2019 EAT-Lancet report presented the inextricable link between human and planetary health. The best weapon against not only most death and disability in the world but also the ongoing destruction of the planet is food. Greenhouse gas emissions from livestock, for example, are one of the biggest contributors to global warming. Animal agriculture is also a significant driver of deforestation, which forces animals to migrate due to loss of habitat. This may expose them to other humans and animals with whom they can share germs. Along with an elimination of fossil fuels, swift changes in the way food is produced and consumed are essential to keeping the earth from spinning into chaos in the coming decades, with rampant preventable diseases and malnutrition afflicting its inhabitants.

Broadly, the report called for a radical decrease in the intake of animal foods and an increase in plant foods. Half of the "planetary health plate" consisted of vegetables and fruits, while whole grains, legumes, nuts, and other plant fats filled the rest of the plate. The option to include animal foods was limited by what the planet could withstand, down to specific portion sizes. Those eating this way would fill their plates with whole plant foods. They might elect to have a small cup of plain yogurt some mornings at breakfast, an egg or two during a weekly brunch, and perhaps a serving of fish or poultry once or twice a week.

The similarities between a diet for planetary health and one that helps to manage inflammation, promoting human health, are striking. Indeed, the allowances for animal flesh, Willett noticed, were just a little higher than what Ancel Keys had recorded during his earliest observations in the Mediterranean. Willett coauthored a white paper,* an evidence-based response correcting revisionist histories of Keys's work and subsequent nutrition research, exploring archival records and primary source materials—a complex task. If history, as Julian Barnes writes, is that "certainty produced at the point where the imperfections of memory meet the inadequacies of documentation," this particular certainty may hold the key to our future.

Eating and living to avert hidden inflammation helps to restore equanimity in the ecosystems within and around us, forestalling chronic inflammatory diseases—which are responsible for most diseases and deaths that plague modern humans—as well as catastrophic events that threaten our existence on earth. It balances the immune response, boosting immunity while preventing distorted reactions to germs or other triggers, so that inflammation is duly wielded and speedily resolved. It helps us to embrace the unprecedented potential of the modern world to maximize human health and longevity.

With proper care, many children born in the twenty-first century may celebrate their one-hundredth birthday. Once fatal illnesses are now treatable due to remarkable medical advances that continue to accrue. Multiple organ transplants may one day become commonplace, and human organs may eventually yield to prosthetic ones, much like prosthetic heart valves or knees. Classic drugs that dampen the immune system—with varying vigor and precision—are making room for those that manipulate the immune response in other ways: drugs that promote resolution pathways, for example, bringing little risk of side effects from immune suppression, or those that incite immune cells to target malignant tumors.

If we aim to survive—and *thrive*—in this new environment, adapting to its strengths and weaknesses, reaping its benefits while tolerating its burdens, we must eat and live in ways that respect the

---

* The white paper, released in 2017, was commissioned by the True Health Initiative, a nonprofit organization founded by David Katz, MD, MPH. The True Health Initiative is made up of a global coalition of experts advocating for evidence-based advice on diet and lifestyle. The lead author on the white paper was Katherine Pett, MS, MEd, RDN, and other writers included Joel Khan, MD, Walter Willett, MD, DrPH, and Katz.

language of the immune system—an endeavor guided not only by the rich history and culture of our past but also the promise of an awe-inspiring future.

The summer of my first pandemic was beset by savage weather all over the world. In the United States, it was one of the hottest, driest summers on record, fraught with heat waves that cooked the country. Hurricanes struck the South before hastily moving east, bringing widespread damage and power outages. Severe thunderstorms with devastating winds raced through the Midwest, from Iowa to Ohio, destroying crops and infrastructure. Numerous wildfires, fanned and fueled by tempestuous winds, dry terrains, and lightning from storms, burned millions of acres across California, Oregon, Washington State, and other parts of the West, setting unprecedented records and flinging fine airborne particles deep into the lungs.

But as I walked the streets of Manhattan that summer, the ashes of a battered city promised restitution. Struggling restaurants began to reinvent themselves, offering takeout, freezer, and ready-to-cook meals, or even socially distanced picnics in the park. People explored their kitchens, tossing a few seeds into a mason jar—for pennies a day—to sprout broccoli. Or they baked a weekly bread. In homes across the city, foods especially adept at boosting immunity—like berries, greens, legumes, mushrooms, tomatoes, onions, carrots, crucifers, garlic, nuts, and seeds—strove to shield individuals from the wreckage of a new germ.

Parks and gardens swelled with motion, as many discovered new, secret retreats in familiar green spaces. They ran, biked, hiked, and walked in the open air as never before. In a place that had always fostered a meeting of minds from all walks of life and as much human connection in the flesh as one desired, the dearth of these tangible encounters—be they serendipitous or scheduled, casual or profound—only underscored their essential roles in human health and pleasure, which technology could never dare to entirely replace. A subtle shift in the flavor of the city's noises man-ifested. Alongside the sounds of sickness and yearning, of misfor-tune and monotony, arose others. The calming hum of a pop-up jazz band in Central Park, its members standing 6 feet apart in damp grass. Impromptu concerts at the hospital, set in spaces meant for lectures and conferences, as health-care workers sang and played instruments in homage to ill and deceased colleagues. Mellow notes from a woman's violin in the apartment across from mine, the rosin between bow and strings allowing me to hear her well on so many nights. And the blues notes Jay wrote when he sought respite from work, neck bent over the shaft of a beloved childhood guitar. The music, a plea of sorts from a city and its people, a mutation of its pain, was so much more. It seeped into our selves and spurred our worn nerves to make resolvins and other tiny molecules, unseen and unheard, that made their way around our minds and bodies, striving to tame the flame within and beyond.

Acknowledgments

In the years I spent writing *A Silent Fire*, I received unwavering support from my partner, Vikram, who reviewed the manuscript and provided sharp and meaningful criticism. Our daughter Fiona, who was born midway through the draft, was a welcome distraction from its contents. I had an outstanding team of editors and agents. Jessica Yao's diligence, and vital insight from both her and Will Hammond, helped to shape this story. Alison Lewis and Zoë Pagnamenta provided keen guidance all along the way, as did Sally Holloway.

This book was inspired by the work of many dedicated physicians and scientists. I had the privilege of interviewing some of them directly, including T. Colin Campbell, Benoit Chassaing, Harold Dvorak, Caldwell Esselstyn, Luigi Ferrucci, Gokhan Hotamisligil, Tomoaki Kato, Peter Libby, Paul Ridker, Charles Serhan, Justin Sonnenburg, and Walter Willett.

I appreciate the early encouragement from Ethan Schmidt, Lauren Sandler, and Timothy Williams, as well as the feedback on select portions of the text from several other friends, family, and colleagues, including Shalini Ravella, Krishna Ravella, Shoma Brahmanandam, Stephen Morse, Ruslan Medzhitov, Katherine Pett, and Henry Blackburn—who, well into his tenth decade, provided generous commentary.

This book would not exist without my parents, Raja and Vijaya Ravella, to whom I will always remain indebted. I thank Maddikunta and the late Geeta Brahmanandam for welcoming me into their lives many years ago. And I am grateful for the support from members of my extended family, including Paula Acosta, Christine Hsu, Yvonne Parris, Mana Raval, Ricardo Rubi, Avni Shah, Alice Wang, and Neha Wattas.

Notes

General Notes

These notes are not meant to be exhaustive. Rather, they aim to provide broad guidance should the reader wish to further explore some of the topics I have discussed in this book. My foray into the vast history of immunology began with an extensive exploration of Rudolf Virchow's intriguing life and work. Erwin H. Ackerknecht, *Rudolf Virchow: Doctor, Statesman, Anthropologist* (Madison: University of Wisconsin Press, 1953) remains a relevant and thorough biography of Virchow in the English language. I also learned from Byron A. Boyd, *Rudolf Virchow: The Scientist as Citizen* (New York: Garland Publishing, 1991), and Brian L. D. Coghlan and Leon P. Bignold, *Virchow's Eulogies: Rudolf Virchow in Tribute to His Fellow Scientists* (Basel: Birkhäuser, 2008), which includes translations of eulogies written by Virchow in honor of a number of his famous teachers, colleagues, and pupils. Virchow's medical writings and lectures helped to inform my work, including several documents translated and edited by L. J. Rather: *Disease, Life and Man: Selected Essays by Rudolf Virchow*, *Collected Essays on Public Health and Epidemiology* (Stanford, CA: Stanford University Press, 1958) and *A Commentary on the Medical Writings of Rudolf Virchow* (San Francisco: Norman Publishing, 1990). Virchow's *Thrombosis and Emboli* (Canton, MA: Science History Publications, 1998), translated by Axel C. Matzdorff and William R. Bell, presents a clear window into his systemic approaches as he accurately describes the formation of blood clots and embolization. Twenty significant lectures delivered by Virchow at the Pathological Institute of Berlin in 1858 are translated by Frank Chance in *Cellular Pathology as Based upon Physiological and Pathological Histology* (Whitefish, MO: Kessinger Publishing, 2008). *Letters to His Parents, 1839 to 1864* (Canton, MA: Science History Publications, 1990), a collection of personal communications by Virchow translated by L. J. Rather and edited by Marie Rabl, divulges many of his early ambitions and personal characteristics, as does an essay written in his high school years, which is translated by Karel B. Absolon in *Virchow on Virchow* (Rockville, MD: Kabel Publishers, 2000).

In *Immunity: How Elie Metchnikoff Changed the Course of Modern Medicine* (Chicago: Chicago Review Press, 2016), Luba Vikhanski provides an excellent modern biography of the scientist whose personal and scientific life weaves through this book. As Metchnikoff is a complex and at times contradictory figure, his depiction is best served by taking into consideration viewpoints from multiple authors, including Alfred I. Tauber and Leon Chernyak, *Metchnikoff and the Origins of Immunology: From Metaphor to Theory* (Oxford: Oxford University Press, 1991), and his wife Olga Metchnikoff, *Life of Elie Metchnikoff, 1845–1916* (London: Constable, 1921). I also looked to Alexandre Besredka, *The Story of an Idea: E. Metchnikoff's Work, Embryogenesis, Inflammation, Immunity, Aging, Pathology, Philosophy* (Bend, OR: Maverick Publications, 1979), translated by Abraham Rivenson and Rolf Oestreicher; Elaine Mardus, *Man with a Microscope: Elie Metchnikoff* (New York: J. Messner, 1968); Charles Dawbarn, *Makers of a New France* (London: Mills and Boon, 1915); and Herman Bernstein, *The Celebrities of Our Time* (London: Hutchinson, 1924). "My Stay in Messina (Memories of the Past, 1908)," an essay in a collection of memoirs by Metchnikoff called *Souvenirs* (Moscow: En Langues Étrangères, 1959), translated by Claudine Neyen, includes a description of his discovery of phagocytosis. Other of Metchnikoff's writings provide an engrossing window into his keen explorations of inflammation and disease, including *Immunity in Infective Diseases*, translated by F. G. Binnie (Cambridge: Cambridge University Press, 1905); *The New Hygiene: Three Lectures on the Prevention of Infectious Diseases* (Chicago: W. T. Keener, 1910); *The Prolongation of Life: Optimistic*

Studies, translated by P. C. Mitchell (New York: The Knickerbocker Press, 1908); *Founders of Modern Medicine: Pasteur, Lister, Koch*, translated by David Berger (New York: Walden Publications, 1939); and *The Nature of Man: Studies in Optimistic Philosophy*, translated by P. Chalmers Mitchell (London: G. P. Putnam's Sons, 1903). In his *Lectures on the Comparative Pathology of Inflammation*, delivered at the Pasteur Institute in 1891 and translated by F. A. Starling and E. H. Starling (New York: Dover Publications, 1968), he examines the mechanisms of inflammation along the hierarchy of animal life with characteristic enthusiasm for the subject.

Several accounts of Paul Ehrlich are illuminating, including Martha Marquardt, *Paul Ehrlich* (New York: Henry Schuman, 1951); Ernst Baumler, *Paul Ehrlich: Scientist for Life* (New York: Holmes & Meier, 1984); Arthur M. Silverstein, *Paul Ehrlich's Receptor Immunology: The Magnificent Obsession* (San Diego: Academic Press, 2002); Herman Goodman, *Paul Ehrlich: A Man of Genius, and an Inspiration to Humanitarians* (New York: reprint from *The Medical Times*, 1924); and Luba Vikhanski's *Immunity*. For a general history of immunology, Silverstein's expansive *A History of Immunology* (San Diego: Academic Press, 1989) is perhaps one of the best resources that I have encountered. Narrower in scope, yet still worth perusing, are Pauline M. H. Mazumdar, *Immunology 1930–1980: Essays on the History of Immunology* (Toronto: Wall and Thompson, 1989); Edward J. Moticka, *A Historical Perspective on Evidence-Based Immunology* (Amsterdam: Elsevier, 2016); Domenico Ribatti, *Milestones in Immunology: Based on Collective Papers* (London: Academic Press, 2017); and Wolfgang Schirmacher, *German Essays on Science in the 19th Century: Paul Ehrlich, Alexander von Humboldt, Werner von Sieme* (New York: Bloomsbury Academic, 1996). I enjoyed, too, William Addison, "Gulstonian Lectures on Fever and Inflammation," *British Medical Journal*, nos. 121–128 (April 23–June 1, 1859), and J. Burdon Sanderson, "Lumleian Lectures on Inflammation," *Lancet* 1, nos. 3057–3061 (April 1–29, 1882).

I obtained reliable insights into the work of Ancel Keys from Todd Tucker, *The Great Starvation Experiment: Ancel Keys and the Men Who Starved for Science* (Minneapolis: University of Minnesota

Press, 2007); Joseph L. Dixon, *Genius and Partnership: Ancel and Margaret Keys and the Discovery of the Mediterranean Diet* (New Brunswick, NJ: Joseph L. Dixon Publishing, 2015); Ancel Keys, Josef Brozek, and Austin Henschel, *The Biology of Human Starvation* (Minneapolis: University of Minnesota Press, 1950); Ancel Keys and Margaret Keys, *Eat Well and Stay Well* (Garden City, NY: Doubleday, 1963); Katherine Pett, Joel Kahn, Walter Willett, and David Katz, *Ancel Keys And The Seven Countries Study: An Evidence-Based Response to Revisionist Histories* (Tulsa, OK: True Health Initiative, 2017); and through communications with Henry Blackburn, an esteemed colleague of Keys. I appreciate the breadth of several evidence-based works on food and health, including T. Colin Campbell, *The China Study* (Dallas, TX: BenBella Books, 2016) and Michael Greger and Gene Stone, *How Not to Die: Discover the Foods Scientifically Proven to Prevent and Reverse Disease* (New York: Flatiron Books, 2015). Greger's voluminous online repository of collected nutrition studies is also a valuable—and easily navigable—resource.

Inflammation is mentioned in nearly every medical textbook and conveyed piecemeal through various intersecting disciplines. *Fundamentals of Inflammation* (Cambridge: Cambridge University Press, 2010), edited by Charles N. Serhan, Peter A. Ward, and Derek W. Gilroy, is a noteworthy general reference on its current cellular and molecular mechanisms. Other informative texts on inflammation and disease include *Inflammation, Lifestyle and Chronic Disease: The Silent Link* (Boca Raton, FL: CRC Press, 2012), edited by Bharat B. Aggarwal, Sunil Krishnan, and Sushovan Guha; *Inflammation and Atherosclerosis* (Berlin: Springer-Verlag Wein, 2012), edited by George Wick and Cecilia Grundtman; and Caleb E. Finch, *The Biology of Human Longevity: Inflammation, Nutrition and Aging in the Evolution of Lifespans* (Burlington, MA: Academic Press, 2007).

Introduction

6   **Only twenty examples:** Jeff Aronson, "When I Use a Word . . . Is It Inflammation? It Is!," *QJM: An International Journal of Medicine* 102 (2009).

Chapter 1: Metamorphosis

11   **had written to his father asking for money:** Byron A. Boyd, *Rudolf Virchow: Scientist as Citizen* (New York: Garland, 1991), 9.

12   **"Did you hear it? We do not know anything anymore":** Erwin H. Ackerknecht, *Rudolf Virchow: Doctor, Statesman, Anthropologist* (Madison: University of Wisconsin Press, 1953), 10.

13   **simply a "vital substance," the "stuff of life":** John Simmons, *Doctors and Discoveries: Lives That Created Today's Medicine* (New York: Houghton Mifflin Harcourt, 2002).

13   **writings on Egyptian papyri:** John F. Nunn, *Ancient Egyptian Medicine* (Norman: University of Oklahoma Press, 2002).

13   **recorded his findings:** A. Cornelius Celsus, *On Medicine*, vol. 3, trans. W. G. Spencer (Cambridge, MA: Harvard University Press, 1938); Russell P. Tracy, "The Five Cardinal Signs of Inflammation: Calor, Dolor, Rubor, Tumor . . . and Penuria (Apologies to Aulus Cornelius Celsus, De Medicina, C. A.D. 25," *Journals of Gerontology: Series A* 61, no. 10 (Oct. 2006).

13   **believed that a noxious buildup:** John Redman Coxe, *The Writings of Hippocrates and Galen. Epitomised from the Original Latin Translations* (Philadelphia: Lindsay and Blakiston, 1846); Vivian Nutton, "The Chronology of Galen's Early Career," *The Classical Quarterly* 23, no. 1 (1973).

14   **developed new methods:** Laura J. Snyder, *Eye of the Beholder: Johannes Vermeer, Antoni Van Leeuwenhoek, and the Reinvention of Seeing* (New York: W. W. Norton, 2015), 6–7, 103.

14   **"If medicine is to fulfill its vast duty":** August Heidland et al., "The Contribution of Rudolf Virchow to the Concept of Inflammation: What Is Still of Importance?," *Journal of Nephrology* 19 Suppl 10 (May–June 2006).

15   **"better than anybody else":** Ackerknecht, *Doctor, Statesman, Anthropologist*. All the quotations that appear in this paragraph are taken from this source.

15   **Virchow added a fifth cardinal sign:** Guido Manjo, *The Healing Hand: Man and Wound in the Ancient World* (Cambridge, MA: Harvard University Press, 1975).

15   **"Nobody would expect a muscle which is inflamed":** Rudolf Virchow, *Cellular Pathology as Based upon Physiological and Pathological Histology* (Whitefish, MO: Kessinger Publishing, 2008). All the quotations that appear in this paragraph are taken from this source.

16   **"Inflammation is thus an active and passive process":** Ackerknecht, *Doctor, Statesman, Anthropologist*.

16   **phenomenal level of productivity:** Carl Vernon Weller, "Rudolf Virchow—Pathologist," *The Scientific Monthly* 13, no. 1 (1921).

17   **tarnished Virchow's reputation after his death:** Ackerknecht, *Doctor, Statesman, Anthropologist*; Boyd, *Scientist as Citizen*.

18   **established germ theory:** Stefan H. E. Kaufman and Florian Winau, "From Bacteriology to Immunology: The Dualism of Specificity," *Nature Immunology* 6 (2005).

18   **went on to create:** Patrice Debre, *Louis Pasteur*, trans. Elborg Forster (Baltimore: Johns Hopkins University Press, 1998).

18 **as Eula Biss writes:** Eula Biss, *On Immunity: An Inoculation* (Minneapolis: Graywolf Press, 2004).

18 **miasma theory:** Jacques Jouanna, *Greek Medicine from Hippocrates to Galen: Selected Papers* (Leiden: Brill, 2012).

20 **but rather a deadly struggle:** The text draws upon Elie Metchnikoff's writings on Messinia, namely a description of his discovery of phagocytosis that was published in a collection of memoirs entitled *Souvenirs*, which was a source for a chapter about the discovery in Olga's biography. Metchnikoff wrote this description some thirty years after he had experienced it, in December of 1908, weeks after winning the Nobel Prize. His account may indeed be true to his emotions as he remembered them. However, the seeds leading up to his epiphany in Messinia may have been sown and smoldering for years during his earlier studies in invertebrates, as argued by Luba Vikhanski and Alfred Tauber, among others. In addition, earlier accounts detailing the process of phagocytosis exist, as Metchnikoff himself humbly acknowledged. Unlike his forebears, however, Metchnikoff studied phagocytosis in great detail. He followed his initial observations with many experiments, giving rise to his classic studies on inflammation, which he detailed in a monograph—*Lectures on the Comparative Pathology of Inflammation*—that remains relevant today. See: Charles T. Ambrose, "The Osler Slide, a Demonstration of Phagocytosis from 1876 Reports of Phagocytosis before Metchnikoff's 1880 Paper," *Cellular Immunology* 240, no. 1 (2006); Siamon Gordon, "Elie Metchnikoff, the Man and the Myth," *Journal of Innate Immunity* 8 (2016).

21 **"We pathologists think and teach the exact opposite":** Luba Vikhanski, *Immunity: How Elie Metchnikoff Changed the Course of Modern Medicine* (Chicago: Chicago Review Press, 2016).

21 **arguing the opposite:** British surgeon John Hunter was another early advocate of inflammation as a natural and, at times, beneficial force. In 1973, he appreciated that inflammation may contribute to host defense rather than to a disease process. See Helene F. Rosenberg and John I. Gallin, "Inflammation," in *Fundamental Immunology*, ed. William E. Paul (Philadelphia: Wolters Kluwer, 2008).

22 **"The larger and less mobile macrophages":** Vikhanski, *Immunity*.

22 **"Protection against disease":** Vikhanski, *Immunity*. All the quotations that appear in this paragraph are taken from this source.

23 **"peaceful little university town":** Olga Metchnikoff, *Life of Elie Metchnikoff, 1845–1916* (London: Constable, 1921).

23 **"I saw a frail elderly man of a short stature":** Elie Metchnikoff, *Founders of Modern Medicine: Pasteur, Lister, Koch* (New York: Walden Publications, 1939).

24 **"most original and so creative":** Vikhanski, *Immunity*.

24 **"I at once placed myself on your side":** Metchnikoff, *Life of Elie Metchnikoff, 1845–1916.*

25 **"an oriental fairy tale":** Vikhanski, *Immunity*.

25 **"Blood is a very special juice":** Vikhanski, *Immunity*.

25 **a "metaphysical speculation":** Daniel P. Todes, *Darwin without Malthus: The Struggle for Existence in Russian Evolutionary Thought* (New York: Oxford University Press, 1989).

26 **"degree of vilification unknown almost in present-day science":** Arthur M. Silverstein, *A History of Immunology*, 2nd ed. (Cambridge, MA: Academic Press, 2009).

26 **"Were I small as a snail, I would hide myself in my shell":** Paul de Kruif, *Microbe Hunters* (San Diego: Harcourt, 2002).

27 **"When this theory was attacked":** Vikhanski, *Immunity*.

27 **a threat to the organism:** The few historic contrarians were ignored. Poet and scientist Erasmus Darwin, in his *Zoonomia* of 1801, had repeatedly referred to a "laudable pus."

27 **"The curative force of nature":** Vikhanski, *Immunity.*

28 **"brilliant eccentric":** Ernst Bäumler, *Paul Ehrlich: Scientist for Life*, trans. Grant Edwards (New York: Holmes and Meier, 1984).

28 **He introduced the term *antikorper*:** Paul Ehrlich, "Croonian Lecture—On Immunity with Special Reference to Cell Life," *Proceedings of the Royal Society of London* 66 (Dec. 31, 1900).

29 **secret was a tiny latch:** Ehrlich, "Croonian Lecture."

29 **"grasping arms" of a sundew's tentacles:** Vikhanski, *Immunity.*

29 **many of the particulars:** Yvonne Bordon, "The Many Sides of Paul Ehrlich," *Nature Immunology* 7, S6 (2016).

30 **"If there was ever a romantic chapter":** Vikhanski, *Immunity.*

Chapter 2: Horror Autotoxicus

37 **"The organism possesses certain contrivances":** Silverstein, *History of Immunology.*

38 **proposed a theory of "autointoxication":** Manon Mathias, "Autointoxication and Historical Precursors of the Microbiome-Gut-Brain Axis," *Microbial Ecology in Health and Disease* 29, no. 2 (2018).

38 **"An autotoxin . . . one that destroys":** Silverstein, *History of Immunology.*

40 **a "paradoxical reaction":** Silverstein, *History of Immunology.*

40 **first described anaphylaxis:** Murray Dworetzky, Sheldon Cohen, and Myrna Zelaya-Quesada, "Portier, Richet, and the Discovery of Anaphylaxis: A Centennial," *Journal of Allergy and Clinical Immunology* 110, no. 2 (2002).

42 **strange constellation of symptoms:** Rober A. Bridges, Heinz Berendes, and Robert A. Good, "A Fatal Granulomatous Disease of Childhood; The Clinical, Pathological, and Laboratory Features of a New Syndrome," *American Journal of Diseases of Children* 97, no. 4 (1959).

42 **presented reports of babies and young children:** Tracy Assari, "Chronic Granulomatous Disease; Fundamental Stages in Our Understanding of CGD," *Medical Immunology* 5 (Sept. 21, 2006).

42 **dared to guess:** Vikhanski, *Immunity.*

43 **leaned toward the podium:** Jules Hoffmann, email to author, May 2021.

45 **"Generations of students will memorize your theory":** Elaine Mardus, *Man with a Microscope: Elie Metchnikoff* (New York: Messner, 1968).

45 **first double hand transplant:** Jean Michel Dubernard et al., "Functional Results of the First Human Double-Hand Transplantation," *Annals of Surgery* 238, no. 1 (2003).

45 **seeded into tissues before birth:** Guillame Hoeffel and Florent Ginhoux, "Fetal Monocytes and the Origins of Tissue-Resident Macrophages," *Cellular Immunology* 330 (August 2018): 5–15.

Chapter 3: A Sense of Strangling

50 **fell dramatically in ensuing decades:** James E. Dalen et al., "The Epidemic of the 20th Century: Coronary Heart Disease," *American Journal of Medicine* 127, no. 9 (2014).

50 **witness to these major advances:** Thomas H. Lee, *Eugene Braunwald and the Rise of Modern Medicine* (Cambridge, MA: Harvard University Press, 2013).

51 **but its beginnings:** Peter Libby, interview with author, February 2019.

51 **struggled with a question similar to Libby's:** Michael E. Silverman, "William Heberden and *Some Account of a Disorder of the Breast*," *Cllinical Cardiology* 10 (1987); Joshua O. Leibowitz, *The History of Coronary Heart Disease* (Berkeley: University of California Press, 1970).

52 **carefully dissected a patient's heart:** O. F. Hedley, "Contributions of Edward Jenner to Modern Concepts of Heart Disease," *American Journal of Public Health* 28 (1938); Silverman, "William Heberden."

52 **"How much the heart must suffer":** John Baron, "Review of the Life of Edward Jenner, M.D., LL.D., F.R.S.," *The Medico-Chirurgical Review and Journal of Medical Science* 33 (Oct. 1, 1838): 497.

52 **"a yellowish matter, comparable to a pea puree":** Georg Wick and Cecilia Grundtman, eds., *Inflammation and Atherosclerosis* (New York: Springer, 2012).

53 **"a vascular question, well expressed in the axiom":** William Osler, *The Principles and Practice of Medicine*, 6th ed. (New York: Appleton, 1906).

53 **would spark a serendipitous discovery:** Gilbert Thompson, ed., *Pioneers of Medicine without a Nobel Prize* (London: Imperial College Press, 2014).

53 **some years earlier:** Thompson, *Pioneers of Medicine.*

53 **excited to see plaques:** Alexander Ignatowski, "Changes in Parenchymatous Organs and in the Aorta of Rabbits under the Influence of Animal Protein [in Russian]," *Izvestia Imperatorskoi Voenno-Medicinskoi Akademii* 18 (1908).

54 **much higher concentrations of cholesterol:** A. Windaus, "Ueber Der Gehalt Normaler Und Atheromatoser Aorten an Cholesterol Und Cholesterinester," *Zeitschrift für Physiologische Chemie* 67 (1910).

54 **sprung into view:** Nikolai N. Anitschkow and S. Chalatov, "Ueber Experimentelle Cholesterinsteatose Und Ihre Bedeutung Fur Die Entstehung Einiger Pathologischer Prozesse," *Zentralblatt für allgemeine Pathologie und pathologische Anatomie* 24 (1913); Nikolai N. Anitschkow and S. Chalatov, "Classics in Arteriosclerosis Research: On Experimental Cholesterin Steatosis and Its Significance in the Origin of Some Pathological Processes," *Arteriosclerosis* 3 (1983).

54 **And with time, effort:** Daniel Steinberg, "In Celebration of the 100th Anniversary of the Lipid Hypothesis of Atherosclerosis," *Journal of Lipid Research* 54 (2013).

54 **sought to identify risk factors:** *Framingham Heart Study: Laying the Foundation for Preventive Health Care*, National Institutes of Health, https://framingham-heartstudy.org/fhs-about/.

55 **decreased heart attacks:** "The Lipid Research Clinics Coronary Primary Prevention Trial Results. I. Reduction in Incidence of Coronary Heart Disease," *Journal of the American Medical Association* 251, no. 3 (Jan. 20 1984).

55 **with the mission:** Daniel Steinberg, *The Cholesterol Wars: The Skeptics vs. the Preponderance of Evidence* (San Diego: Academic Press, 2007).

56 **urging scientists to look beyond lipids:** Russell Ross and John A Glomset, "The Pathogenesis of Atherosclerosis: (First of Two Parts)," *New England Journal of Medicine* 295, no. 7 (1976); Russell Ross and John A Glomset, "The Pathogenesis of Atherosclerosis: (Second of Two Parts)," *New England Journal of Medicine* 295, no. 8 (1976).

58 **musings on Bach as a medical healer:** Peter Libby, "Johann Sebastian Bach: A Healer in His Time," *Circulation Research* 124, no. 9 (2019).

58 **Virchow had hypothesized that inflammation played a key role in heart dis-ease:** Virchow, *Cellular Pathology*.

58 **speculated on the link:** Joseph Hodgson, *A Treatise on the Diseases of Arteries and Veins, Containing the Pathology and Treatment of Aneurisms and Wounded Arteries* (London: Underwood, 1815).

58 **"a depot which lies deep beneath the comparatively normal surface":** Wick and Grundtman, *Inflammation and Atherosclerosis*.

59 **He did not believe:** J. B. Duguid, "Pathogenesis of Atherosclerosis," *Lancet* 2, no. 6586 (Nov. 19, 1949).

63 **"clearly an inflammatory disease":** Russell Ross, "Atherosclerosis—An Inflammatory Disease," *New England Journal of Medicine* 340, no. 2 (Jan. 14, 1999): 115–26.

64 **heart disease occurs more often:** Peter Libby, "Role of Inflammation in Atherosclerosis Associated with Rheumatoid Arthritis," *American Journal of Medicine* 121, no. 10, Suppl 1 (Oct. 2008).

67 **initial 1997 paper on CRP:** Paul M. Ridker et al., "Inflammation, Aspirin, and the Risk of Cardiovascular Disease in Apparently Healthy Men," *New England Journal of Medicine* 336, no. 14 (April 3, 1997).

68 **in 2001 he launched JUPITER:** Paul M. Ridker et al., "Rosuvastatin to Prevent Vascular Events in Men and Women with Elevated C-Reactive Protein," *New England Journal of Medicine* 359, no. 21 (Nov. 20, 2008).

69 **upstream cytokines like IL-1β:** Paul M. Ridker, "From C-Reactive Protein to Interleukin-6 to Interleukin-1: Moving Upstream to Identify Novel Targets for Atheroprotection," *Circulation Research* 118, no. 1 (Jan. 8, 2016).

70 **asking the essential questions:** Paul Ridker, interview with author, Feb. 2019. Also see Paul M. Ridker, "Closing the Loop on Inflammation and Atherothrombosis: Why Perform the Cirt and Cantos Trials?," *Transactions of the American Clinical and Climatological Association* 124 (2013).

70 **showed the world:** Paul M. Ridker et al., "Antiinflammatory Therapy with Canakinumab for Atherosclerotic Disease," *New England Journal of Medicine* 377, no. 12 (2017).

71 **Other trials followed suit:** Jean-Claude Tardif et al., "Efficacy and Safety of Low-Dose Colchicine after Myocardial Infarction," *New England Journal of Medicine* 381, no. 26 (2019); Stefan M. Nidorf et al., "Colchicine in Patients with Chronic Coronary Disease," *New England Journal of Medicine* 383, no. 19 (2020).

Chapter 4: Open Wounds

73 **"the larynx was completely destroyed through cancer":** Rudolf Virchow, "Professor Virchow's Report on the Portion of Growth Removed from the Larynx of H.I.H. The Crown Prince of Germany by Dr. M. Mackenzie on June 28th," *British Medical Journal* 2, no. 1386 (1887).

74 **"blacker in color than inflammation":** Jeremiah Reedy, "Galen on Cancer and Related Diseases," in *Clio Medica. Acta Academiae Internationalis Historiae Medicinae*, ed. Lester S. King (Leiden, The Netherlands: Brill | Rodopi, 1975).

74 **same set of functional capabilities:** Douglas Hanahan and Robert A. Weinberg, "The Hallmarks of Cancer," *Cell* 100, no. 1 (2000).

75 **"immunologic surveillance":** Bäumler, *Paul Ehrlich: Scientist for Life*.

75 **the mice did not develop tumors:** Robert J. Moore et al., "Mice Deficient in Tumor Necrosis Factor-Alpha Are Resistant to Skin Carcinogenesis," *Nature Medicine* 5, no. 7 (July 1999).

76 **published an essay:** Harold F. Dvorak, "Tumors: Wounds That Do Not Heal," *New England Journal of Medicine* 315, no. 26 (Dec. 25, 1986).

76 **identified an angiogenic factor and inspired other labs:** Patricia K. Donahoe, *Judah Folkman, 1933–2008: A Biographical Memoir*, National Academy of Sciences (2014), http://www.nasonline.org/publications/biographical-memoirs/memoir-pdfs/folkman-judah.pdf.

77 **firm molecular basis:** Harold F. Dvorak, "Tumors: Wounds That Do Not Heal—Redux," *Cancer Immunology Research* 3, no. 1 (2015).

77 **not the first one:** Dvorak, "Tumors: Wounds That Do Not Heal—Redux."

77 **developed tumors only at the injection sites:** David S. Dolberg et al., "Wounding and Its Role in RSV-Mediated Tumor Formation," *Science* 230, no. 4726 (Nov. 8, 1985).

77 **"Cancer is not creative":** Harold Dvorak, email to author, Feb. 2019.

78 **dubbed the "first violin":** Ben-Neriah Yinon and Michael Karin, "Inflammation Meets Cancer, with Nf-κb as the Matchmaker," *Nature Immunology* 12 (2011).

78 **turn into traitors:** See, for example, Maria Rosaria Galdiero, Gianni Marone, and Alberto Mantovani, "Cancer Inflammation and Cytokines," *Cold Spring Harbor Perspectives in Biology* (2017); Mingen Liu, Anusha Kalbasi, and Gregory L. Beatty, "Functio Laesa: Cancer Inflammation and Therapeutic Resistance," *Journal of Oncology Practice* 13, no. 3 (2017); Shanthini M. Crusz and Frances R. Balkwill, "Inflammation and Cancer: Advances and New Agents," *Nature Reviews Clinical Oncology* 12 (Oct. 2015).

78 **showed that macrophages:** Elaine Y. Lin et al., "Colony-Stimulating Factor 1 Promotes Progression of Mammary Tumors to Malignancy," *Journal of Experimental Medicine* 193, no. 6 (March 19 2001).

78 **are also emerging:** Galdiero, Marone, and Mantovani, "Cancer Inflammation and Cytokines."

79 **added two additional hallmarks of cancer:** "Enabling hallmarks" are mechanisms that allow cancer cells to acquire core hallmarks. Enabling hallmarks include inflammation as well as mutable genomes (whose mutability greatly accelerates the pace of cancer formation). Immune evasion, a trait of many cancers, was also elevated to the level of a core hallmark. See Douglas Hanahan and Robert A. Weinberg, "Hallmarks of Cancer: The Next Generation," *Cell* 144, no. 5 (2011).

79 **can affect all its life stages:** See, for example, Bharat B. Aggarwal, Bokyung Sung, and Subash Chandra Gupta, eds., *Inflammation and Cancer* (Basel: Springer, 2014); Crusz and Balkwill, "Inflammation and Cancer: Advances and New Agents"; Shabnam Shalapour and Michael Karin, "Immunity, Inflammation, and Cancer: An Eternal Fight between Good and Evil," *Journal of Clinical Investigation* 125, no. 9 (2015).

79 **from the initial genetic:** Dawit Kidane et al., "Interplay between DNA Repair and Inflammation, and the Link to Cancer," *Critical Reviews in Biochemistry and Molecular Biology* 4, no. 9 (2014).

79 **At least a quarter:** Sergei I. Grivennikov, Florian R. Greten, and Michael Karin, "Immunity, Inflammation, and Cancer," *Cell* 140 (March 19, 2010); Hugo Gonzalez, Catharina Hagerling, and Zena Werb, "Roles of the Immune System in Cancer: From Tumor Initiation to Metastatic Progression," *Genes & Development* 32 (2018).

80 **might not be as prone to developing:** Eran Elinav et al., "Inflammation-Induced Cancer: Crosstalk between Tumours, Immune Cells and Microorganisms," *Nature Reviews Cancer* 13, no. 11 (Nov. 2013).

81 **burns beneath many developing tumors:** Ruslan Medzhitov, email to author, June 2021.

82 **They "taste" their surroundings:** Hoeffel and Ginhoux, "Fetal Monocytes and the Origins of Tissue-Resident Macrophages."

82 **lies what Medzhitov calls *parainflammation*:** Ruslan Medzhitov, "Origin and Physiological Roles of Inflammation," *Nature* 454, no. 7203 (2008).

82 **Several studies link:** Bharat B. Aggarwal et al., "Inflammation and Cancer: How Hot Is the Link?," *Biochemical Pharmacology* 72, no. 11 (Nov. 30, 2006); Joydeb Kumar Kundu and Young-Joon Surh, "Inflammation: Gearing the Journey to Cancer," *Mutation Research/Reviews in Mutation Research* 659, no. 1 (July 2008).

83 **could be that parainflammation interacted:** Audrey Lasry et al., "Cancer Cell-Autonomous Parainflammation Mimics Immune Cell Infiltration," *Cancer Research* 77, no. 14 (2017).

Chapter 5: Anatomical Intimacies

86 **Obesity rose steadily:** Garabed Eknoyan, "A History of Obesity, or How What Was Good Became Ugly and Then Bad," *Advances in Chronic Kidney Disease* 13, no. 4 (Oct. 2006).

86 **since the eighteenth century:** Eknoyan, "A History of Obesity."

86 **obesity research began in earnest:** Eknoyan, "A History of Obesity."

87 **not a satisfying answer:** Robert W. O'Rourke, "Inflammation, Obesity, and the Promise of Immunotherapy for Metabolic Disease," *Surgery for Obesity and Related Diseases* 9, no. 5 (Sept.–Oct. 2013).

87 **found TNF-α in fat tissue:** Gökhan S. Hotamisligil, Narinder S. Shargill, and Bruce M. Spiegelman, "Adipose Expression of Tumor Necrosis Factor-Alpha: Direct Role in Obesity-Linked Insulin Resistance," *Science* 259, no. 5091 (Jan. 1, 1993).

88 **true for humans as well:** Gökhan S. Hotamisligil et al., "Increased Adipose Tissue Expression of Tumor Necrosis Factor-Alpha in Human Obesity and Insulin Resistance," *Journal of Clinical Investigation* 95, no. 5 (May 1995).

88 **but many others as well:** S. K. Garg et al., "Diabetes and Cancer: Two Diseases with Obesity as a Common Risk Factor," *Diabetes, Obesity, and Metabolism: A Journal of Pharmacology and Therapeutics* 16, no. 2 (Feb. 2014); Maximilian Zeyda and Thomas M. Stulnig, "Obesity, Inflammation, and Insulin Resistance —a Mini-Review," *Gerontology* 55, no. 4 (2009).

89 **most of the inflammatory cytokines:** Stuart P. Weisberg et al., "Obesity Is Associated with Macrophage Accumulation in Adipose Tissue," *Journal of Clinical Investigation* 112, no. 12 (Dec. 2003); Haiyan Xu et al., "Chronic Inflammation in Fat Plays a Crucial Role in the Development of Obesity-Related Insulin Resistance," *Journal of Clinical Investigation* 112, no. 12 (Dec. 2003).

89 **much like an infection does:** Marc Y. Donath et al., "Inflammation in Obesity and Diabetes: Islet Dysfunction and Therapeutic Opportunity," *Cell Metabolism* 17, no. 6 (June 4, 2013).

89 **were actually immune cells:** Anthony Ferrante, email to author, May 2021.

90 **as well as adaptive immune cells:** A. W. Ferrante, Jr., "The Immune Cells in Adipose Tissue," *Diabetes, Obesity, and Metabolism: A Journal of Pharmacology and Therapeutics* 15, Suppl. 3 (Sept. 2013).

90 **immunometabolism was emerging:** Diane Mathis and Steven E. Shoelson, "Immunometabolism: An Emerging Frontier," *Nature Reviews Immunology* 11, no. 2 (Feb. 2011).

90 **interdependence of the immune system and metabolism:** Justin I. Odegaard and Ajay Chawla, "Pleiotropic Actions of Insulin Resistance and Inflammation in Metabolic Homeostasis," *Science* 339, no. 6116 (Jan. 11, 2013); Gökhan S. Hotamisligil, "Inflammation and Metabolic Disorders," *Nature* 444, no. 7121 (Dec. 14, 2006).

91 **same ancestral cell:** O'Rourke, "Inflammation, Obesity, and the Promise."

91 **embracing them tightly:** O'Rourke, "Inflammation, Obesity, and the Promise."

91 **a potential mechanistic link:** See, for example, Carey N. Lumeng and Alan R. Saltiel, "Inflammatory Links between Obesity and Metabolic Disease," *Journal of Clinical Investigation* 121, no. 6 (June 2011); Margaret F. Gregor and Gökhan S. Hotamisligil, "Inflammatory Mechanisms in Obesity," *Annual Review of Immunology* 29 (2011); F. Tona et al., "Systemic Inflammation Is Related to Coronary Microvascular Dysfunction in Obese Patients without Obstructive Coronary Disease," *Nutrition, Metabolism, and Cardiovascular Diseases* 24, no. 4 (April 2014); Fátima Pérez de Heredia, Sonia Gómez-Martinez, and Ascensión Marcos, "Obesity, Inflammation and the Immune System," *Proceedings of the Nutrition Society* 71, no. 2 (May 2012).

91 **observed that body fat distribution:** P. Mathieu, I. Lemieux, and J. P. Després, "Obesity, Inflammation, and Cardiovascular Risk," *Clinical Pharmacology & Therapeutics* 87, no. 4 (April 1, 2010).

92 **higher risk of early death:** Ahmad Jayedi et al., "Central Fatness and Risk of All Cause Mortality: Systematic Review and Dose-Response Meta-Analysis of 72 Prospective Cohort Studies," *British Medical Journal* 370 (Sept. 23, 2020).

92 **like the heart and its vessels:** Zdenek Matloch et al., "The Role of Inflammation in Epicardial Adipose Tissue in Heart Diseases," *Current Pharmaceutical Design* 24, no. 3 (2018); M. Iantorno et al., "Obesity, Inflammation and Endothelial Dysfunction," *Journal of Biological Regulators and Homeostatic Agents* 28, no. 2 (April–June 2014).

93 **a metabolic and an immunological disease:** See, for example, Steven E. Shoelson, Laura Herrero, and Afia Naaz, "Obesity, Inflammation, and Insulin Resistance," *Gastroenterology* 132, no. 6 (May 2007); Jongsoon Lee, "Adipose Tissue Macrophages in the Development of Obesity-Induced Inflammation, Insulin Resistance and Type 2 Diabetes," *Archives of Pharmacal Research* 36, no. 2 (Feb. 2013); Joanne C. McNelis and Jerrold M. Olefsky, "Macrophages, Immunity, and Metabolic Disease," *Immunity* 41, no. 1 (July 17, 2014); Marc Y. Donath, "Targeting Inflammation in the Treatment of Type 2 Diabetes: Time to Start," *Nature Reviews Drug Discovery* 13, no. 6 (June 2014).

93 **eating themselves to death:** Odegaard and Chawla, "Pleiotropic Actions of Insulin Resistance."

94 **Wilhelm Ebstein wrote:** Steven Shoelson, "JMM—Past and Present," *Journal of Molecular Medicine* 80 (2002).

94 **chanced upon the same finding:** Steven E. Shoelson, Jongsoon Lee, and Allison B. Goldfine, "Inflammation and Insulin Resistance," *Journal of Clinical Investigation* 116, no. 7 (July 2006).

94 **with a few exceptions:** Robert B. Supernaw, "Chapter 111— Simple Analgesics," in *Pain Management*, ed. Steven D. Waldman and Joseph I. Bloch (Philadelphia: W. B. Saunders, 2007).

95 **TNF-α disrupts insulin signaling:** Gökhan S. Hotamisligil et al., "IRS-1-Mediated Inhibition of Insulin Receptor Tyrosine Kinase Activity in TNF-αlpha- and Obesity-Induced Insulin Resistance," *Science* 271, no. 5249 (Feb. 2, 1996).

95 **found to dampen insulin secretion:** Michael L. McDaniel et al., "Cytokines and

Nitric Oxide in Islet Inflammation and Diabetes," *Proceedings of the Society for Experimental Biology and Medicine* 211, no. 1 (Jan. 1996).

95 **but also the pancreas:** K. Eguchi and I. Manabe, "Macrophages and Islet Inflammation in Type 2 Diabetes," *Diabetes, Obesity, and Metabolism: A Journal of Pharmacology and Therapeutics* 15, Suppl. 3 (Sept. 2013).

95 **much more likely to be diagnosed:** Aruna D. Pradhan et al., "C-Reactive Protein, Interleukin 6, and Risk of Developing Type 2 Diabetes Mellitus," *Journal of the American Medical Association* 286, no. 3 (July 18, 2001).

95 **some critics wonder:** Shannon M. Reilly and Alan R. Saltiel, "Adapting to Obesity with Adipose Tissue Inflammation," *Nature Reviews Endocrinology* 13, no. 11 (Nov. 2017).

96 **damaging and absolutely causal to the process:** Gokhan Hotamisligil, interview with author, Feb. 2019.

97 **feed liver disease:** Giovanni Tarantino, "Gut Microbiome, Obesity-Related Comorbidities, and Low-Grade Chronic Inflammation," *Journal of Clinical Endocrinology and Metabolism* 99, no. 7 (July 2014); Anne M. Minihane et al., "Low-Grade Inflammation, Diet Composition and Health: Current Research Evidence and Its Translation," *British Journal of Nutrition* 114, no. 7 (Oct. 14, 2015).

97 **confers an additional hefty risk:** See, for example, Andrew J. Dannenberg and Nathan A. Berger, *Obesity, Inflammation and Cancer*, 7 (New York: Springer, 2013); Ryan Kolb, Fayyaz S. Sutterwala, and Weizhou Zhang, "Obesity and Cancer: Inflammation Bridges the Two," *Current Opinion in Pharmacology* 29 (Aug. 2016); Marek Wagner, Eli Sihn Samdal Steinskog, and Helge Wiig, "Adipose Tissue Macrophages: The Inflammatory Link between Obesity and Cancer?," *Expert Opinion on Therapeutic Targets* 19, no. 4 (April 2015); Tuo Deng et al., "Obesity, Inflammation, and Cancer," *Annual Review of Pathology: Mechanisms of Disease* 11 (May 23, 2016).

Chapter 6: Gray Matter

99 **Journalists commented:** Vikhanski, *Immunity*.

100 **focus on the cytokine IL-6:** Luigi Ferrucci et al., "Serum Il-6 Level and the Development of Disability in Older Persons," *Journal of the American Geriatrics Society* 47, no. 6 (1999); Tamara B. Harris et al., "Associations of Elevated Interleukin-6 and C-Reactive Protein Levels with Mortality in the Elderly," *American Journal of Medicine* 106, no. 5 (May 1999).

101 **became increasingly important:** See, for example, Claudio Franceschi and Judith Campisi, "Chronic Inflammation (Inflammaging) and Its Potential Contribution to Age-Associated Diseases," *Journals of Gerontology: Series A* 69, Suppl. 1 (2014); Claudio Franceschi et al., "Inflammaging," in *Handbook of Immunosenescence: Basic Understanding and Clinical Implications*, ed. Tamas Fulop et al. (Cham: Springer International Publishing, 2019); Luigi Ferrucci and Elisa Fabbri, "Inflammageing: Chronic Inflammation in Ageing, Cardiovascular Disease, and Frailty," *Nature Reviews Cardiology* 15, no. 9 (Sept. 2018); Yumiko Oishi and Ichiro Manabe, "Macrophages in Age-Related Chronic Inflammatory Diseases," *NPJ Aging and Mechanisms of Disease* 2, no. 1 (2016).

101 **recognized as one of the "hallmarks of aging":** The basic hallmarks of biological aging were first published in 2013. In 2014, inflammaging was recognized as one of the hallmarks of the aging process. See Carlos López-Otín et al., "The Hallmarks of Aging," *Cell* 153, no. 6 (June 6, 2013); Brian K. Kennedy et al., "Geroscience: Linking Aging to Chronic Disease," *Cell* 159, no. 4 (Nov. 6, 2014).

102   **excites macrophages:** Claudio Franceschi et al., "Inflammaging and 'Garb-Aging'," *Trends in Endocrinology & Metabolism* 28, no. 3 (2017).

102   **can firmly impinge:** Arsun Bektas et al., "Aging, Inflammation and the Environment," *Experimental Gerontology* 105 (May 1, 2018).

102   **surrender to time:** Darren J. Baker et al., "Naturally Occurring P16ink4a-Positive Cells Shorten Healthy Lifespan," *Nature* 530, no. 7589 (Feb. 1, 2016); Francesco Prattichizzo et al., "Senescence Associated Macrophages and "Macroph-Aging": Are They Pieces of the Same Puzzle?," *Aging (Albany NY)* 8, no. 12 (Dec. 7, 2016).

103   **animal studies suggest:** Melissa L. Harris et al., "A Direct Link between Mitf, Innate Immunity, and Hair Graying," *PLOS Biology* 16, no. 5 (2018).

104   **of developing kidney disease:** Richard L. Amdur et al., "Inflammation and Progression of Ckd: The Cric Study," *Clinical Journal of the American Society of Nephrology* 11, no. 9 (Sept. 7, 2016).

104   **builds up inflammatory triggers:** Oleh M. Akchurin and Frederick Kaskel, "Update on Inflammation in Chronic Kidney Disease," *Blood Purification* 39, nos. 1–3 (2015); Dominic S. Raj, Roberto Pecoits-Filho, and Paul L. Kimmel, "Chapter 17: Inflammation in Chronic Kidney Disease," in *Chronic Renal Disease*, ed. Paul L. Kimmel and Mark E. Rosenberg (San Diego: Academic Press, 2015); Simona Mihai et al., "Inflammation-Related Mechanisms in Chronic Kidney Disease Prediction, Progression, and Outcome," *Journal of Immunology Research* 2018 (Sept. 6, 2018); Gabriela Cobo, Bengt Lindholm, and Peter Stenvinkel, "Chronic Inflammation in End-Stage Renal Disease and Dialysis," *Nephrology Dialysis Transplantation* 33, Suppl. 3 (Oct. 1, 2018).

104   **in part, an inflammatory one:** F. Berenbaum, "Osteoarthritis as an Inflammatory Disease (Osteoarthritis Is Not Osteoarthrosis!)," *Osteoarthritis Cartilage* 21, no. 1 (Jan. 2013).

104   **around 60 percent of the inflammatory cells:** Yu-sheng Li et al., "T Cells in Osteoarthritis: Alterations and Beyond," Review, *Frontiers in Immunology* 8, no. 356 (March 30, 2017).

105   **hidden inflammation upon muscles:** Sebastiaan Dalle, Lenka Rossmeislova, and Katrien Koppo, "The Role of Inflammation in Age-Related Sarcopenia," Review, *Frontiers in Physiology* 8, no. 1045 (Dec. 12, 2017); Giulia Bano et al., "Inflammation and Sarcopenia: A Systematic Review and Meta-Analysis," *Maturitas* 96 (Feb. 1, 2017).

105   **play a major role:** Chang-Yi Cui and Luigi Ferrucci, "Macrophages in Skeletal Muscle Aging," *Aging* 12, no. 1 (2020).

105   **wrote of his findings:** Alison Abbott, "Is 'Friendly Fire' in the Brain Provoking Alzheimer's Disease?," *Nature* 556 (April 26, 2018); Edward Bullmore, *The Inflamed Mind: A Radical New Approach to Depression* (New York: Picador, 2018).

106   **at least a third of cases:** Gill Livingston et al., "Dementia Prevention, Intervention, and Care," *Lancet* 390, no. 10113 (Dec .16, 2017).

106   **closely mirrored modern microglia:** Helmut Kettenmann and Alexei Verkhratsky, "Neuroglia: The 150 Years After," *Trends in Neurosciences* 31, no. 12 (2008).

106   **Heneka was incredulous:** Abbott, "Is 'Friendly Fire' in the Brain "; Michael T. Heneka et al., "Nlrp3 Is Activated in Alzheimer's Disease and Contributes to Pathology in App/Ps1 Mice," *Nature* 493, no. 7434 (Jan. 1, 2013).

106   **Nearly all are related:** Brian W. Kunkle et al., "Genetic Meta-Analysis of Diagnosed Alzheimer's Disease Identifies New Risk Loci and Implicates Aβ, Tau, Immunity and Lipid Processing," *Nature Genetics* 51, no. 3 (March 1, 2019); Rita Guerreiro et al., "Trem2 Variants in Alzheimer's Disease," *New England Journal of*

Medicine 368, no. 2 (2012); Thorlakur Jonsson et al., "Variant of Trem2 Associated with the Risk of Alzheimer's Disease," *New England Journal of Medicine* 368, no. 2 (2012).

107 **central to the link:** Julia Marschallinger, Kira Irving Mosher, and Tony Wyss-Coray, "Microglial Dysfunction in Brain Aging and Neurodegeneration," in *Handbook of Immunosenescence: Basic Understanding and Clinical Implications*, ed. Tamas Fulop et al. (London: Springer Nature, 2018).

107 **fuel the growth:** Ji-Yeun Hur et al., "The Innate Immunity Protein Ifitm3 Modulates Γ-Secretase in Alzheimer's Disease," *Nature* 586, no. 7831 (Oct. 1, 2020); Christina Ising et al., "Nlrp3 Inflammasome Activation Drives Tau Pathology," *Nature* 575, no. 7784 (Nov. 2019).

107 **brains filled with amyloid plaques:** Beatriz G. Perez-Nievas et al., "Dissecting Phenotypic Traits Linked to Human Resilience to Alzheimer's Pathology," *Brain* 136, Pt. 8 (Aug. 2013).

107 **might need inflammation:** See, for example, Robert Moir and Rudolph E. Tanzi, "The Innate Immune Protection Hypothesis of Alzheimer's Disease," *Advances in Motion* (Feb. 4, 2020); Hur et al., " Innate Immunity Protein "; Michael T Heneka et al., "Neuroinflammation in Alzheimer's Disease," *The Lancet Neurology* 14, no. 4 (2015).

108 **is linked to mental decline in later years:** Keenan A. Walker et al., "Systemic Inflammation During Midlife and Cognitive Change over 20 Years," *Neurology* 92, no. 11 (2019).

108 **smaller brain volumes:** Keenan A. Walker et al., "Midlife Systemic Inflammatory Markers Are Associated with Late-Life Brain Volume," *Neurology* 89, no. 22 (2017).

108 **cold sores or gum disease:** Angela R. Kamer et al., "Inflammation and Alzheimer's Disease: Possible Role of Periodontal Diseases," *Alzheimer's & Dementia* 4, no. 4 (July 1, 2008).

108 **buildup of tau and amyloid:** Robert A. Stern et al., "Tau Positron-Emission Tomography in Former National Football League Players," *New England Journal of Medicine* 380, no. 18 (May 2, 2019); Thor D. Stein et al., "Beta-Amyloid Deposition in Chronic Traumatic Encephalopathy," *Acta Neuropathologica* 130, no. 1 (July 1, 2015).

109 **neurodegenerative and psychiatric health:** Megan E. Renna et al., "The Association between Anxiety, Traumatic Stress, and Obsessive-Compulsive Disorders and Chronic Inflammation: A Systematic Review and Meta-Analysis," *Depression & Anxiety* 35, no. 11 (2018); Heeok Hong, Byung Sun Kim, and Heh-In Im, "Pathophysiological Role of Neuroinflammation in Neurodegenerative Diseases and Psychiatric Disorders," *International Neurourology Journal* 20, Suppl. 1 (2016).

109 **a scandalous paper:** R. S. Smith, "The Macrophage Theory of Depression," *Medical Hypotheses* 35, no. 4 (Aug. 1991).

110 **including depression:** Bullmore, *The Inflamed Mind*; Beurel, Toups, and Nemeroff, "Bidirectional Relationship of Depression and Inflammation"; Chieh-Hsin Lee and Fabrizio Giuliani, "The Role of Inflammation in Depression and Fatigue," Review, *Frontiers in Immunology* 10, no. 1696 (July 19, 2019).

110 **One large 2013 study:** Marie Kim Wium-Andersen et al., "Elevated C-Reactive Protein Levels, Psychological Distress, and Depression in 73,131 Individuals," *JAMA Psychiatry* 70, no. 2 (2013).

110 **directly affect the brain and mood:** Neil A. Harrison et al., "Inflammation Causes Mood Changes through Alterations in Subgenual Cingulate Activity and Mesolimbic Connectivity," *Biological Psychiatry* 66, no. 5 (Sept. 1, 2009); Thomas

E. Kraynak et al., "Functional Neuroanatomy of Peripheral Inflammatory Physiology: A Meta-Analysis of Human Neuroimaging Studies," *Neuroscience and Biobehavioral Reviews* 94 (Nov. 2018).

110 **of both innate and adaptive immune responses:** Eléonore Beurel, Marisa Toups, and Charles B. Nemeroff, "The Bidirectional Relationship of Depression and Inflammation: Double Trouble," *Neuron* 107, no. 2 (July 22, 2020); George M. Slavich and Michael R. Irwin, "From Stress to Inflammation and Major Depressive Disorder: A Social Signal Transduction Theory of Depression," *Psychological Bulletin* 140, no. 3 (May 2014).

111 **Even a vaccine can elicit:** Harrison et al., "Inflammation Causes Mood Changes."

112 **the most common causes of death:** David S. Jones, Scott H. Podolsky, and Jeremy A. Greene, "The Burden of Disease and the Changing Task of Medicine," *New England Journal of Medicine* 366, no. 25 (June 21, 2012).

114 **new identities with alternate names:** Rehan P. Visser, "Fernando Pessoa's Art of Living: Ironic Multiples, Multiple Ironies," *The Philosophical Forum* 50, no. 4 (Dec. 1, 2019).

114 **on his deathbed at the Hospital Sao Luis:** Carmela Ciuraru, *Nom De Plume: A (Secret) History of Pseudonyms* (New York: HarperCollins, 2011).

Chapter 7: Resolution

115 **among the first reported drugs:** Nicolâas G. Bazâan, Jack H. Botting, and John R. Vane, *New Targets in Inflammation: Inhibitors of Cox-2 or Adhesion Molecules* (Dordrecht: Springer Netherlands, 1996); J. R. Vane and R. M. Botting, "The History of Anti-Inflammatory Drugs and Their Mechanism of Action," in *New Targets in Inflammation: Inhibitors of Cox-2 or Adhesion Molecules Proceedings of a Conference Held on April 15–16, 1996, in New Orleans, USA, Supported by an Educational Grant from Boehringer Ingelheim,* ed. Nicolas Bazan, Jack Botting, and John Vane (Dordrecht: Springer Netherlands, 1996).

116 **accidentally tasted the willow bark:** Vane and Botting, "History of Anti-Inflammatory Drugs."

116 **succeeded in extracting:** Alan Jones, "Terminology and Processes Used in Drug Manufacture," in *Chemistry: An Introduction for Medical and Health Sciences* (Chichester, UK: John Wiley & Sons, 2005).

116 **possibly working under:** Walter Sneader, "The Discovery of Aspirin: A Reappraisal," *British Medical Journal* 321, no. 7276 (Dec. 23–30, 2000).

117 **thousands of Americans die:** Michael Fine, "Quantifying the Impact of Nsaid-Associated Adverse Events," *American Journal of Managed Care* 19, no. 14, Suppl. (Nov. 2013).

117 **hailed as miracle drugs:** Kay Brune and Burkhard Hinz, "The Discovery and Development of Antiinflammatory Drugs," *Arthritis & Rheumatism* 50, no. 8 (2004).

118 **named the hypothetical drug *Zauberkugel*, or "magic bullet":** Frank Heynick, "The Original 'Magic Bullet' Is 100 Years Old," *British Journal of Psychiatry* 195, no. 5 (2009).

123 **is indeed an active process:** See, for example, James N. Fullerton and Derek W. Gilroy, "Resolution of Inflammation: A New Therapeutic Frontier," *Nature Reviews Drug Discovery* 15, no. 8 (Aug. 1, 2016); I. Tabas and C. K. Glass, "Anti-Inflammatory Therapy in Chronic Disease: Challenges and Opportunities," *Science* 339, no. 6116 (Jan. 11, 2013).

123 **produce most of the specialized pro-resolving mediators:** Charles Serhan, interview with author, Feb. 2019.
125 **as one scientist said:** Caroline Williams, "How to Extinguish the Inflammation Epidemic," *The New Scientist* (June 14, 2017).
125 **modifies its function:** Charles N. Serhan, Stephanie Yacoubian, and Rong Yang, "Anti-Inflammatory and Proresolving Lipid Mediators," *Annual Review of Pathology: Mechanisms of Disease* 3 (2008).
126 **in response to inflammatory challenges:** Minihane et al., "Low-Grade Inflammation."
128 **cannot alone explain:** Stephen M. Rappaport, "Genetic Factors Are Not the Major Causes of Chronic Diseases," *PLOS ONE* 11, no. 4 (2016).
128 **are a product of modern life:** Grivennikov, Greten, and Karin, "Immunity, Inflammation, and Cancer"; Haijiang Dai et al., "Global, Regional, and National Burden of Ischaemic Heart Disease and Its Attributable Risk Factors, 1990–2017: Results from the Global Burden of Disease Study 2017," *European Heart Journal— Quality of Care and Clinical Outcomes* (2020).
129 **the deadliest factors:** Carine Lenders et al., "A Novel Nutrition Medicine Education Model: The Boston University Experience," *Advances in Nutrition* 4, no. 1 (Jan. 1, 2013); GBD 2017 Diet Collaborators, "Health Effects of Dietary Risks in 195 Countries, 1990–2017: A Systematic Analysis for the Global Burden of Disease Study 2017," *Lancet* 393, no. 10184 (May 11, 2019); U.S. Burden of Disease Collaborators, "The State of US Health, 1990–2016: Burden of Diseases, Injuries, and Risk Factors among US States," *Journal of the American Medical Association* 319, no. 14 (2018).

Chapter 8: Quiet Conversations

133 **prepared to fight food as it would a germ:** See, for example, Marc Veldhoen and Verena Brucklacher-Waldert, "Dietary Influences on Intestinal Immunity," *Nature Reviews Immunology* 12, no. 10 (Oct. 2012); Ling Zhao, Joo Y. Lee, and Daniel H. Hwang, "Inhibition of Pattern Recognition Receptor-Mediated Inflammation by Bioactive Phytochemicals," *Nutrition Reviews* 69, no. 6 (June 2011); Lili Yu et al., "Pattern Recognition Receptor-Mediated Chronic Inflammation in the Development and Progression of Obesity-Related Metabolic Diseases," *Mediators of Inflammation* 2019 (2019).
133 **almost all cells:** Daniel M. Davis, *The Beautiful Cure: The Revolution in Immunology and What It Means for Your Health* (Chicago: University of Chicago Press, 2018).
133 **also by noninfectious matter:** Tao Gong et al., "Damp-Sensing Receptors in Sterile Inflammation and Inflammatory Diseases," *Nature Reviews Immunology* 20, no. 2 (Feb. 2020).
134 **"The flora of the human stomach":** Vikhanski, *Immunity.*
134 **"The idea is to define precisely":** Vikhanski, *Immunity.*
134 **"The intestinal flora," he wrote:** Metchnikoff, *Life of Elie Metchnikoff, 1845–1916.*
136 **contain the largest reservoir:** Lesley E. Smythies, Larry M. Wahl, and Phillip D. Smith, "Isolation and Purification of Human Intestinal Macrophages," *Current Protocols in Immunology* 70 (Jan. 2006).
137 **are trained to tolerate:** Lesley E. Smythies et al., "Human Intestinal Macrophages Display Profound Inflammatory Anergy Despite Avid Phagocytic and Bacteriocidal Activity," *Journal of Clinical Investigation* 115, no. 1 (Jan. 3, 2005).

138 **showed that a common gut bacterium:** S. K. Mazmanian et al., "An Immunomodulatory Molecule of Symbiotic Bacteria Directs Maturation of the Host Immune System," *Cell* 122, no. 1 (July 15, 2005).

138 **dramatically increased:** Ivaylo I. Ivanov et al., "Induction of Intestinal Th17 Cells by Segmented Filamentous Bacteria," *Cell* 139, no. 3 (Oct. 30, 2009).

138 **is one of the underlying causes:** Keiko Yasuda et al., "Satb1 Regulates the Effector Program of Encephalitogenic Tissue Th17 Cells in Chronic Inflammation," *Nature Communications* 10, no. 1 (2019).

139 **the first human study of its kind:** Jonas Schluter et al., "The Gut Microbiota Is Associated with Immune Cell Dynamics in Humans," *Nature* 588, no. 7837 (Dec. 1, 2020).

140 **calibrate the immune system:** Jun Sun and Ikuko Kato, "Gut Microbiota, Inflammation and Colorectal Cancer," *Genes & Diseases* 3, no. 2 (June 1, 2016).

140 **first fecal transplant:** B. Eiseman et al., "Fecal Enema as an Adjunct in the Treatment of Pseudomembranous Enterocolitis," *Surgery* 44, no. 5 (Nov. 1958).

141 **transferred gut microbes:** Fredrik Bäckhed et al., "The Gut Microbiota as an Environmental Factor That Regulates Fat Storage," *Proceedings of the National Academy of Sciences of the USA* 101, no. 44 (2004).

141 **infused them into skinny, germ-free mice:** K. Ridaura Vanessa et al., "Gut Microbiota from Twins Discordant for Obesity Modulate Metabolism in Mice," *Science* 341, no. 6150 (Sept. 6, 2013).

141 **often confined to describing microbiomes:** H. Renz et al., "An Exposome Perspective: Early-Life Events and Immune Development in a Changing World," *Journal of Allergy and Clinical Immunology* 140, no. 1 (July 2017).

142 **part of a sinister loop:** H. K. Somineni and S. Kugathasan, "The Microbiome in Patients with Inflammatory Diseases," *Clinical Gastroenterology and Hepatology* 17, no. 2 (Jan. 2019).

142 **one important mechanism:** See, for example, Benoit Chassaing and Andrew T. Gewirtz, "Gut Microbiota, Low-Grade Inflammation, and Metabolic Syndrome," *Toxicologic Pathology* 42, no. 1 (Jan. 2014).

142 **plays an important role:** Judith Aron-Wisnewsky et al., "Metabolism and Metabolic Disorders and the Microbiome: The Intestinal Microbiota Associated with Obesity, Lipid Metabolism, and Metabolic Health—Pathophysiology and Therapeutic Strategies," *Gastroenterology* 160, no. 2 (2021).

143 **helps to regulate the immune system:** Martin J. Blaser, *Missing Microbes: How the Overuse of Antibiotics Is Fueling Our Modern Plagues* (New York: Holt, 2014).

143 **even appears to protect:** Fergus Shanahan, Tarini S. Ghosh, and Paul W. O'Toole, "The Healthy Microbiome—What Is the Definition of a Healthy Gut Microbiome?" *Gastroenterology* 160, no. 2 (Jan. 2021).

145 **may be filled with metabolites:** Stephan J. Ott et al., "Efficacy of Sterile Fecal Filtrate Transfer for Treating Patients with *Clostridium difficile* Infection," *Gastroenterology* 152, no. 4 (March 1, 2017); D. H. Kao et al., "A51 Effect of Lyophilized Sterile Fecal Filtrate Vs Lyophilized Donor Stool on Recurrent *Clostridium difficile* Infection (Rcdi): Preliminary Results from a Randomized, Double-Blind Pilot Study," *Journal of the Canadian Association of Gastroenterology* 2, Suppl. 2 (2019).

145 **on a high-fat diet:** Patrice D. Cani et al., "Metabolic Endotoxemia Initiates Obesity and Insulin Resistance," *Diabetes* 56, no. 7 (2007).

146 **a similar experiment conducted in humans:** Yi Wan et al., "Effects of Dietary Fat on Gut Microbiota and Faecal Metabolites, and Their Relationship with Cardiometabolic Risk Factors: A 6-Month Randomised Controlled-Feeding Trial," *Gut* 68, no. 8 (2019).

Chapter 9: Fat Wars

150 **only space available:** Henry Blackburn, email to author, May 2021.

150 **a select group of Minneapolis men:** Ancel Keys et al., "Coronary Heart Disease among Minnesota Business and Professional Men Followed Fifteen Years," *Circulation* 28 (1963). Although the Minnesota Business and Professional Men's Study was launched in 1947 as one of the first prospective studies of coronary heart disease and followed up five hundred men annually for fifteen years, this article was the study's first of several subsequent follow-up study publications.

151 **at a frightening rate by mid-century:** David S. Jones and Jeremy A. Greene, "The Decline and Rise of Coronary Heart Disease: Understanding Public Health Catastrophism," *American Journal of Public Health* 103, no. 7 (July 2013).

151 **"We felt warm all over":** Ancel Keys and Margaret H. Keys, *Eat Well and Stay Well* (Garden City, NY: Doubleday, 1963).

152 **"never more than a few hours from the oven":** Keys and Keys, *Eat Well and Stay Well.*

153 **"the prolonged influence of the war and its aftermath":** Katherine D. Pett et al., *Ancel Keys and the Seven Countries Study: An Evidence-Based Response to Revisionist Histories* (Tulsa, OK: The True Health Initiative, 2017).

153 **led to planning:** Henry Blackburn, email to author, May 2021.

154 **"of the highest caliber":** Pett et al., *Ancel Keys and the Seven Countries Study.*

154 **provided an account:** Ancel Keys et al., "The Diet and 15-Year Death Rate in the Seven Countries Study," *American Journal of Epidemiology* 124, no. 6 (1986).

154 **as Henry Blackburn writes:** Henry Blackburn, "Invited Commentary: 30-Year Perspective on the Seven Countries Study," *American Journal of Epidemiology* 185, no. 11 (June 1, 2017).

155 **"The fat in the land":** Keys and Keys, *Eat Well and Stay Well.*

155 **"The important peculiarity of the American diet":** Keys and Keys, *Eat Well and Stay Well.*

156 **first five- and ten-year follow-up data:** "Study Findings," accessed November 8, 2021, https://www.sevencountriesstudy.com/study-findings/.

156 **an "important and remarkable substance":** Keys and Keys, *Eat Well and Stay Well.*

156 **"for the purposes of controlling the serum level":** A. Keys, J. T. Anderson, and F. Grande, "Serum Cholesterol Response to Changes in the Diet: II. The Effect of Cholesterol in the Diet," *Metabolism* 14, no. 7 (July 1965).

157 **the Seven Countries Study faced limitations:** Pett et al., *Ancel Keys and the Seven Countries Study.*

157 **can dictate the behavior of macrophages and other immune cells:** See, for example, Jose E. Galgani and Diego García, "Chapter 25: Role of Saturated and Polyunsaturated Fat in Obesity-Related Inflammation," in *Inflammation, Advancing Age and Nutrition*, ed. Irfan Rahman and Debasis Bagchi (San Diego: Academic Press, 2014); Robert Caesar et al., "Crosstalk between Gut Microbiota and Dietary Lipids Aggravates WAT Inflammation through Tlr Signaling," *Cell Metabolism* 22, no. 4 (2015); Yan Y. Lam et al., "Effects of Dietary Fat Profile on Gut Permeability and Microbiota and Their Relationships with Metabolic Changes in Mice," *Obesity* 23, no. 7 (July 1, 2015); Christopher K. Glass and Jerrold M. Olefsky, "Inflammation and Lipid Signaling in the Etiology of Insulin Resistance," *Cell Metabolism* 15 (May 2, 2012).

157 **saturated fats activate:** See, for example, C. Lawrence Kien et al., "Lipidomic Evidence That Lowering the Typical Dietary Palmitate to Oleate Ratio in Humans Decreases the Leukocyte Production of Proinflammatory Cytokines and Muscle

Expression of Redox-Sensitive Genes," *Journal of Nutritional Biochemistry* 26, no. 12 (Dec. 1, 2015); David L. Katz, Rachel S. C. Friedman, and Sean C. Lucan, *Nutrition in Clinical Practice: A Comprehensive, Evidence-Based Manual for the Practitioner* (Philadelphia: Wolters Kluwer, 2015); Rupali Deopurkar et al., "Differential Effects of Cream, Glucose, and Orange Juice on Inflammation, Endotoxin, and the Expression of Toll-Like Receptor-4 and Suppressor of Cytokine Signaling-3," *Diabetes Care* 33, no. 5 (2010).

157   **drive macrophages to inflame:** Megan M. Robblee et al., "Saturated Fatty Acids Engage an Ire1–Dependent Pathway to Activate the Nlrp3 Inflammasome in Myeloid Cells," *Cell Reports* 14, no. 11 (2016); Suzanne Devkota et al., "Dietary-Fat-Induced Taurocholic Acid Promotes Pathobiont Expansion and Colitis in Il-10 Mice," *Nature* 487, no. 7405 (July 1, 2012); Haitao Wen et al., "Fatty Acid–Induced Nlrp3-Asc Inflammasome Activation Interferes with Insulin Signaling," *Nature Immunology* 12, no. 5 (May 1, 2011).

157   **rendering it dysfunctional and inflammatory:** Stephen J. Nicholls et al., "Consumption of Saturated Fat Impairs the Anti-Inflammatory Properties of High-Density Lipoproteins and Endothelial Function," *Journal of the American College of Cardiology* 48, no. 4 (Aug. 15, 2006).

158   **breeds inflammatory bacteria:** Justin Sonnenburg and Erica Sonnenburg, *The Good Gut: Taking Control of Your Weight, Your Mood, and Your Long-Term Health* (New York: Penguin, 2016); Melisa A. Bailey and Hannah D. Holscher, "Microbiome-Mediated Effects of the Mediterranean Diet on Inflammation," *Advances in Nutrition* 9, no. 3 (2018); Tien S. Dong and Arpana Gupta, "Influence of Early Life, Diet, and the Environment on the Microbiome," *Clinical Gastroenterology and Hepatology* 17, no. 2 (Jan. 2019); Suzanne Devkota and Eugene B. Chang, "Nutrition, Microbiomes, and Intestinal Inflammation," *Current Opinion in Gastroenterology* 29, no. 6 (2013).

158   **is less incendiary:** David L. Katz, Kim Doughty, and Ather Ali, "Cocoa and Chocolate in Human Health and Disease," *Antioxidants & Redox Signaling* 15, no. 10 (Nov. 15, 2011).

158   **helped to confirm this notion:** Frank M. Sacks et al., "Dietary Fats and Cardiovascular Disease: A Presidential Advisory from the American Heart Association," *Circulation* 136, no. 3 (July 18, 2017).

159   **In 1980, Willett began collecting data:** Walter Willett, interview with author, Feb. 2020.

159   **biggest epidemiological undertakings:** "Nurses' Health Study," accessed Nov. 8, 2021, https://nurseshealthstudy.org/.

159   **captures the most attention:** See, for example, Cristina Nocella et al., "Extra Virgin Olive Oil and Cardiovascular Diseases: Benefits for Human Health," *Endocrine, Metabolic & Immune Disorders—Drug Targets* 18, no. 1 (2018); Lukas Schwingshackl, Marina Christoph, and Georg Hoffmann, "Effects of Olive Oil on Markers of Inflammation and Endothelial Function—A Systematic Review and Meta-Analysis," *Nutrients* 7, no. 9 (Sept. 11, 2015).

159   **against many chronic inflammatory diseases:** Cheng Luo et al., "Nut Consumption and Risk of Type 2 Diabetes, Cardiovascular Disease, and All-Cause Mortality: A Systematic Review and Meta-Analysis," *American Journal of Clinical Nutrition* 100, no. 1 (2014); Marta Guasch-Ferré et al., "Frequency of Nut Consumption and Mortality Risk in the Predimed Nutrition Intervention Trial," *BMC Medicine* 11 (July 16, 2013); Cyril W. C. Kendall et al., "Nuts, Metabolic Syndrome and Diabetes," *British Journal of Nutrition* 104, no. 4 (2010).

160   **lower cholesterol levels and inflammation:** Emilio Ros, "Nuts and Novel Bio-

markers of Cardiovascular Disease," *American Journal of Clinical Nutrition* 89, no. 5 (May 2009); Bamini Gopinath et al., "Consumption of Polyunsaturated Fatty Acids, Fish, and Nuts and Risk of Inflammatory Disease Mortality," *American Journal of Clinical Nutrition* 93, no. 5 (2011); Zhi Yu et al., "Associations between Nut Consumption and Inflammatory Biomarkers," *American Journal of Clinical Nutrition* 104, no. 3 (2016).

160 **contribute to problems:** Yu-Shian Cheng et al., "Supplementation of Omega 3 Fatty Acids May Improve Hyperactivity, Lethargy, and Stereotypy in Children with Autism Spectrum Disorders: A Meta-Analysis of Randomized Controlled Trials," *Neuropsychiatric Disease and Treatment* 13 (2017); Alexandra J. Richardson, "Omega-3 Fatty Acids in ADHD and Related Neurodevelopmental Disorders," *International Review of Psychiatry* 18, no. 2 (April 2006).

160 **prevent heart attacks, strokes, and even death:** Deepak L. Bhatt et al., "Reduce-It USA: Results from the 3146 Patients Randomized in the United States," *Circulation* 141, no. 5 (Feb. 4, 2020).

160 **shrink atherosclerotic plaques:** Matthew J. Budoff et al., "Effect of Icosapent Ethyl on Progression of Coronary Atherosclerosis in Patients with Elevated Triglycerides on Statin Therapy: Final Results of the Evaporate Trial," *European Heart Journal* 41, no. 40 (Oct. 21, 2020).

161 **potent effects on the immune system:** See, for example, Janice K. Kiecolt-Glaser et al., "Omega-3 Supplementation Lowers Inflammation and Anxiety in Medical Students: A Randomized Controlled Trial," *Brain, Behavior, and Immunity* 25, no. 8 (Nov. 2011); Artemis P. Simopoulos, "Omega-3 Fatty Acids in Inflammation and Autoimmune Diseases," *Journal of the American College of Nutrition* 21, no. 6 (Dec. 2002); Phillip C. Calder, "Omega-3 Fatty Acids and Inflammatory Processes: From Molecules to Man," *Biochemical Society Transactions* 45, no. 5 (Oct. 15, 2017); Seyedeh Parisa Moosavian et al., "The Effect of Omega-3 and Vitamin E on Oxidative Stress and Inflammation: Systematic Review and Meta-Analysis of Randomized Controlled Trials," *International Journal for Vitamin and Nutrition Research* 90, no. 5–6 (Oct. 2020).

162 **promote microbial diversity:** Henry Watson et al., "A Randomised Trial of the Effect of Omega-3 Polyunsaturated Fatty Acid Supplements on the Human Intestinal Microbiota," *Gut* 67, no. 11 (2018); Mingyang Song and Andrew T. Chan, "Environmental Factors, Gut Microbiota, and Colorectal Cancer Prevention," *Clinical Gastroenterology and Hepatology* 17, no. 2 (Jan. 2019).

162 **adapted from the phrase "crystallized cottonseed oil":** Gary List and Michael Jackson, "Giants of the Past: The Battle over Hydrogenation " *Inform* 18 (2007).

163 **"déclassé Southern aristocrat with a boyish charm.":** Lauren Coodley, *Upton Sinclair: California Socialist, Celebrity Intellectual* (Lincoln: University of Nebraska Press, 2013).

163 **including "knockers," "rippers," "leg breakers," and "gutters.":** Constitutional Rights Foundation, "Upton Sinclair's *the Jungle*: Muckraking the Meat-Packing Industry," *Bill of Rights in Action* 24, no. 1 (2008).

163 **still fed into the assembly lines:** Coodley, *Upton Sinclair.*

163 **"there were things that went into the sausage":** Upton Sinclair, *The Jungle*, ed. Harold Bloom (New York: Chelsea House , 2002).

163 **"overlooked for days, till all but the bones of them":** Sinclair, *The Jungle.*

163 **"It seemed to me I was confronting a veritable fortress":** Upton Sinclair, *The Autobiography of Upton Sinclair* (New York: Harcourt Brace and World, 1962).

163 **"It is alive and warm":** Sinclair, *The Jungle.*

164 **"artifact of a culture in the making.":** Susan Strasser, *Satisfaction Guaranteed: The Making of the American Mass Market* (Washington, DC: Smithsonian Books, 2004).

164 **"America has been termed a country of dyspeptics":** Procter & Gamble Co., *The Story of Crisco* (Cincinnatti, OH: Procter & Gamble , 1913).

165 **as research on the role of diet:** Dariush Mozaffarian, Irwin Rosenberg, and Ricardo Uauy, "History of Modern Nutrition Science—Implications for Current Research, Dietary Guidelines, and Food Policy," *British Medical Journal* 361 (2018).

165 **are tied to chronic, low-level inflammation:** Fred A. Kummerow, "The Negative Effects of Hydrogenated Trans Fats and What To Do about Them," *Atherosclerosis* 205 (2009).

166 **inflame the endothelial cells:** Naomi G. Iwata et al., "Trans Fatty Acids Induce Vascular Inflammation and Reduce Vascular Nitric Oxide Production in Endothelial Cells," *PLOS ONE* 6, no. 12 (2011).

Chapter 10: *Sweet, Salty, Deadly*

168 **John Harvey Kellogg was the fifth son:** Howard Markel, *The Kelloggs: The Battling Brothers of Battle Creek* (New York: Vintage Books, 2018); Richard W. Schwarz, *John Harvey Kellogg, M.D.: Pioneering Health Reformer* (Hagerstown, MD: Review and Herald Publishing Association, 2006); John Harvey Kellogg, *Shall We Slay to Eat?* (Battle Creek, MI: Good Health Publishing Company, 1906).

169 **"Hot beefsteak," food historian Abigail Carroll wrote:** Abigail Carroll, *Three Squares: The Invention of the American Meal* (New York: Basic Books, 2013).

169 **"the great American evil":** Walt Whitman, "Manly Health and Training," *New York Atlas*, Sept. 26, 1858.

170 **among the top killers of humankind:** Jones, Podolsky, and Greene, "The Burden of Disease."

170 **"When I was a boy, we knew nothing about diet":** Markel, *The Kelloggs: The Battling Brothers.*

170 **"Eat what the monkey eats":** Kellogg, *Shall We Slay to Eat?*

171 **potential to alter immune function:** See, for example, Xin Zhou et al., "Variation in Dietary Salt Intake Induces Coordinated Dynamics of Monocyte Subsets and Monocyte-Platelet Aggregates in Humans: Implications in End Organ Inflammation," *PLOS One* 8, no. 4 (2013); Johanna Sigaux et al., "Salt, Inflammatory Joint Disease, and Autoimmunity," *Joint Bone Spine* 85, no. 4 (July 1, 2018); Markus Kleinewietfeld et al., "Sodium Chloride Drives Autoimmune Disease by the Induction of Pathogenic Th17 Cells," *Nature* 496, no. 7446 (April 25, 2013).

171 **disables Tregs:** Tomokazu Sumida et al., "Activated B-Catenin in Foxp3 Regulatory T Cells Links Inflammatory Environments to Autoimmunity," *Nature Immunology* 19, no. 12 (Dec. 2018).

172 **hypertension is, in part, an inflammatory disease:** Jason D. Foss, Annet Kirabo, and David G. Harrison, "Do High-Salt Microenvironments Drive Hypertensive Inflammation?," *American Journal of Physiology—Regulatory, Integrative and Comparative Physiology* 312, no. 1 (Jan. 2017): R1–R4; Natalia R. Barbaro et al., "Dendritic Cell Amiloride-Sensitive Channels Mediate Sodium-Induced Inflammation and Hypertension," *Cell Reports* 21, no. 4 (Oct. 24, 2017).

172 **is implicated in arterial stiffening:** Panagiota Pietri and Christodoulos Stefanadis, "Cardiovascular Aging and Longevity: JACC State-of-the-Art Review," *Journal of the American College of Cardiology* 77, no. 2 (Jan. 19, 2021).

172 **as one researcher found:** Julia Matthias et al., "Sodium Chloride Is an Ionic Checkpoint for Human Th2 Cells and Shapes the Atopic Skin Microenvironment," *Science Translational Medicine* 11, no. 480 (Feb. 20, 2019).

173 **tasted like "cold roast mutton":** William Shurtleff and Akiko Aoyagi, *History of Meat Alternatives (965 CE to 2014): Extensively Annotated Bibliography and Sourcebook* (Lafayette, CA: Soyinfo Center, 2014).

174 **"There is no physiological requirement for sugar":** John Yudkin, *Pure, White and Deadly: How Sugar Is Killing Us and What We Can Do to Stop It* (New York: Viking, 2012).

174 **came under attack:** Harvey Levenstein, *Fear of Food: A History of Why We Worry about What We Eat* (Chicago: University of Chicago Press, 2012).

176 **stressing the body:** Scott Dickinson et al., "High-Glycemic Index Carbohydrate Increases Nuclear Factor-Kappab Activation in Mononuclear Cells of Young, Lean Healthy Subjects," *American Journal of Clinical Nutrition* 87, no. 5 (May 2008); Katherine Esposito and Dario Giugliano, "Diet and Inflammation: A Link to Metabolic and Cardiovascular Diseases," *European Heart Journal* 27, no. 1 (2006).

176 **converts dietary carbohydrates to fatty acids:** Thomas Jensen et al., "Fructose and Sugar: A Major Mediator of Non-Alcoholic Fatty Liver Disease," *Journal of Hepatology* 68, no. 5 (May 1, 2018).

177 **heighten our risk:** Quanhe Yang et al., "Added Sugar Intake and Cardiovascular Diseases Mortality among US Adults," *JAMA Internal Medicine* 174, no. 4 (April 2014).

177 **daily can of soda:** Isabelle Aeberli et al., "Low to Moderate Sugar-Sweetened Beverage Consumption Impairs Glucose and Lipid Metabolism and Promotes Inflammation in Healthy Young Men: A Randomized Controlled Trial," *American Journal of Clinical Nutrition* 94, no. 2 (Aug. 2011); J. M. Bruun et al., "Consumption of Sucrose-Sweetened Soft Drinks Increases Plasma Levels of Uric Acid in Overweight and Obese Subjects: A 6-Month Randomised Controlled Trial," *European Journal of Clinical Nutrition* 69, no. 8 (Aug. 2015).

177 **consumed with impunity:** Allan S. Christensen et al., "Effect of Fruit Restriction on Glycemic Control in Patients with Type 2 Diabetes—A Randomized Trial," *Nutrition Journal* 12 (March 5, 2013); B. J. Meyer et al., "Some Biochemical Effects of a Mainly Fruit Diet in Man," *South African Medical Journal* 45, no. 10 (March 6, 1971); David J. A. Jenkins et al., "Effect of a Very-High-Fiber Vegetable, Fruit, and Nut Diet on Serum Lipids and Colonic Function," *Metabolism* 50, no. 4 (April 2001).

Chapter 11: Feeding Germs

178 **had developed unruly features:** Yi Rang Na et al., "Macrophages in Intestinal Inflammation and Resolution: A Potential Therapeutic Target in IBD," *Nature Reviews Gastroenterology & Hepatology* 16, no. 9 (Sept. 1, 2019).

179 **"What is the difference":** Markel, *The Kelloggs.*

179 **is also an important inducer:** Alessandra Geremia et al., "Innate and Adaptive Immunity in Inflammatory Bowel Disease," *Autoimmunity Reviews* 13, no. 1 (Jan. 2014).

180 **"it is no wonder the human gastric machine":** John Harvey Kellogg, *The New Dietetics: What to Eat and How: A Guide to Scientific Feeding in Health and Disease* (Battle Creek, MI: Modern Medicine Publishing, 1923).

180 **could harm the body:** Karen Windey, Vicky De Preter, and Kristin Verbeke, "Relevance of Protein Fermentation to Gut Health," *Molecular Nutrition & Food Research* 56, no. 1 (Jan. 2012); Stephen J. D. O'Keefe et al., "Fat, Fibre and Cancer Risk in Afri-

can Americans and Rural Africans," *Nature Communications* 6 (2015); Devkota and Chang, "Nutrition, Microbiomes, and Intestinal Inflammation."

181 **all kinds of human studies:** Patricia Lopez-Legarrea et al., "The Protein Type within a Hypocaloric Diet Affects Obesity-Related Inflammation: The Resmena Project," *Nutrition* 30, no. 4 (April 2014); Monique van Nielen et al., "Dietary Protein Intake and Incidence of Type 2 Diabetes in Europe: The Epic-Interact Case-Cohort Study," *Diabetes Care* 37, no. 7 (July 2014); Nathalie Bergeron et al., "Effects of Red Meat, White Meat, and Nonmeat Protein Sources on Atherogenic Lipoprotein Measures in the Context of Low Compared with High Saturated Fat Intake: A Randomized Controlled Trial," *The American Journal of Clinical Nutrition* 110, no. 1 (2019); Heli E. K. Virtanen et al., "Dietary Proteins and Protein Sources and Risk of Death: The Kuopio Ischaemic Heart Disease Risk Factor Study," *American Journal of Clinical Nutrition* 109, no. 5 (2019).

182 **including hydrogen sulfide:** Jose C. Clemente, Julia Manasson, and Jose U. Scher, "The Role of the Gut Microbiome in Systemic Inflammatory Disease," *British Medical Journal* 360 (2018); Song and Chan, "Environmental Factors, Gut Microbiota"; E. Magee, "A Nutritional Component to Inflammatory Bowel Disease: The Contribution of Meat to Fecal Sulfide Excretion," *Nutrition* 15, no. 3 (March 1999); S. L. Jowett et al., "Influence of Dietary Factors on the Clinical Course of Ulcerative Colitis: A Prospective Cohort Study," *Gut* 53, no. 10 (Oct. 2004).

182 **continue to point out:** Dong and Gupta, "Influence of Early Life."

182 **a sugar in red meat:** Maria Hedlund et al., "Evidence for a Human-Specific Mechanism for Diet and Antibody-Mediated Inflammation in Carcinoma Progression," *Proceedings of the National Academy of Sciences of the USA* 105, no. 48 (Dec. 2, 2008).

182 **Or heme iron:** Lu Qi et al., "Heme Iron from Diet as a Risk Factor for Coronary Heart Disease in Women with Type 2 Diabetes," *Diabetes Care* 30, no. 1 (2007).

183 **and found massive levels:** Clett Erridge, "The Capacity of Foodstuffs to Induce Innate Immune Activation of Human Monocytes in Vitro Is Dependent on Food Content of Stimulants of Toll-Like Receptors 2 and 4," *British Journal of Nutrition* 105, no. 1 (2011).

183 **emit enough endotoxins to inflame the whole body:** Robert A. Vogel, Mary C. Corretti, and Gary D. Plotnick, "Effect of a Single High-Fat Meal on Endothelial Function in Healthy Subjects," *American Journal of Cardiology* 79, no. 3 (Feb. 1, 1997).

183 **where he began a medical practice:** Stephen O'Keefe, email to author, May 2021.

183 **who had made similar observations:** Brian Kellock, *Fiber Man: The Life Story of Dr. Denis Burkit* (Tring: Lion Publishing, 1985).

185 **enrolled twenty African Americans:** O'Keefe et al., "Fat, Fibre and Cancer Risk."

185 **including one comparing children:** Carlotta De Filippo et al., "Impact of Diet in Shaping Gut Microbiota Revealed by a Comparative Study in Children from Europe and Rural Africa," *Proceedings of the National Academy of Sciences of the USA* 107, no. 33 (2010); H. L. Simpson and B. J. Campbell, "Review Article: Dietary Fibre-Microbiota Interactions," *Alimentary Pharmacology and Therapeutics* 42 (2015).

186 **on the innate and adaptive:** Melanie Schirmer et al., "Linking the Human Gut Microbiome to Inflammatory Cytokine Production Capacity," *Cell* 167, no. 4 (2016); Lisa Rizzetto et al., "Connecting the Immune System, Systemic Chronic Inflammation and the Gut Microbiome: The Role of Sex," *Journal of Autoimmunity* 92 (Aug. 2018).

186   **the ability of fat cells to take up fatty acids:** Rizzetto et al., "Connecting the Immune System."

187   **induce Tregs:** Herbert Tilg and Alexander R. Moschen, "Food, Immunity, and the Microbiome," *Gastroenterology* 148, no. 6 (May 2015); James L. Richards et al., "Dietary Metabolites and the Gut Microbiota: An Alternative Approach to Control Inflammatory and Autoimmune Diseases," *Clinical & Translational Immunology* 5, no. 5 (May 2016).

187   **help to breed calm macrophages:** See, for example, Shuai Wang et al., "Functions of Macrophages in the Maintenance of Intestinal Homeostasis," *Journal of Immunology Research* 2019 (March 2019); Na et al., "Macrophages in Intestinal Inflammation and Resolution"; Song and Chan, "Environmental Factors, Gut Microbiota"; Hideo Ohira, Wao Tsutsui, and Yoshio Fujioka, "Are Short Chain Fatty Acids in Gut Microbiota Defensive Players for Inflammation and Atherosclerosis?," *Journal of Atherosclerosis and Thrombosis* 24, no. 7 (2017).

187   **weave through the placenta:** Yu Anne Yap and Eliana Mariño, "An Insight into the Intestinal Web of Mucosal Immunity, Microbiota, and Diet in Inflammation," *Frontiers in Immunology* 9 (2018).

188   **inflammation within the gut:** Huawei Zeng et al., "Secondary Bile Acids and Short Chain Fatty Acids in the Colon: A Focus on Colonic Microbiome, Cell Proliferation, Inflammation, and Cancer," *International Journal of Molecular Sciences* 20, no. 5 (2019).

188   **correlate with the size:** J. E. Park et al., "Differential Effect of Short-Term Popular Diets on TMAO and Other Cardio-Metabolic Risk Markers," *Nutrition, Metabolism and Cardiovascular Diseases* 29, no. 5 (May 1, 2019).

188   **processed mostly plants:** Joanna E. Lambert, "Primate Nutritional Ecology: Feeding Biology and Diet at Ecological and Evolutionary Scales," in *Primates in Perspective*, ed. Christina Campbell et al. (Oxford: Oxford University Press, 2010); Joanna E. Lambert, "Primate Digestion: Interactions among Anatomy, Physiology, and Feeding Ecology," *Evolutionary Anthropology: Issues, News, and Reviews* 7, no. 1 (Jan. 1, 1998).

188   **but subsisted largely on leaves:** Peter S. Ungar, *Evolution of the Human Diet: The Known, the Unknown, and the Unknowable*, Human Evolution Series (Oxford: Oxford University Press, 2007); Lisa Ringhofer, *Fishing, Foraging and Farming in the Bolivian Amazon: On a Local Society in Transition* (Dordrecht: Springer, 2010).

189   **also feasted largely on plant foods:** Ringhofer, *Fishing, Foraging and Farming*; H. J. Challa, M. Bandlamudi, and K. R. Uppaluri, "Paleolithic Diet," in *Statpearls* (2021).

189   **fossilized dental plaques:** Jessica Hendy et al., "Proteomic Evidence of Dietary Sources in Ancient Dental Calculus," *Proceedings of the Royal Society B: Biological Sciences* 285, no. 1883 (July 25, 2018).

189   **triggers less inflammation:** Fatemeh Arya et al., "Differences in Postprandial Inflammatory Responses to a 'Modern' V. Traditional Meat Meal: A Preliminary Study," *British Journal of Nutrition* 104, no. 5 (Sept. 2010).

189   **are particularly insidious:** Shivam Joshi, Robert J. Ostfeld, and Michelle McMacken, "The Ketogenic Diet for Obesity and Diabetes—Enthusiasm Outpaces Evidence," *JAMA Internal Medicine* 179, no. 9 (2019); Steven R. Smith, "A Look at the Low-Carbohydrate Diet," *New England Journal of Medicine* 361 (Dec. 3, 2009); Ilana M. Bank et al., "Sudden Cardiac Death in Association with the Ketogenic Diet," *Pediatric Neurology* 39, no. 6 (Dec. 1, 2008); Beth Zupec-Kania and Mary L. Zupanc, "Long-Term Management of the Ketogenic Diet: Seizure

Monitoring, Nutrition, and Supplementation," *Epilepsia* 49, Suppl. 8 (Nov. 2008).

190   **are tied to an increased risk:** Jason K. Hou, Bincy Abraham, and Hashem El-Serag, "Dietary Intake and Risk of Developing Inflammatory Bowel Disease: A Systematic Review of the Literature," *American Journal of Gastroenterology* 106, no. 4 (April 2011).

190   **while fiber intake:** Mitsuro Chiba et al., "Lifestyle-Related Disease in Crohn's Disease: Relapse Prevention by a Semi-Vegetarian Diet," *World Journal of Gastroenterology* 16, no. 20 (2010); Ashwin N. Ananthakrishnan et al., "A Prospective Study of Long-Term Intake of Dietary Fiber and Risk of Crohn's Disease and Ulcerative Colitis," *Gastroenterology* 145, no. 5 (Nov. 2013).

190   **In the small intestine:** Carol L. Roberts et al., "Translocation of Crohn's Disease *Escherichia Coli* across M-Cells: Contrasting Effects of Soluble Plant Fibres and Emulsifiers," *Gut* 59, no. 10 (Oct. 2010); Isobel Franks, "Crohn's Disease: Soluble Plant Fibers May Protect against *E. Coli* Translocation," *Nature Reviews Gastroenterology & Hepatology* 7, no. 12 (Dec. 2010).

Chapter 12: *Farm Country*

191   **hidden scourge of hunger:** Marion Nestle, *Food Politics: How the Food Industry Influences Nutrition and Health* (Berkeley: University of California Press, 2007); Robert Sam Anson, *McGovern: A Biography* (New York: Holt, Rinehart and Winston, 1972).

191   **science was controversial:** William J. Broad, "NIH Deals Gingerly with Diet-Disease Link: Federal Dietary Guidelines for Disease Prevention Have Scant Support from NIH, but Pressure to Take a Stand Is Building," *Science* 204, no. 4398 (1979); William J. Broad, "Jump in Funding Feeds Research on Nutrition: But the Dollars Also Fuel a Departmental Turf War That Threatens to Sap the Field of Its Newfound Nourishment," *Science* 204, no. 4397 (1979); George McGovern, "Statement of Senator George McGovern on the Publication of Dietary Goals for the United States," in *Dietary Goals for the United States* (Washington, DC: US Government Printing Office, 1977).

191   **released the country's first dietary guidelines:** "The McGovern Report," Nutrition Facts, April 12, 2013, https://nutritionfacts.org/video/the-mcgovern-report/. All the quotations regarding the country's first dietary guidelines that appear in this paragraph and the ones that follow are taken from this source.

192   **sending a warning:** Michael Pollan, *In Defense of Food: An Eater's Manifesto* (New York: Penguin Press, 2008).

193   **spoke of nutrients rather than whole foods:** Pollan, *In Defense of Food.*

193   **objected to this approach:** Pollan, *In Defense of Food.* All the quotations that appear in this paragraph are taken from this source.

193   **called this diet "plant-based":** T. Colin Campbell, interview with author, Feb. 2020.

194   **are prone to building up:** See, for example, Song and Chan, "Environmental Factors, Gut Microbiota"; Mari Anoushka Ricker and William Christian Haas, "Anti-Inflammatory Diet in Clinical Practice: A Review," *Nutrition in Clinical Practice* 32, no. 3 (June 1, 2017); Franceschi et al., "Inflammaging and 'Garb-Aging'"; Adriaan A. van Beek et al., "Metabolic Alterations in Aging Macrophages: Ingredients for Inflammaging?," *Trends in Immunology* 40, no. 2 (Feb. 1, 2019).

195 "'Immunologists are forced to use unusual expressions'": Biss, *On Immunity: An Inoculation.*

197 **increase our risk of getting sick or dying:** Anaïs Rico-Campà et al., "Association between Consumption of Ultra-Processed Foods and All Cause Mortality: Sun Prospective Cohort Study," *British Medical Journal* 365 (2019); Bernard Srour et al., "Ultraprocessed Food Consumption and Risk of Type 2 Diabetes among Participants of the Nutrinet-Santé Prospective Cohort," *JAMA Internal Medicine* 180, no. 2 (2020); Bernard Srour et al., "Ultra-Processed Food Intake and Risk of Cardiovascular Disease: Prospective Cohort Study (Nutrinet-Santé)," *British Medical Journal* 365 (2019).

198 **that promote inflammation:** Jotham Suez et al., "Artificial Sweeteners Induce Glucose Intolerance by Altering the Gut Microbiota," *Nature* 514, no. 7521 (Oct. 1, 2014); Jotham Suez et al., "Non-Caloric Artificial Sweeteners and the Microbiome: Findings and Challenges," *Gut Microbes* 6, no. 2 (March 4, 2015); Iryna Liauchonak et al., "Non-Nutritive Sweeteners and Their Implications on the Development of Metabolic Syndrome," *Nutrients* 11, no. 3 (2019); Stephanie Olivier-Van Stichelen, Kristina I. Rother, and John A. Hanover, "Maternal Exposure to Non-Nutritive Sweeteners Impacts Progeny's Metabolism and Microbiome," *Frontiers in Microbiology* 10, no. 1360 (June 20, 2019).

199 **and insulin resistance:** M. Y. Pepino et al., "Sucralose Affects Glycemic and Hormonal Responses to an Oral Glucose Load," *Diabetes Care* 36, no. 9 (Sept. 2013); Alonso Romo-Romo et al., "Sucralose Decreases Insulin Sensitivity in Healthy Subjects: A Randomized Controlled Trial," *American Journal of Clinical Nutrition* 108, no. 3 (Sept. 1, 2018).

199 **inflamed healthy mice:** Benoit Chassaing et al., "Dietary Emulsifiers Impact the Mouse Gut Microbiota Promoting Colitis and Metabolic Syndrome," *Nature* 519, no. 7541 (March 1, 2015).

199 **some soluble plant fibers:** Roberts et al., "Translocation of Crohn's Disease *Escherichia Coli* across M-Cells."

200 **directly activate the immune system:** See, for example, Anette Christ, Mario Lauterbach, and Eicke Latz, "Western Diet and the Immune System: An Inflammatory Connection," *Immunity* 51 (Nov. 19, 2019); Tilg and Moschen, "Food, Immunity, and the Microbiome"; Janett Barbaresko et al., "Dietary Pattern Analysis and Biomarkers of Low-Grade Inflammation: A Systematic Literature Review," *Nutrition Reviews* 71, no. 8 (Aug. 2013); Dario Giugliano, Antonio Ceriello, and Katherine Esposito, "The Effects of Diet on Inflammation: Emphasis on the Metabolic Syndrome," *Journal of the American College of Cardiology* 48, no. 4 (Aug. 15, 2006).

200 **hasten the degradation:** Sonia García-Calzón et al., "Dietary Inflammatory Index and Telomere Length in Subjects with a High Cardiovascular Disease Risk from the Predimed-Navarra Study: Cross-Sectional and Longitudinal Analyses over 5 Y," *American Journal of Clinical Nutrition* 102, no. 4 (2015).

201 **with longer telomeres:** Marta Crous-Bou et al., "Mediterranean Diet and Telomere Length in Nurses' Health Study: Population Based Cohort Study," *British Medical Journal* 349 (Dec. 2, 2014); Dean Ornish et al., "Effect of Comprehensive Lifestyle Changes on Telomerase Activity and Telomere Length in Men with Biopsy-Proven Low-Risk Prostate Cancer: 5-Year Follow-Up of a Descriptive Pilot Study," *The Lancet Oncology* 14, no. 11 (Oct. 2013).

201 **may *remember* a noxious diet:** Oliver Soehnlein and Peter Libby, "Targeting Inflammation in Atherosclerosis—From Experimental Insights to the Clinic," *Nature Reviews Drug Discovery* 20 (2021).

201   **also wields its influence:** Christ, Lauterbach, and Latz, "Western Diet and the Immune System."

202   **inspired fifteenth-century explorers:** *The Routledge Handbook of Soft Power*, ed. Naren Chitty et al. (New York: Routledge, 2017).

Chapter 13: Mangiafoglia

204   **"a much darker and more solid bread":** Keys and Keys, *Eat Well and Stay Well.*

204   **"still warm and gloriously fragrant":** Keys and Keys, *Eat Well and Stay Well.*

204   **"darkly roasted, newly ground beans":** Keys and Keys, *Eat Well and Stay Well.*

205   **boost salivary levels of IgA:** Sang Chul Jeong, Sundar Rao Koyyalamudi, and Gerald Pang, "Dietary Intake of Agaricus bisporus White Button Mushroom Accelerates Salivary Immunoglobulin A Secretion in Healthy Volunteers," *Nutrition* 28, no. 5 (May 2012).

205   **may lower the rates:** F. Meng, "Baker's Yeast Beta-Glucan Decreases Episodes of Common Childhood Illness in 1 to 4 Year Old Children during Cold Season in China," *Journal of Nutrition and Food Science* 6, no. 4 (2016).

205   **integral to the health of the immune system:** See, for example, Barbara Prietl et al., "Vitamin D and Immune Function," *Nutrients* 5, no. 7 (July 5, 2013); Wei Liu et al., "The Anti-Inflammatory Effects of Vitamin D in Tumorigenesis," *International Journal of Molecular Sciences* 19, no. 9 (Sept. 13, 2018); Neng Chen et al., "Effect of Vitamin D Supplementation on the Level of Circulating High-Sensitivity C-Reactive Protein: A Meta-Analysis of Randomized Controlled Trials," *Nutrients* 6, no. 6 (June 10, 2014); Wang et al., "Functions of Macrophages in the Maintenance of Intestinal Homeostasis."

206   **"No main meal in the Mediterranean":** Ancel Keys, "Mediterranean Diet and Public Health: Personal Reflections," *American Journal of Clinical Nutrition* 61, no. 6 Suppl. (June 1995).

206   **unique class of phytochemicals:** Albena T. Dinkova-Kostova and Rumen V. Kostov, "Glucosinolates and Isothiocyanates in Health and Disease," *Trends in Molecular Medicine* 18, no. 6 (June 1, 2012).

207   **powerful protection against chronic inflammatory diseases:** See, for example, Nagisa Mori et al., "Cruciferous Vegetable Intake and Mortality in Middle-Aged Adults: A Prospective Cohort Study," *Clinical Nutrition* 38, no. 2 (April 1, 2019); Dagfinn Aune, "Plant Foods, Antioxidant Biomarkers, and the Risk of Cardiovascular Disease, Cancer, and Mortality: A Review of the Evidence," *Advances in Nutrition* 10, Suppl. 4 (2019); Patrizia Riso et al., "Effect of 10-Day Broccoli Consumption on Inflammatory Status of Young Healthy Smokers," *International Journal of Food Sciences and Nutrition* 65, no. 1 (Feb. 2014); Yu Jiang et al., "Cruciferous Vegetable Intake Is Inversely Correlated with Circulating Levels of Proinflammatory Markers in Women," *Journal of the Academy of Nutrition and Dietetics* 114, no. 5 (May 2014).

207   **secrete enzymes that activate isothiocyanates:** Sicong Tian et al., "Microbiota: A Mediator to Transform Glucosinolate Precursors in Cruciferous Vegetables to the Active Isothiocyanates," *Journal of the Science of Food and Agriculture* 98, no. 4 (March 2018).

207   **interact with a special molecule:** Tilg and Moschen, "Food, Immunity, and the Microbiome"; Veldhoen and Brucklacher-Waldert, "Dietary Influences on Intestinal Immunity."

207   **influence inflammation through multiple pathways:** Marialaura Bonaccio et al., "Mediterranean Diet, Dietary Polyphenols and Low Grade Inflammation:

Results from the Moli-Sani Study," *British Journal of Clinical Pharmacology* 83, no. 1 (Jan. 2017); Alexa Serino and Gloria Salazar, "Protective Role of Polyphenols against Vascular Inflammation, Aging and Cardiovascular Disease," *Nutrients* 11, no. 1 (2019); Ricker and Haas, "Anti-Inflammatory Diet in Clinical Practice"; Sashwati Roy and Siba P. Raychaudhuri, eds., *Chronic Inflammation: Molecular Pathophysiology, Nutritional and Therapeutic Interventions* (Boca Raton, FL: CRC Press, 2012).

208   **fermented by microbes to yield beneficial metabolites:** Rizzetto et al., "Connecting the Immune System"; Bailey and Holscher, "Microbiome-Mediated Effects of the Mediterranean Diet."

208   **dropped by 80 percent and life expectancy increased:** Pett et al., *Ancel Keys and the Seven Countries Study.*

209   **"My room is filled with books I cannot read":** Todd Tucker, *The Great Starvation Experiment* (Minneapolis: University of Minnesota Press, 2007).

209   **first half of the twentieth century:** Mozaffarian, Rosenberg, and Uauy, "History of Modern Nutrition Science."

210   **devised the randomized controlled trial:** Vern Farewell and Tony Johnson, "Woods and Russell, Hill, and the Emergence of Medical Statistics," *Statistics in Medicine* 29, no. 14 (June 30, 2010).

210   **half a pack of cigarettes a day:** Centers for Disease Control and Prevention, "Tobacco Use—United States, 1900–1999," *Morbidity Mortality Weekly Report* 48, no. 43 (Nov. 5, 1999).

210   **toyed with the idea:** National Tobacco Reform Leadership Team, "Letter to FDA Commissioner Scott Gottlieb, M.D.— Recent FDA Actions to Reduce Adolescent Tobacco Use," (Dec. 7, 2018), https://www.tobaccoreform.org/letter-to-fda-commissioner-scott-gottlieb-m-d-actions-to-reduce-adolescent-tobacco-use/.

210   **Many doctors smoked:** Martha N. Gardner and Allan M. Brandt, "'The Doctors' Choice Is America's Choice': The Physician in US Cigarette Advertisements, 1930–1953," *American Journal of Public Health* 96, no. 2 (2006).

210   **intuitively making connections:** Richard Doll, "The First Report on Smoking and Lung Cancer," in *Ashes to Ashes: The History of Smoking and Health*, ed. Stephen Lock, Lois A. Reynolds, and E. M. Tansey (Amsterdam: Editions Rodopi B. V., 1998).

211   **tracked their health:** Richard Doll and A. Bradford Hill, "Lung Cancer and Other Causes of Death in Relation to Smoking; A Second Report on the Mortality of British Doctors," *British Medical Journal* 2, no. 5001 (Nov. 10, 1956); Richard Doll and A. Bradford Hill, "The Mortality of Doctors in Relation to Their Smoking Habits: A Preliminary Report," *British Medical Journal* 1, no. 4877 (1954).

211   **needed a revamping of the word *cause*:** A. Bradford Hill, "The Environment and Disease: Association or Causation?," *Proceedings of the Royal Society of Medicine* 58, no. 5 (1965).

211   **designed to assess:** Julia A. Segre, "What Does It Take to Satisfy Koch's Postulates Two Centuries Later? Microbial Genomics and Propionibacteria Acnes," *Journal of Investigative Dermatology* 133, no. 9 (Sept. 2013).

212   **are not ideal:** Ambika Satija et al., "Understanding Nutritional Epidemiology and Its Role in Policy," *Advances in Nutrition* 6, no. 1 (2015).

214   **many of which derived from plants:** Mouhssen Lahlou, "The Success of Natural Products in Drug Discovery," *Pharmacology & Pharmacy* 4 (2013).

215   **best fight the "cauldron of inflammation":** Caldwell Esselstyn, interview with author, Feb. 2020.

216 **like vitamin D, is especially important:** Veldhoen and Brucklacher-Waldert, "Dietary Influences on Intestinal Immunity."

216 **eating a purple potato:** Kerrie L Kaspar et al., "Pigmented Potato Consumption Alters Oxidative Stress and Inflammatory Damage in Men," *Journal of Nutrition* 141, no. 1 (Jan. 2011); Joe A. Vinson et al., "High-Antioxidant Potatoes: Acute in Vivo Antioxidant Source and Hypotensive Agent in Humans after Supplementation to Hypertensive Subjects," *Journal of Agricultural and Food Chemistry* 60, no. 27 (July 11, 2012).

216 **dietary intervention data:** Caldwell B. Esselstyn Jr. et al., "A Way to Reverse CAD?," *Journal of Family Practice* 63, no. 7 (July 2014).

217 **low levels of salicylic acid:** C. J. Blacklock et al., "Salicylic Acid in the Serum of Subjects Not Taking Aspirin: Comparison of Salicylic Acid Concentrations in the Serum of Vegetarians, Non-Vegetarians, and Patients Taking Low Dose Aspirin," *Journal of Clinical Pathology* 54, no. 7 (July 2001).

217 **show that amla:** Pingali Usharani, Nishat Fatima, and Nizampatnam Muralidhar, "Effects of Phyllanthus emblica Extract on Endothelial Dysfunction and Biomarkers of Oxidative Stress in Patients with Type 2 Diabetes Mellitus: A Randomized, Double-Blind, Controlled Study," *Diabetes, Metabolic Syndrome and Obesity: Targets and Therapy* 6 (2013); Pingali Usharani, Padma Latha Merugu, and Chandrasekhar Nutalapati, "Evaluation of the Effects of a Standardized Aqueous Extract of Phyllanthus emblica Fruits on Endothelial Dysfunction, Oxidative Stress, Systemic Inflammation and Lipid Profile in Subjects with Metabolic Syndrome: A Randomised, Double Blind, Placebo Controlled Clinical Study," *BMC Complementary and Alternative Medicine* 19, no. 1 (May 6, 2019).

218 **favorably modulate the immune system:** See, for example, Susan S. Percival et al., "Bioavailability of Herbs and Spices in Humans as Determined by Ex Vivo Inflammatory Suppression and DNA Strand Breaks," *Journal of the American College of Nutrition* 31, no. 4 (Aug. 2012); Ricker and Haas, "Anti-Inflammatory Diet in Clinical Practice"; Changyou Zhu et al., "Impact of Cinnamon Supplementation on Cardiometabolic Biomarkers of Inflammation and Oxidative Stress: A Systematic Review and Meta-Analysis of Randomized Controlled Trials," *Complementary Therapies in Medicine* 53 (Sept. 1, 2020); Shiva Kazemi et al., "Cardamom Supplementation Improves Inflammatory and Oxidative Stress Biomarkers in Hyperlipidemic, Overweight, and Obese Pre-Diabetic Women: A Randomized Double-Blind Clinical Trial," *Journal of the Science of Food and Agriculture* 97, no. 15 (Dec. 2017).

218 **point to its utility:** See, for example, Yasmin Anum Mohd Yusof, "Gingerol and Its Role in Chronic Diseases," in *Drug Discovery from Mother Nature*, ed. Subash Chandra Gupta, Sahdeo Prasad, and Bharat B. Aggarwal (Cham, Switzerland: Springer, 2016); Jing Wang et al., "Beneficial Effects of Ginger Zingiber officinale Roscoe on Obesity and Metabolic Syndrome: A Review," *Annals of the New York Academy of Sciences* 1398, no. 1 (June 1, 2017); Hassan Mozaffari-Khosravi et al., "The Effect of Ginger Powder Supplementation on Insulin Resistance and Glycemic Indices in Patients with Type 2 Diabetes: A Randomized, Double-Blind, Placebo-Controlled Trial," *Complementary Therapies in Medicine* 22, no. 1 (Feb. 2014); Mehran Rahimlou et al., "Ginger Supplementation in Nonalcoholic Fatty Liver Disease: A Randomized, Double-Blind, Placebo-Controlled Pilot Study," *Hepatitis Monthly* 16, no. 1 (Jan. 2016); E. M. Bartels et al., "Efficacy and Safety of Ginger in Osteoarthritis Patients: A Meta-Analysis of Randomized Placebo-Controlled Trials," *Osteoarthritis Cartilage* 23, no. 1 (Jan. 2015).

218 **even painful menses and migraine headaches:** Mehdi Maghbooli et al., "Com-

parison between the Efficacy of Ginger and Sumatriptan in the Ablative Treatment of the Common Migraine," *Phytotherapy Research* 28, no. 3 (March 2014); James W. Daily et al., "Efficacy of Ginger for Alleviating the Symptoms of Primary Dysmenorrhea: A Systematic Review and Meta-Analysis of Randomized Clinical Trials," *Pain Medicine* 16, no. 12 (Dec. 2015).

218 **dozens of human clinical trials:** See, for example, Subash C. Gupta, Sridevi Patchva, and Bharat B. Aggarwal, "Therapeutic Roles of Curcumin: Lessons Learned from Clinical Trials," *The AAPS Journal* 15, no. 1 (Jan. 2013); Binu Chandran and Ajay Goel, "A Randomized, Pilot Study to Assess the Efficacy and Safety of Curcumin in Patients with Active Rheumatoid Arthritis," *Phytotherapy Research* 26, no. 11 (Nov. 2012); Vilai Kuptniratsaikul et al., "Efficacy and Safety of Curcuma domestica Extracts in Patients with Knee Osteoarthritis," *Journal of Alternative and Complementary Medicine* 15, no. 8 (Aug. 2009); Krishna Adit Agarwal et al., "Efficacy of Turmeric (Curcumin) in Pain and Postoperative Fatigue after Laparoscopic Cholecystectomy: A Double-Blind, Randomized Placebo-Controlled Study," *Surgical Endoscopy* 25, no. 12 (Dec. 2011).

218 **inhibits many inflammatory pathways:** S. Prasad and B. B. Aggarwal, "Turmeric, the Gold Spice: From Traditional to Modern Medicine," in *Herbal Medicine: Biomolecular and Clinical Aspects*, ed. Iris F. F. Benzie and Sissi Wachtel-Galor (Boca Raton, FL: CRC Press, 2011); Roy and Raychaudhuri, *Molecular Pathophysiology, Nutritional and Therapeutic Interventions*.

218 **In inflammatory bowel disease patients:** Hiroyuki Hanai et al., "Curcumin Maintenance Therapy for Ulcerative Colitis: Randomized, Multicenter, Double-Blind, Placebo-Controlled Trial," *Clinical Gastroenterology and Hepatology* 4, no. 12 (Dec. 2006); David L. Suskind et al., "Tolerability of Curcumin in Pediatric Inflammatory Bowel Disease: A Forced-Dose Titration Study," *Journal of Pediatric Gastroenterology and Nutrition* 56, no. 3 (March 2013).

218 **discourages infectious germs and tumors:** Caleb E. Finch, *The Biology of Human Longevity: Inflammation, Nutrition, and Aging in the Evolution of Life Spans* (Burlington, MA: Academic Press, 2007).

219 **continues to exhibit potent anti-inflammatory activity:** Bharat B Aggarwal et al., "Curcumin-Free Turmeric Exhibits Anti-Inflammatory and Anticancer Activities: Identification of Novel Components of Turmeric," *Molecular Nutrition & Food Research* 57, no. 9 (Sept. 2013).

219 **boosts the bioavailability of curcumin:** Guido Shoba et al., "Influence of Piperine on the Pharmacokinetics of Curcumin in Animals and Human Volunteers," *Planta Medica* 64, no. 4 (May 1998); Preetha Anand et al., "Bioavailability of Curcumin: Problems and Promises," *Molecular Pharmaceutics* 4, no. 6 (Nov.–Dec. 2007).

220 **hundreds of varieties:** Samuel A. Smits et al., "Seasonal Cycling in the Gut Microbiome of the Hadza Hunter-Gatherers of Tanzania," *Science* 357, no. 6353 (2017).

220 **prevent—or, in many cases, to treat:** See, for example, Michael Greger and Gene Stone, *How Not to Die: Discover the Foods Scientifically Proven to Prevent and Reverse Disease* (New York: Flatiron Books, 2015); T. Colin Campbell and Thomas M. Campbell II, *The China Study: Revised and Expanded Edition: The Most Comprehensive Study of Nutrition Ever Conducted and the Startling Implications for Diet, Weight Loss, and Long-Term Health* (Dallas, TX: BenBella Books, 2016); David L. Katz, *The Truth about Food: Why Pandas Eat Bamboo and People Get Bamboozled* (independently published, 2018).

220 **drawn on this large body of literature:** See, for example, Barbaresko et al., "Dietary Pattern Analysis and Biomarkers of Low-Grade Inflammation"; Wolfgang Marx et al., "The Dietary Inflammatory Index and Human Health: An

Umbrella Review of Meta-Analyses of Observational Studies," *Advances in Nutrition* (2021); Fred K. Tabung et al., "Development and Validation of an Empirical Dietary Inflammatory Index," *Journal of Nutrition* 146, no. 8 (2016); Nitin Shivappa et al., "Designing and Developing a Literature-Derived, Population-Based Dietary Inflammatory Index," *Public Health Nutrition* 17, no. 8 (2014).

220 **which have been tied to:** See, for example, Fred K. Tabung et al., "Association of Dietary Inflammatory Potential with Colorectal Cancer Risk in Men and Women," *JAMA Oncology* 4, no. 3 (2018); Chun-Han Lo et al., "Dietary Inflammatory Potential and Risk of Crohn's Disease and Ulcerative Colitis," *Gastroenterology* 159, no. 3 (Sept. 1, 2020); Jun Li et al., "Dietary Inflammatory Potential and Risk of Cardiovascular Disease among Men and Women in the U.S," *Journal of the American College of Cardiology* 76, no. 19 (Nov. 10, 2020).

221 **Randomized controlled trials point out:** Richard Rosenfeld and Megan Hall, *Evidence Summary for Plant-Based Diets: Reviews, Trials, Large Cohort, and Landmark Observational Studies*, SUNY Downstate Committee on Plant-Based Health and Nutrition (Aug. 2021), https://www.downstate.edu/about/community-impact/plant-based/evidence.html.

221 **Observational studies that have enrolled:** Rosenfeld and Hall, *Evidence Summary for Plant-Based Diets.*

Chapter 14: Shaping Sustenance

223 **"in man, as well as in several other mammals":** Elie Metchnikoff, *The New Hygiene: Three Lectures on the Prevention of Infectious Diseases* (Chicago: W. T. Keener, 1910).

223 **isolated, excessive doses:** Vincent J. van Buul and Fred J. P. H. Brouns, "Health Effects of Wheat Lectins: A Review," *Journal of Cereal Science* 59, no. 2 (March 1, 2014).

223 **may hold promise:** A. Pusztai, "Dietary Lectins Are Metabolic Signals for the Gut and Modulate Immune and Hormone Functions," *European Journal of Clinical Nutrition* 47, no. 10 (Oct. 1993); Ram Sarup Singh, Hemant Preet Kaur, and Jagat Rakesh Kanwar, "Mushroom Lectins as Promising Anticancer Substances," *Current Protein & Peptide Science* 17, no. 8 (2016); Jasminka Giacometti, "Plant Lectins in Cancer Prevention and Treatment," *Medicina* 51, no. 2 (2015).

224 **cause CRP and other markers:** See, for example, Parisa Hajihashemi and Fahimeh Haghighatdoost, "Effects of Whole-Grain Consumption on Selected Biomarkers of Systematic Inflammation: A Systematic Review and Meta-Analysis of Randomized Controlled Trials," *Journal of the American College of Nutrition* 38, no. 3 (April 3, 2019); Yujie Xu et al., "Whole Grain Diet Reduces Systemic Inflammation: A Meta-Analysis of 9 Randomized Trials," *Medicine* 97, no. 43 (2018); Abdolrasoul Safaeiyan et al., "Randomized Controlled Trial on the Effects of Legumes on Cardiovascular Risk Factors in Women with Abdominal Obesity," *ARYA Atherosclerosis* 11, no. 2 (2015).

225 **reduce low-level inflammation:** Mahsa Ghavipour et al., "Tomato Juice Consumption Reduces Systemic Inflammation in Overweight and Obese Females," *British Journal of Nutrition* 109, no. 11 (June 2013); Young-il Kim et al., "Tomato Extract Suppresses the Production of Proinflammatory Mediators Induced by Interaction between Adipocytes and Macrophages," *Bioscience, Biotechnology, and Biochemistry* 79, no. 1 (2015); Helena Hermana M. Hermsdorff et al., "Fruit and Vegetable Consumption and Proinflammatory Gene Expression from Peripheral

Blood Mononuclear Cells in Young Adults: A Translational Study," *Nutrition & Metabolism* 7 (May 13, 2010): 42.

225 **variable magnitude and duration:** Husam Ghanim et al., "Increase in Plasma Endotoxin Concentrations and the Expression of Toll-Like Receptors and Suppressor of Cytokine Signaling-3 in Mononuclear Cells after a High-Fat, High-Carbohydrate Meal," *Diabetes Care* 32, no. 12 (2009).

225 **sears through our body:** Vogel, Corretti, and Plotnick, "Effect of a Single High-Fat Meal"; Sara K. Rosenkranz et al., "Effects of a High-Fat Meal on Pulmonary Function in Healthy Subjects," *European Journal of Applied Physiology* 109, no. 3 (June 2010); Deopurkar et al., "Differential Effects of Cream, Glucose, and Orange Juice."

225 **and chronic disease:** James H. O'Keefe and David S. H. Bell, "Postprandial Hyperglycemia/Hyperlipidemia (Postprandial Dysmetabolism) Is a Cardiovascular Risk Factor," *American Journal of Cardiology* 100, no. 5 (2007); F. Cavalot et al., "Postprandial Blood Glucose Is a Stronger Predictor of Cardiovascular Events Than Fasting Blood Glucose in Type 2 Diabetes Mellitus, Particularly in Women: Lessons from the San Luigi Gonzaga Diabetes Study," *Journal of Clinical Endocrinology & Metabolism* 91, no. 3 (2006).

225 **temper the inflammation created by other foods:** See, for example, Esposito and Giugliano, "A Link to Metabolic and Cardiovascular Diseases"; Paresh Dandona et al., "Macronutrient Intake Induces Oxidative and Inflammatory Stress: Potential Relevance to Atherosclerosis and Insulin Resistance," *Experimental & Molecular Medicine* 42, no. 4 (April 1, 2010); James H. O'Keefe, Neil M. Gheewala, and Joan O. O'Keefe, "Dietary Strategies for Improving Post-Prandial Glucose, Lipids, Inflammation, and Cardiovascular Health," *Journal of the American College of Cardiology* 51, no. 3 (2008); Janice K. Kiecolt-Glaser, "Stress, Food, and Inflammation: Psychoneuroimmunology and Nutrition at the Cutting Edge," *Psychosomatic Medicine* 72, no. 4 (2010).

225 **load of vegetables:** Katherine Esposito et al., "Effect of Dietary Antioxidants on Postprandial Endothelial Dysfunction Induced by a High-Fat Meal in Healthy Subjects," *American Journal of Clinical Nutrition* 77, no. 1 (2003).

225 **half an avocado atop a burger:** Zhaoping Li et al., "Hass Avocado Modulates Postprandial Vascular Reactivity and Postprandial Inflammatory Responses to a Hamburger Meal in Healthy Volunteers," *Food & Function* 4, no. 3 (Feb. 26, 2013).

225 **blend of spices:** Ester S. Oh et al., "Spices in a High-Saturated-Fat, High-Carbohydrate Meal Reduce Postprandial Proinflammatory Cytokine Secretion in Men with Overweight or Obesity: A 3-Period, Crossover, Randomized Controlled Trial," *Journal of Nutrition* 150, no. 6 (2020).

225 **handful of berries or nuts:** See, for example, Britt Burton-Freeman et al., "Strawberry Modulates LDL Oxidation and Postprandial Lipemia in Response to High-Fat Meal in Overweight Hyperlipidemic Men and Women," *Journal of the American College of Nutrition* 29, no. 1 (2010); Riitta Törrönen et al., "Berries Reduce Postprandial Insulin Responses to Wheat and Rye Breads in Healthy Women," *Journal of Nutrition* 143, no. 4 (2013); Bryan C. Blacker et al., "Consumption of Blueberries with a High-Carbohydrate, Low-Fat Breakfast Decreases Postprandial Serum Markers of Oxidation," *British Journal of Nutrition* 109, no. 9 (2013); David J. A. Jenkins et al., "Almonds Decrease Postprandial Glycemia, Insulinemia, and Oxidative Damage in Healthy Individuals," *Journal of Nutrition* 136, no. 12 (2006).

225 **carry over to the next meal:** T. M. Wolever et al., "Second-Meal Effect: Low-

Glycemic-Index Foods Eaten at Dinner Improve Subsequent Breakfast Glycemic Response," *American Journal of Clinical Nutrition* 48, no. 4 (1988); Rebecca C. Mollard et al., "First and Second Meal Effects of Pulses on Blood Glucose, Appetite, and Food Intake at a Later Meal," *Applied Physiology, Nutrition, and Metabolism* 36, no. 5 (2011).

226   **eat a rare slab:** Tilg and Moschen, "Food, Immunity, and the Microbiome."

226   **transformed sugars into alcohol:** Allison Lassieur, *Louis Pasteur: Revolutionary Scientist* (New York: Franklin Watts, 2005); Luisa Alba-Lois and Claudia Segal-Kischinevsky, "Yeast Fermentation and the Making of Beer and Wine," *Nature Education* 3, no. 9 (2010).

227   **"respiration without air":** Fermentation does not always occur in the absence of oxygen. To capture the wide variety of reactions and pathways that occur in fermented foods and beverages, the 2019 International Scientific Association for Probiotics and Prebiotics consensus statement on fermented foods defines these foods and beverages as "foods made through desired microbial growth and enzymatic conversions of food components." See Maria L. Marco et al., "The International Scientific Association for Probiotics and Prebiotics (ISAPP) Consensus Statement on Fermented Foods," *Nature Reviews Gastroenterology & Hepatology* 18, no. 3 (March 1, 2021).

228   **"A reader may be surprised by my recommendation":** Elie Metchnikoff, *The Prolongation of Life: Optimistic Studies* (New York: G. P. Putnam's Sons, 1908).

228   **"placed the whole world under obligation to him":** John Harvey Kellogg, *Autointoxication or Intestinal Toxemia* (Whitefish, MT: Kessinger Publishing, 1922; repr., 2010).

229   **"a quirky, beautiful thing in biology":** Justin Sonnenburg, interview with author, May 2020.

229   **In a 2021 study:** Hannah C. Wastyk et al., "Gut-Microbiota-Targeted Diets Modulate Human Immune Status," *Cell* 184, no. 16 (2021).

230   **as Metchnikoff once proposed:** Elie Metchnikoff, *The Nature of Man: Studies in Optimistic Philosophy* (London: Heinemann, 1906).

230   **due to poor transport mechanisms:** Peter R. Gibson and Susan J. Shepherd, "Evidence-Based Dietary Management of Functional Gastrointestinal Symptoms: The FODMAP Approach," *Journal of Gastroenterology and Hepatology* 25, no. 2 (Feb. 1, 2010).

230   **decreases after weaning:** Benjamin Misselwitz et al., "Lactose Malabsorption and Intolerance: Pathogenesis, Diagnosis and Treatment," *United European Gastroenterology Journal* 1, no. 3 (2013).

231   **rare yet serious health risks:** Ailsa Holbourn and Judith Hurdman, "Kombucha: Is a Cup of Tea Good for You?," *Case Reports* 2017 (2017); Maheedhar Gedela et al., "A Case of Hepatotoxicity Related to Kombucha Tea Consumption," *South Dakota Journal of Medicine* 69, no. 1 (2016).

232   **with a variety of specific conditions:** Eamonn M. M. Quigley, "Prebiotics and Probiotics in Digestive Health," *Clinical Gastroenterology and Hepatology* 17, no. 2 (2019).

232   **in petri dishes and in people:** Veronica Valli et al., "Health Benefits of Ancient Grains: Comparison among Bread Made with Ancient, Heritage and Modern Grain Flours in Human Cultured Cells," *Food Research International* 107 (May 1, 2018); Monica Dinu et al., "Ancient Wheat Species and Human Health: Biochemical and Clinical Implications," *Journal of Nutritional Biochemistry* 52 (Feb. 1, 2018).

232 **may suppress inflammation more easily:** Francesco Sofi et al., "Effect of *Triticum turgidum* Subsp. *turanicum* Wheat on Irritable Bowel Syndrome: A Double-Blinded Randomised Dietary Intervention Trial," *British Journal of Nutrition* 111, no. 11 (June 14, 2014); Anne Whittaker et al., "A Khorasan Wheat-Based Replacement Diet Improves Risk Profile of Patients with Type 2 Diabetes Mellitus (T2DM): A Randomized Crossover Trial," *European Journal of Nutrition* 56, no. 3 (April 2017); Anne Whittaker et al., "An Organic Khorasan Wheat-Based Replacement Diet Improves Risk Profile of Patients with Acute Coronary Syndrome: A Randomized Crossover Trial," *Nutrients* 7, no. 5 (May 11, 2015); Monica Dinu et al., "A Khorasan Wheat-Based Replacement Diet Improves Risk Profile of Patients with Nonalcoholic Fatty Liver Disease (NAFLD): A Randomized Clinical Trial," *Journal of the American College of Nutrition* 37, no. 6 (Aug. 2018).

233 **counter the inflammatory effects of air pollutants:** Patricia A Egner et al., "Rapid and Sustainable Detoxication of Airborne Pollutants by Broccoli Sprout Beverage: Results of a Randomized Clinical Trial in China," *Cancer Prevention Research* 7, no. 8 (2014); Stacey A Ritz, Junxiang Wan, and David Diaz-Sanchez, "Sulforaphane-Stimulated Phase II Enzyme Induction Inhibits Cytokine Production by Airway Epithelial Cells Stimulated with Diesel Extract," *American Journal of Physiology—Lung Cellular and Molecular Physiology* 292, no. 1 (2007); Marc A Riedl, Andrew Saxon, and David Diaz-Sanchez, "Oral Sulforaphane Increases Phase II Antioxidant Enzymes in the Human Upper Airway," *Clinical Immunology* 130, no. 3 (2009); David Heber et al., "Sulforaphane-Rich Broccoli Sprout Extract Attenuates Nasal Allergic Response to Diesel Exhaust Particles," *Food & Function* 5, no. 1 (2014).

233 **increase immunity and dampen inflammation:** Terry L Noah et al., "Effect of Broccoli Sprouts on Nasal Response to Live Attenuated Influenza Virus in Smokers: A Randomized, Double-Blind Study," *PLOS ONE* 9, no. 6 (2014).

233 **can be profitable:** Max Leenders et al., "Fruit and Vegetable Intake and Cause-Specific Mortality in the Epic Study," *European Journal of Epidemiology* 29, no. 9 (Sept. 2014).

233 **symptomatic improvement:** Osmo Hänninen et al., "Antioxidants in Vegan Diet and Rheumatic Disorders," *Toxicology* 155, nos. 1–3 (2000); Mikko T. Nenonen et al., "Uncooked, Lactobacilli-Rich, Vegan Food and Rheumatoid Arthritis," *British Journal of Rheumatology* 37, no. 3 (1998); R. Peltonen et al., "Faecal Microbial Flora and Disease Activity in Rheumatoid Arthritis During a Vegan Diet," *British Journal of Rheumatology* 36, no. 1 (1997).

234 **data suggest that the innate immune response:** Melanie Uhde et al., "Intestinal Cell Damage and Systemic Immune Activation in Individuals Reporting Sensitivity to Wheat in the Absence of Coeliac Disease," *Gut* 65, no. 12 (Dec. 2016).

234 **and even sensitivities:** Stuart M. Brierley, "Food for Thought about the Immune Drivers of Gut Pain," *Nature* 590 (Feb. 4, 2021).

234 **is different from that found in celiac disease:** Melanie Uhde et al., "Subclass Profile of IgG Antibody Response to Gluten Differentiates Nonceliac Gluten Sensitivity from Celiac Disease," *Gastroenterology* 159, no. 5 (2020).

234 **just as much of an increase:** Uhde et al., "Intestinal Cell Damage and Systemic Immune Activation."

237 **cost for immune health:** Jo Robinson, *Eating on the Wild Side: The Missing Link to Optimum Health* (New York: Little, Brown, 2013).

238 **as British philosopher A. C. Grayling writes:** A. C. Grayling, *The Reason of Things: Living with Philosophy* (London: Phoenix, 2007).

Chapter 15: Dirty Cures

240   **"is older than history, older than tradition"**: Mark Twain, *Following the Equator: A Journey around the World* (New York: Dover Publications, 1897; repr., 1989).

240   **deemed it the "water of immortality"**: Richard Garbe, *Akbar, Emperor of India: A Picture of Life and Customs from the Sixteenth Century*, trans. Lydia Gillingham Robinson (Chicago: The Open Court Publishing Company, 1909).

241   **pioneered antiseptic surgery:** Lindsey Fitzharris, *The Butchering Art: Joseph Lister's Quest to Transform the Grisly World of Victorian Medicine* (New York: Farrar, Straus and Giroux, 2017).

242   **"the higher life is everywhere inter-penetrated"**: Nancy Tomes, "The Private Side of Public Health: Sanitary Science, Domestic Hygiene, and the Germ Theory, 1870–1900," *Bulletin of the History of Medicine* 64, no. 4 (1990).

242   **fated fall day:** Eric Lax, *The Mold in Dr. Florey's Coat: The Story of Penicillin and the Modern Age of Medical Miracles* (New York: Henry Holt, 2004).

243   **exposes her fetus:** Andrew J. Macpherson, Mercedes Gomez de Agüero, and Stephanie C. Ganal-Vonarburg, "How Nutrition and the Maternal Microbiota Shape the Neonatal Immune System," *Nature Reviews Immunology* 17, no. 8 (Aug. 1, 2017).

243   **others disagree:** Graham Rook et al., "Evolution, Human-Microbe Interactions, and Life History Plasticity," *Lancet* 390 (July 29, 2017).

244   **"hygiene hypothesis"**: David P. Strachan, "Hay Fever, Hygiene, and Household Size," *British Medical Journal* 299, no. 6710 (1989).

245   **"of no obvious utility for either biological survival":** Dr. Lagerspetz makes this comment in reference to Sigmund Freud's views. Olli Lagerspetz, *A Philosophy of Dirt* (Chicago: University of Chicago Press, 2018).

245   **in the first few days and years:** F. Shanahan, T. S. Ghosh, and P. W. O'Toole, "The Healthy Microbiome—What Is the Definition of a Healthy Gut Microbiome?," *Gastroenterology* 160, no. 2 (Jan. 2021); Mirae Lee and Eugene B. Chang, "Inflammatory Bowel Diseases (IBD) and the Microbiome—Searching the Crime Scene for Clues," *Gastroenterology* 160, no. 2 (Jan. 1, 2021).

245   **greater numbers of microbes during infancy:** Thomas W. McDade, "Early Environments and the Ecology of Inflammation," *Proceedings of the National Academy of Sciences* 109, Suppl. 2 (2012); Thomas W. McDade et al., "Early Origins of Inflammation: Microbial Exposures in Infancy Predict Lower Levels of C-Reactive Protein in Adulthood," *Proceedings of the Royal Society B: Biological Sciences* 277, no. 1684 (2010).

246   **considered a new hypothesis:** Graham A. W. Rook, "Hygiene Hypothesis and Autoimmune Diseases," *Clinical Reviews in Allergy & Immunology* 42 (2012).

246   **a deep, gnawing hunger for solitude:** Michael Finkel, *The Stranger in the Woods: The Extraordinary Story of the Last True Hermit* (New York: Alfred A. Knopf, 2017).

247   **lower the risk of allergies and asthma:** Bill Hesselmar et al., "Pet-Keeping in Early Life Reduces the Risk of Allergy in a Dose-Dependent Fashion," *PLOS ONE* 13, no. 12 (2018); Tove Fall et al., "Early Exposure to Dogs and Farm Animals and the Risk of Childhood Asthma," *JAMA Pediatrics* 169, no. 11 (2015).

248   **mimic traditional farm environments:** Michelle M. Stein et al., "Innate Immunity and Asthma Risk in Amish and Hutterite Farm Children," *New England Journal of Medicine* 375, no. 5 (Aug. 4, 2016).

248   **asthma, allergies, and autoimmune diseases:** Michael Elten et al., "Residential Greenspace in Childhood Reduces Risk of Pediatric Inflammatory Bowel Disease: A Population-Based Cohort Study," *American Journal of Gastroenterology* 116, no. 2 (2021).

250 **also *defined* it:** Eileen Crist and Alfred I. Tauber, "Selfhood, Immunity, and the Biological Imagination: The Thought of Frank Macfarlane Burnet," *Biology and Philosophy* 15 (1999); Alfred I. Tauber, "Moving Beyond the Immune Self?," *Seminars in Immunology* 12 (2000).

251 **expanded the common view:** F. Macfarlane Burnet, *Biological Aspects of Infectious Disease* (Cambridge: Cambridge University Press, 1940); F. Macfarlane Burnet, *The Virus as Organism* (Cambridge: Cambridge University Press, 1946).

251 **does not suffice:** Tauber, "Moving Beyond the Immune Self?"

Chapter 16: Easter Island

254 **published a paper:** Michel Poulain et al., "Identification of a Geographic Area Characterized by Extreme Longevity in the Sardinia Island: The AKEA Study," *Experimental Gerontology* 39, no. 9 (Sept. 1, 2004).

254 **set out to identify:** Dan Buettner, email to author, May 2021.

256 **may slow aging and help to prevent or even treat:** See, for example, Rafael de Cabo and Mark P. Mattson, "Effects of Intermittent Fasting on Health, Aging, and Disease," *New England Journal of Medicine* 381, no. 26 (Dec. 26, 2019); Elizabeth F. Sutton et al., "Early Time-Restricted Feeding Improves Insulin Sensitivity, Blood Pressure, and Oxidative Stress Even without Weight Loss in Men with Prediabetes," *Cell Metabolism* 27, no. 6 (June 5, 2018); Michael J. Wilkinson et al., "Ten-Hour Time-Restricted Eating Reduces Weight, Blood Pressure, and Atherogenic Lipids in Patients with Metabolic Syndrome," *Cell Metabolism* 31, no. 1 (2020); Adrienne R Barnosky et al., "Intermittent Fasting vs Daily Calorie Restriction for Type 2 Diabetes Prevention: A Review of Human Findings," *Translational Research: The Journal of Laboratory and Clinical Medicine* 164, no. 4 (Oct. 2014).

256 **begin to back down:** Chia-Wei Cheng et al., "Prolonged Fasting Reduces IGF-1/PKA to Promote Hematopoietic-Stem-Cell-Based Regeneration and Reverse Immunosuppression," *Cell Stem Cell* 14, no. 6 (June 5, 2014).

257 **allaying the inflammatory actions of macrophages:** Stefan Jordan et al., "Dietary Intake Regulates the Circulating Inflammatory Monocyte Pool," *Cell* 178, no. 5 (2019).

257 **across age-groups:** Jeffrey A. Woods et al., "Exercise, Inflammation and Aging," *Aging and Disease* 3, no. 1 (2012); Kaleen M. Lavin et al., "Effects of Aging and Lifelong Aerobic Exercise on Basal and Exercise-Induced Inflammation," *Journal of Applied Physiology* 128, no. 1 (Jan. 2020).

257 **tones down chronic, low-level inflammation:** See, for example, Stoyan Dimitrov, Elaine Hulteng, and Suzi Hong, "Inflammation and Exercise: Inhibition of Monocytic Intracellular TNF Production by Acute Exercise Via β_2-Adrenergic Activation," *Brain, Behavior, and Immunity* 61 (March 1, 2017); Earl S. Ford, "Does Exercise Reduce Inflammation? Physical Activity and C-Reactive Protein among U.S. Adults," *Epidemiology* 13, no. 5 (Sept. 2002); Peter T. Campbell et al., "A Yearlong Exercise Intervention Decreases CRP among Obese Postmenopausal Women," *Medicine & Science in Sports & Exercise* 41, no. 8 (2009); Laura A Daray et al., "Endurance and Resistance Training Lowers C-Reactive Protein in Young, Healthy Females," *Applied Physiology, Nutrition, and Metabolism* 36, no. 5 (Oct. 2011).

258 **influences most of the hallmarks:** Nuria Garatachea et al., "Exercise Attenuates the Major Hallmarks of Aging," *Rejuvenation Research* 18, no. 1 (Feb. 2015).

258 **lowers the numbers of the macrophages:** Dannenberg and Berger, *Obesity, Inflammation and Cancer.*

258   **manipulates microglial behavior:** Onanong Mee-inta, Zi-Wei Zhao, and Yu-Min Kuo, "Physical Exercise Inhibits Inflammation and Microglial Activation," *Cells* 8, no. 7 (2019).

258   **shrinks inflammatory fat around blood vessels:** Mee-inta, Zhao, and Kuo, "Physical Exercise Inhibits Inflammation."

258   **have experimented with "rat yoga":** Lisbeth Berrueta et al., "Stretching Impacts Inflammation Resolution in Connective Tissue," *Journal of Cellular Physiology* 231, no. 7 (July 1, 2016).

259   **central embodiment of chronic stress:** See, for example, Kimberley J. Smith et al., "The Association between Loneliness, Social Isolation and Inflammation: A Systematic Review and Meta-Analysis," *Neuroscience & Biobehavioral Reviews* 112 (May 1, 2020); Naomi I. Eisenberger et al., "In Sickness and in Health: The Co-Regulation of Inflammation and Social Behavior," *Neuropsychopharmacology* 42, no. 1 (Jan. 1, 2017); Janice K. Kiecolt-Glaser, Jean-Philippe Gouin, and Liisa Hantsoo, "Close Relationships, Inflammation, and Health," *Neuroscience & Biobehavioral Reviews* 35, no. 1 (Sept. 1, 2010); Paula V. Nersesian et al., "Loneliness in Middle Age and Biomarkers of Systemic Inflammation: Findings from Midlife in the United States," *Social Science & Medicine* 209 (July 1, 2018); Bert N. Uchino et al., "Social Support, Social Integration, and Inflammatory Cytokines: A Meta-Analysis," *Health Psychology* 37, no. 5 (2018).

259   **perhaps one mechanistic link:** See, for example, Slavich and Irwin, "From Stress to Inflammation and Major Depressive Disorder."

259   **like public speaking:** Bullmore, *The Inflamed Mind*.

259   **can feed hidden inflammation:** Janet M. Mullington et al., "Sleep Loss and Inflammation," *Best Practice & Research. Clinical Endocrinology & Metabolism* 24, no. 5 (2010); Michael R. Irwin et al., "Sleep Loss Activates Cellular Inflammatory Signaling," *Biological Psychiatry* 64, no. 6 (2008); Michael R. Irwin, "Sleep and Inflammation: Partners in Sickness and in Health," *Nature Reviews Immunology* 19, no. 11 (Nov. 1, 2019); Michael R. Irwin, Richard Olmstead, and Judith E. Carroll, "Sleep Disturbance, Sleep Duration, and Inflammation: A Systematic Review and Meta-Analysis of Cohort Studies and Experimental Sleep Deprivation," *Biological Psychiatry* 80, no. 1 (2016).

259   **and disease:** Raphael Vallat et al., "Broken Sleep Predicts Hardened Blood Vessels," *PLOS Biology* 18, no. 6 (2020); Tabitha R. F. Green et al., "The Bidirectional Relationship between Sleep and Inflammation Links Traumatic Brain Injury and Alzheimer's Disease," Review, *Frontiers in Neuroscience* 14, no. 894 (Aug. 25, 2020).

259   **disruption of circadian rhythms:** Maria Comas et al., "A Circadian Based Inflammatory Response—Implications for Respiratory Disease and Treatment," *Sleep Science and Practice* 1, no. 1 (Sept. 25, 2017).

260   **have been shown to lower inflammation:** See, for example, Paula R. Pullen et al., "Effects of Yoga on Inflammation and Exercise Capacity in Patients with Chronic Heart Failure," *Journal of Cardiac Failure* 14, no. 5 (June 2008); David S. Black and George M. Slavich, "Mindfulness Meditation and the Immune System: A Systematic Review of Randomized Controlled Trials," *Annals of the New York Academy of Sciences* 1373, no. 1 (June 2016); J. D. Creswell et al., "Mindfulness-Based Stress Reduction Training Reduces Loneliness and Pro-Inflammatory Gene Expression in Older Adults: A Small Randomized Controlled Trial," *Brain, Behavior, and Immunity* 26, no. 7 (Oct. 2012); David S Black et al., "Yogic Meditation Reverses NF-κB and IRF-Related Transcriptome Dynamics in Leukocytes of Family Dementia Caregivers in a Randomized Controlled Trial," *Psychoneuroendocrinology* 38, no. 3 (2013).

260 **attempted to capture their essence:** Panagiota Pietri, Theodore Papaioannou, and Christodoulos Stefanadis, "Environment: An Old Clue to the Secret of Longevity," *Nature* 544 (April 27, 2017).

260 **from pollutants:** See, for example, Dean E. Schraufnagel et al., "Air Pollution and Noncommunicable Diseases: A Review by the Forum of International Respiratory Societies' Environmental Committee, Part 1: The Damaging Effects of Air Pollution," *CHEST* 155, no. 2 (2019); Hector A. Olvera Alvarez et al., "Early Life Stress, Air Pollution, Inflammation, and Disease: An Integrative Review and Immunologic Model of Social-Environmental Adversity and Lifespan Health," *Neuroscience & Biobehavioral Reviews* 92 (Sept. 1, 2018); C. Arden Pope et al., "Exposure to Fine Particulate Air Pollution Is Associated with Endothelial Injury and Systemic Inflammation," *Circulation Research* 119, no. 11 (Nov. 11, 2016); Weidong Wu, Yuefei Jin, and Chris Carlsten, "Inflammatory Health Effects of Indoor and Outdoor Particulate Matter," *Journal of Allergy and Clinical Immunology* 141, no. 3 (2018).

260 **and smoking:** Jennifer O'Loughlin et al., "Association between Cigarette Smoking and C-Reactive Protein in a Representative, Population-Based Sample of Adolescents," *Nicotine & Tobacco Research* 10, no. 3 (March 2008); Russell P. Tracy et al., "Lifetime Smoking Exposure Affects the Association of C-Reactive Protein with Cardiovascular Disease Risk Factors and Subclinical Disease in Healthy Elderly Subjects," *Arteriosclerosis, Thrombosis, and Vascular Biology* 17, no. 10 (Oct. 1997); Ritienne Attard et al., "The Impact of Passive and Active Smoking on Inflammation, Lipid Profile and the Risk of Myocardial Infarction," *Open Heart* 4, no. 2 (2017).

260 **onslaught of chemicals:** Jinghua Yuan et al., "Long-Term Persistent Organic Pollutants Exposure Induced Telomere Dysfunction and Senescence-Associated Secretary Phenotype," *Journals of Gerontology: Series A* 73, no. 8 (2018).

260 **fumes from biomass fuels:** Jamie Rylance et al., "The Global Burden of Air Pollution on Mortality: The Need to Include Exposure to Household Biomass Fuel–Derived Particulates," *Environmental Health Perspectives* 118, no. 10 (Oct. 1, 2010).

261 **can still be found inside these macrophages:** Yoav Arnson, Yehuda Shoenfeld, and Howard Amital, "Effects of Tobacco Smoke on Immunity, Inflammation and Autoimmunity," *Journal of Autoimmunity* 34, no. 3 (May 1, 2010).

261 **manage to retain an abundance of factors:** Daniela Monti et al., "Inflammaging and Human Longevity in the Omics Era," *Mechanisms of Ageing and Development* 165 (July 1, 2017).

261 **develop hidden inflammation that may persist:** Christopher P. Fagundes and Baldwin Way, "Early-Life Stress and Adult Inflammation," *Current Directions in Psychological Science* 23, no. 4 (Aug. 1, 2014); Olvera Alvarez et al., "Early Life Stress, Air Pollution, Inflammation, and Disease."

261 **naturally decrease with age:** Charles Serhan, interview with author, Feb. 2019.

262 **story starts even earlier:** Renz et al., "An Exposome Perspective"; Kanakadurga Singer and Carey N. Lumeng, "The Initiation of Metabolic Inflammation in Childhood Obesity," *Journal of Clinical Investigation* 127, no. 1 (Jan. 3, 2017).

Chapter 17: Human Chimeras

265 **with a genetic mutation:** Florian J. Clemente et al., "A Selective Sweep on a Deleterious Mutation in CPT1A in Arctic Populations," *American Journal of Human Genetics* 95, no. 5 (Nov. 6, 2014); Melanie B. Gillingham et al., "Impaired Fasting Tolerance among Alaska Native Children with a Common Carnitine Palmitoyltransferase 1A Sequence Variant," *Molecular Genetics and Metabolism* 104, no. 3 (2011).

265 **from deeply embedded evolutionary vulnerabilities:** Gokhan Hotamisligil, interview with author, Feb. 2019. See also Daniel Okin and Ruslan Medzhitov, "Evolution of Inflammatory Diseases," *Current Biology* 22, no. 17 (Sept. 11, 2012).

266 **wrote of organ transplantation:** Silverstein, *History of Immunology.*

268 **also plays a role in organ rejection:** Daniel N. Mori et al., "Inflammatory Triggers of Acute Rejection of Organ Allografts," *Immunological Reviews* 258, no. 1 (2014).

268 **from an irked innate immune system:** Faouzi Braza et al., "Role of TLRs and DAMPs in Allograft Inflammation and Transplant Outcomes," *Nature Reviews Nephrology* 12, no. 5 (May 1, 2016); Dag Olav Dahle et al., "Inflammation-Associated Graft Loss in Renal Transplant Recipients," *Nephrology Dialysis Transplantation* 26, no. 11 (2011); Daniel Kreisel and Daniel R. Goldstein, "Innate Immunity and Organ Transplantation: Focus on Lung Transplantation," *Transplant International* 26, no. 1 (Jan. 1, 2013).

268 **cutting its blood supply:** Karsten Bartels, Almut Grenz, and Holger K. Eltzschig, "Hypoxia and Inflammation Are Two Sides of the Same Coin," *Proceedings of the National Academy of Sciences of the USA* 110, no. 46 (Nov. 12, 2013).

269 **special immune cells called Tregs:** Muhammad Atif et al., "Regulatory T Cells in Solid Organ Transplantation," *Clinical & Translational Immunology* 9, no. 2 (2020).

269 **can help to enhance a unique state:** Jun-Feng Du et al., "Treg-Based Therapy and Mixed Chimerism in Small Intestinal Transplantation: Does Treg + BMT Equal Intestine Allograft Tolerance?," *Medical Hypotheses* 76, no. 1 (Jan. 2011).

270 **lower rates of organ rejection:** J. Zuber et al., "Macrochimerism in Intestinal Transplantation: Association with Lower Rejection Rates and Multivisceral Transplants, without GVHD," *American Journal of Transplantation* 15, no. 10 (Oct. 1, 2015).

270 **entirely wean off:** Joseph Leventhal et al., "Chimerism and Tolerance without GVHD or Engraftment Syndrome in HLA-Mismatched Combined Kidney and Hematopoietic Stem Cell Transplantation," *Science Translational Medicine* 4, no. 124 (March 7, 2012).

270 **may naturally shift with lifestyle changes:** See, for example, Wooki Kim and Hyungjae Lee, "Advances in Nutritional Research on Regulatory T-Cells," *Nutrients* 5, no. 11 (Oct. 28, 2013); Shohreh Issazadeh-Navikas, Roman Teimer, and Robert Bockermann, "Influence of Dietary Components on Regulatory T Cells," *Molecular Medicine* 18, no. 1 (2012); Rebeca Arroyo Hornero et al., "The Impact of Dietary Components on Regulatory T Cells and Disease," *Frontiers in Immunology* 11 (2020); J. A. Fishman and A. W. Thomson, "Immune Homeostasis and the Microbiome—Dietary and Therapeutic Modulation and Implications for Transplantation," *American Journal of Transplantation* 15, no. 7 (2015).

270 **lower rates of graft failure and loss:** António W. Gomes-Neto et al., "Mediterranean Style Diet and Kidney Function Loss in Kidney Transplant Recipients," *Clinical Journal of the American Society of Nephrology* 15, no. 2 (Feb. 7, 2020).

271 **is linked to worse outcomes:** Maral Baghai Arassi et al., "The Gut Microbiome in Solid Organ Transplantation," *Pediatric Transplantation* 24, no. 7 (Nov. 1, 2020).

272 **noted in 1962:** Felissa R. Lashley and Jerry D. Durham, eds., *Emerging Infectious Diseases: Trends and Issues,* 2nd ed. (New York: Springer, 2007).

272 **How much of the germ:** Jesse Fajnzylber et al., "SARS-CoV-2 Viral Load Is Associated with Increased Disease Severity and Mortality," *Nature Communica-*

tions 11, no. 1 (Oct. 30, 2020); Elisabet Pujadas et al., "SARS-CoV-2 Viral Load Predicts COVID-19 Mortality," *The Lancet Respiratory Medicine* 8, no. 9 (Sept. 1, 2020).

273   **foil the immune system:** Yoriyuki Konno et al., "SARS-CoV-2 ORF3b Is a Potent Interferon Antagonist Whose Activity Is Further Increased by a Naturally Occurring Elongation Variant," *Cell Reports* 32, no. 12 (Sept. 2020); John M. Lubinski et al., "Herpes Simplex Virus Type 1 Evades the Effects of Antibody and Complement in Vivo," *Journal of Virology* 76, no. 18 (2002).

273   **can later resort:** Rose H. Manjili et al., "COVID-19 as an Acute Inflammatory Disease," *Journal of Immunology* 205, no. 1 (2020).

273   **plays a role in outcomes:** Stephen Morse, email to author, May 2021.

274   **some of the main culprits:** David C. Fajgenbaum and Carl H. June, "Cytokine Storm," *New England Journal of Medicine* 383 (2020).

274   **has been labeled:** Puja Mehta et al., "COVID-19: Consider Cytokine Storm Syndromes and Immunosuppression," *The Lancet* 395, no. 10229 (2020).

274   **tend to form more easily:** Ricardo J. Jose and Ari Manuel, "COVID-19 Cytokine Storm: The Interplay between Inflammation and Coagulation," *The Lancet Respiratory Medicine* 8, no. 6 (2020).

275   **can also help to explain:** Erola Pairo-Castineira et al., "Genetic Mechanisms of Critical Illness in COVID-19," *Nature* 591, no. 7848 (March 1, 2021).

275   **widely but not universally:** Stephen Morse, email to author, May 2021.

275   **have a nonfunctional *IFITM3* gene:** Davis, *The Beautiful Cure.*

276   **suffer with hidden inflammation:** Alexander Kroemer et al., "Inflammasome Activation and Pyroptosis in Lymphopenic Liver Patients with COVID-19," *Journal of Hepatology* 73, no. 5 (2020); Carolina Lucas et al., "Longitudinal Immunological Analyses Reveal Inflammatory Misfiring in Severe COVID-19 Patients," *medRxiv* (2020); Matthew J. Cummings et al., "Epidemiology, Clinical Course, and Outcomes of Critically Ill Adults with COVID-19 in New York City: A Prospective Cohort Study," *The Lancet* 395, no. 10239 (2020); Alisa A. Mueller et al., "Inflammatory Biomarker Trends Predict Respiratory Decline in COVID-19 Patients," *Cell Reports Medicine* 1, no. 8 (2020).

276   **a major risk factor:** Kaveh Hajifathalian et al., "Obesity Is Associated with Worse Outcomes in COVID-19: Analysis of Early Data from New York City," *Obesity* 28, no. 9 (Sept. 2020).

276   **may overreact to germs:** Mireya G. Ramos Muniz et al., "Obesity Exacerbates the Cytokine Storm Elicited by *Francisella tularensis* Infection of Females and Is Associated with Increased Mortality," *BioMed Research International* 2018 (June 26, 2018); C. Tsatsanis, A. N. Margioris, and D. P. Kontoyiannis, "Association between H1N1 Infection Severity and Obesity—Adiponectin as a Potential Etiologic Factor," *Journal of Infectious Diseases* 202, no. 3 (2010); Gabrielle P. Huizinga, Benjamin H. Singer, and Kanakadurga Singer, "The Collision of Meta-Inflammation and SARS-CoV-2 Pandemic Infection," *Endocrinology* 161, no. 11 (2020).

277   **gains a foothold:** Annsea Park and Akiko Iwasaki, "Type I and Type III Interferons—Induction, Signaling, Evasion, and Application to Combat COVID-19," *Cell Host & Microbe* 27, no. 6 (June 10, 2020).

278   **natural order between human and microbes:** Michael Greger, *How to Survive a Pandemic* (New York: Flatiron Books, 2020).

278   **most of the antibiotics:** Michael J. Martin, Sapna E. Thottathil, and Thomas B. Newman, "Antibiotics Overuse in Animal Agriculture: A Call to Action for Health Care Providers," *American Journal of Public Health* 105, no. 12 (2015).

280 **"Every leaf and every glass blade":** David Wallace-Wells, *The Uninhabitable Earth: Life after Warming* (New York: Tim Duggan Books, 2019).

280 **higher than what Ancel Keys had recorded:** Walter Willett, interview with author, Feb. 2020.

280 **coauthored a white paper:** Pett et al., *Ancel Keys and the Seven Countries Study.*

281 **as Julian Barnes writes:** Julian Barnes, *The Sense of an Ending* (New York: Vintage Books, 2012).

Index

achromatic lens, 14
acute inflammation
  autoimmunity and, 35
  evolution and, 265
  exercise and, 257
  inflammatory markers
    and, 126
  inflammatory mediators
    and, 122
  innate immune system
    and, 31
  phagocytes and, 22, 31, 32,
    122, 123
  resolvins and, 123
adaptation, 264
adaptive immune system,
  30–31
  aging and, 276
  antibodies and, 41
  autoimmune diseases
    and, 34
  branches of, 26, 30
  cancer and, 78
  dendritic cells and, 44
  depression and, 110n
  fiber and, 186
  gluten and, 234
  heart disease and, 63
  hidden inflammation and,
    63, 78, 89–90, 270–71,
    276–77
  high salt intake and, 171
  Hoffman on, 43
  immune system deficien-
    cies and, 42
  inflammatory bowel disease
    and, 179n
  innate immune system
    cooperation, 32, 44,
    273, 274
  intestinal immune system
    and, 136, 137
  lymphocyte functions in,
    31–32

  macrophages and, 113
  microbiome and, 139
  obesity and, 89–90, 270–71
  organ rejection and, 268
  pattern recognition recep-
    tors and, 133n
  research focus on, 42
  rheumatoid arthritis and,
    104–5
  Tregs and, 124
  vaccines and, 277
  See also immune cells; mac-
    rophages; neutrophils;
    phagocytes
adipocytes, 91, 93, 113
adiponectin, 89
adipose tissue. *See* body fat;
  obesity
African Americans, 170, 185
aging, 37
  See also healthy aging;
    inflammaging
agrarian lifestyles, 271–72
agriculture, regenerative,
  248–49
Akbar (emperor, 16th century),
  240
allergic diseases, 39–41, 49, 112
Alzheimer, Alois, 105, 106
Alzheimer's disease, 105–6, 108
American diet. *See* Western
  diet
American Pediatric Society, 42
amylase-trypsin inhibitors,
  226, 230, 235
amyloid plaques, 106, 107, 108
anaphylaxis, 40
angina pectoris, 51–52
angiogenesis, 76, 77
animal foods
  aging and, 256
  animal husbandry and, 271,
    277, 278–79, 280
  blue zone diet and, 255

  cancer and, 193
  chronic inflammatory
    diseases and, 181, 182,
    189, 194
  climate change and, 271, 280
  cooking methods and, 194
  diet for planetary health
    and, 280
  endothelial cells and, 215
  endotoxins and, 182, 183
  fiber and, 179, 182
  food-borne pathogens and,
    182–83
  healthy aging and, 255
  heme iron and, 182
  hidden inflammation and,
    189–90
  hydrogen sulfide and, 182
  immune system and, 200
  inflammatory bowel disease
    and, 190
  inflammatory microbiome
    and, 179, 182
  insulin-like growth factor
    and, 256
  Kellogg diet and, 172, 175
  macrophages and, 183
  mammalian target of rapa-
    mycin and, 256
  meatpacking industry
    and, 163
  Mediterranean diet and,
    280
  metabolites and, 187–88
  microbes and, 277, 278
  microbiota-food inter-
    actions and, 179–83,
    189–90
  N-glycolylneuraminic acid
    and, 182
  19th-century game, 169, 171
  Okinawan diet and, 215
  Paleolithic diets and, 189
  pandemics and, 271, 277–79